Deeley.

# Genetic Engineering

## Principles and Methods

## Volume 2

GENETIC ENGINEERING
Principles and Methods

Advisory Board

Carl W. Anderson
Donald D. Brown
Peter Day
Donald R. Helinski
Tom Maniatis

A Continuation Order Plan is available for this series. A continuation order will bring
delivery of each new volume immediately upon publication. Volumes are billed only upon
actual shipment. For further information please contact the publisher.

# Genetic Engineering

## Principles and Methods

## Volume 2

Edited by

### Jane K. Setlow
Brookhaven National Laboratory
Upton, New York

and

### Alexander Hollaender
Associated Universities, Inc.
Washington, D.C.

Plenum Press · New York and London

Library of Congress Catalog Card Number 79-644807
ISBN 0-306-40447-8


© 1980 Plenum Press, New York
A Division of Plenum Publishing Corporation
227 West 17th Street, New York, N.Y. 10011

Printed in the United States of America

PREFACE TO VOLUME 1


This volume is the first of a series concerning a new technology which is revolutionizing the study of biology, perhaps as profoundly as the discovery of the gene. As pointed out in the introductory chapter, we look forward to the future impact of the technology, but we cannot see where it might take us. The purpose of these volumes is to follow closely the explosion of new techniques and information that is occurring as a result of the newly-acquired ability to make particular kinds of precise cuts in DNA molecules. Thus we are particularly committed to rapid publication.


Jane K. Setlow

Alexander Hollaender

ACKNOWLEDGMENT

Laine McCarthy did all the final processing of the manuscripts, as well as helping with the editing of this volume.  The Editors are profoundly grateful for her skill and patience.

CONTENTS

# CLONING OF REPEATED SEQUENCE DNA FROM CEREAL PLANTS

J.R. Bedbrook* and W.L. Gerlach*

Cytogenetics Department, Plant Breeding Institute
Maris Lane, Trumpington, Cambridge, CB2 2LQ
England

## INTRODUCTION

While the specific functions of repeated sequence DNA remain
enigmatic, several of their properties are clear.  First, they are
ubiquitous among eukaryotes, and second, in the eukaryotes with high
nuclear DNA content they represent the major DNA component of
chromosomes.  Repeated sequence DNA accounts for at least 70% of the
total chromosomal DNA in cereal plants(1).

The structure of repeated DNA sequences has been examined in a
number of animal species.  Much of this work has depended on the fact
that many repeated sequences can be isolated as density satellites
upon equilibrium density centrifugation.  Only a small fraction of
the total repeated DNA in cereals can be purified by density centrifugation (2-4) and, in general, details of the organization of
repeated sequences in cereals has been dependent on renaturation
analysis (5) and enrichment of very highly repeated DNA fractions by
renaturation (6).  We have used molecular cloning to isolate repeated
DNA sequences and repeated genes from cereal plants (7-10).  The
advantage of molecular cloning lies in the purification of large
amounts of specific repeated DNA sequences that cannot otherwise be
isolated as density satellites.  It is also possible, using clones of
repeated sequences, to check individual members of these sequence
families for subtle changes in structure.  They can provide examples
of the types of mutation, rearrangement and amplification events that
occur in chromosome evolution.

---

*Present Address--Division of Plant Industry, CSIRO, P.O. Box 1600,
Canberra City, A.C.T., Australia 2601.

It is the purpose of this chapter to describe some of the
results obtained by studying clones of repeated DNA sequences
from cereal plants. We first describe some of the technical considerations involved in cloning plant DNA in the form of restriction
fragments. This is valid in view of various unpublished reports of
problems with the cloning of plant DNA. Problems with stable cloning
of certain very simple satellite DNA sequences from Drosophila
melanogaster have also been reported (11). We evaluate the stability
of different types of repeated cereal DNA clones in bacterial plasmids and outline the various sequences we have cloned and the information their study is providing on the structure and evolution of
cereal plant genomes.


RESTRICTION ENDONUCLEASE ANALYSIS OF CEREAL DNA

Restriction endonuclease-generated fragments of eukaryotic DNA
have often been used in molecular cloning experiments. A useful
feature of this approach is that the cloned fragments can be easily
identified and physical correspondence with fragments in the genome
can be simply established. In size-fractionated restriction enzyme
digests, many families of repeated sequences are revealed as bands
superimposed on the smear of fluorescence derived from more random
sequences. By determining the fraction of DNA in these bands, we
have been able to quantify the variation in amounts of specific
repeated sequences between closely related species (8). In this way,
we can account for 50% of the difference in DNA content between
Secale cereale and S. silvestre, two species that represent the
extremes in nuclear DNA amount in the rye genus (12).
Restriction endonucleases have also provided methods of enriching for groups of repeated sequence DNA families. Clearly,
enzymes will not cut repeated sequences when the repeat unit does
not contain sites for the enzyme. The spared fraction in complete
restriction endonuclease digests is therefore enriched for such
repeats. Various restriction endonuclease-spared DNA fractions have
been purified from rye by velocity sedimentation. These fractions
have enabled us, in concert with molecular cloning, to identify and
characterize specific repeat sequence families from rye (see below).
In addition to the bands seen in gels of restriction enzyme
digests of cereal DNA, we have attempted to make an overall
characterization of the digestion products, including the background
smear of fluorescence seen in such gels. We have compared the
conformance of the size distribution of the smear of fluorescence
with that predicted by random fragmentation theory and concluded that
most of the visible fluorescence in gels of restriction digests of
this DNA must be derived from repeated sequences of relatively low
complexity (7). This conclusion was based on the fact that the size
distribution of fragments produced by restriction enzyme digestion of
rye DNA did not conform with the predicted size distribution of
fragments for a random sequence. In contrast, the bulk of Drosophila

melanogaster DNA behaves as a random sequence when analyzed in this way (13). This conclusion was further supported by the observation that highly repeated DNA fractions of rye DNA prepared by renaturation fractionation gave a size distribution of fragments indistinguishable from total DNA.

Table 1 gives this comparison of theoretical and empirical frequencies for the occurrence of restriction endonuclease targets in rye DNA. The number average fragment lengths for EcoRI, BamHI, BglII, HindIII and HaeIII enzymes are reasonably close to the theoretical values. The observed disparity in these values probably reflects the influence of the relatively low complexity of the bulk of the assayed DNA (see above). The enzymes HpaII, SmaI and PstI give markedly different values than those predicted. The most probable explanation for the lack of HpaII and SmaI sites is the high level of 5-methyl-cytosine (5MC) in cereal plant DNA (14,15). Restriction activity of HpaII (16), and probably SmaI, is prevented by 5MC. HaeIII cleavage is also eliminated by 5MC (16) in one of the two possible positions in the target sequence and it is interesting that HaeIII sites are present at the expected frequency in rye DNA. Taken together, these results must mean that most 5MC in plant DNA is in the sequence 5MCpG. This is discussed in relation to the ribosomal RNA genes in a later section. We have no simple explanation for the lack of PstI sites in rye DNA.

Restriction analysis of cereal genomes has also been informative for technical reasons. Firstly, restriction enzyme analysis is a useful assay for DNA purity. Plant DNA seems to contain impurities that inhibit the action of restriction endonucleases. Completeness of digestion could be assayed by the relative stoichiometry of bands in the digest. DNA prepared by standard procedures on CsCl gradients required a vast excess of enzyme for complete digestion. We found that CsCl gradients containing ethidium bromide (see Appendix) provided the best purification based on the restriction endonuclease assay.

The second technical way in which restriction enzyme analysis was useful was in giving an estimate of the necessary number of clones for a complete clone library of cereal DNA. This number is of the order of $10^6$ for most restriction endonucleases.


CLONING OF CEREAL NUCLEAR DNA

To ascertain that there were no inherent difficulties in cloning cereal DNA as prepared by our procedure (see Appendix), we compared (10) the cloning efficiency of identically prepared wheat (Triticum aestivum var. Chinese Spring) and Escherichia coli DNA in the plasmid pAC184 (17) using the recA⁻ E. coli host strain HB101. Circular pAC184 vector yielded $10^6$ transformants per μg (= transformation efficiency). This decreased to $10^3$ when the vector was linearized with the restriction endonuclease EcoRI. Self-ligation of this DNA

Table 1

Comparison of Theoretical and Empirical Frequencies for the Occurrence of
Restriction Endonuclease Targets in the DNA of S. cereale

| Restriction enzyme | Probability of occurrence of sequence in genome p | Mean segment length determined from p | Calculated mean segment length determined from restriction enzyme digests | Calculated number of different fragments produced by digestion of S. cereale DNA |
|---|---|---|---|---|
| EcoRI | $2.887 \times 10^{-4}$ | 3464 | 3400 | $2.30 \times 10^{6}$ |
| BamHI | $1.948 \times 10^{-4}$ | 5132 | 3311 | $2.35 \times 10^{6}$ |
| BglII | $2.887 \times 10^{-4}$ | 3464 | 2512 | $3.10 \times 10^{6}$ |
| HaeIII | $25.860 \times 10^{-4}$ | 387 | 400 | $1.95 \times 10^{7}$ |
| HpaII | $25.860 \times 10^{-4}$ | 387 | >10,000 | $<7.80 \times 10^{5}$ |
| HindIII | $2.887 \times 10^{-4}$ | 3464 | 2400 | $3.25 \times 10^{6}$ |
| PstI | $1.948 \times 10^{-4}$ | 5132 | 8128 | $9.50 \times 10^{5}$ |
| SmaI | $1.315 \times 10^{-4}$ | 7605 | >10,000 | $<7.80 \times 10^{5}$ |

Rye DNA was digested with restriction enzymes and fractionated by gel electrophoresis as described in the legend to Figure 1. Traces of photographic negatives of the gel were made with a
scanning densitometer. Cumulative size distributions were obtained by measuring the fluorescence within specific fragment size classes determined from molecular weight markers. The integral of the weight over the range of fragment sizes was plotted against fragment size. The
plots were tested for conformance to the Kuhn approximation ($w_t = tp^2e^{-pt}$) where $w_t$ is the
weight fraction of DNA found in segments t base pairs long. The value for p (probability of a
double-strand cleavage) was determined from the base composition of rye DNA and the base composition of the enzyme recognition sequence.

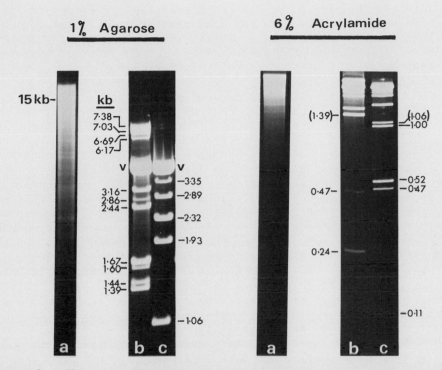

Figure 1.   The size distribution of some EcoRI-generated wheat DNA
fragments cloned in plasmid pAC184 with E. coli host strain HB101.
The tracks show electrophoresis separation patterns of the different
fragment sizes.   (a)  Wheat DNA after digestion with EcoRI.   This was
the DNA used in the cloning experiment.   Note that a proportion of
the DNA is present as relatively long (>15 kilobases) molecules after
digestion.   (b),(c)  EcoRI-digested plasmid DNA from two different
mixed cultures showing an array of wheat DNA inserts cloned into the
pAC184 plasmid vector (V = vector DNA).   (Molecular sizes of bands
were determined from a consideration of the sizes of restriction
endonuclease digested phage lambda and plasmid pBR322 DNAs included
in the gels.)

led to a transformation efficiency of $10^5$, an order of magnitude
lower than that of the original circular DNA.   Ligation of the
linearized plasmid with EcoRI-digested E. coli DNA also resulted in a
transformation efficiency of about $10^5$, similar to the self-ligated
plasmid.   When wheat DNA was ligated to the linearized plasmid the
resultant efficiency was 2.5 to 4 x $10^4$.   Twelve to fifteen percent
of the transformants contained DNA inserts regardless of whether the
ligated DNA originated from wheat or E. coli.

This slight decrease in the transformation efficiency of ligated wheat DNA relative to E. coli DNA is a regular observation in our experiments and we are not sure of its cause. One contributing factor certainly may be the presence of a portion of EcoRI-digested wheat DNA in fragments too long (>15 kilobases, Figure 1) to be recovered efficiently as clones when ligated to pAC184.

In order to characterize the average insert size of the cloned wheat DNA, plasmid DNA was prepared from mixed cultures and the size of the insertions was determined by electrophoresis after digestion with EcoRI (Figure 1). Sixty-one inserts were measured ranking in size from 14 kilobases to <0.1 kilobases, the limit of detection, with a mean of 2.9 kilobases. It is interesting to compare this spectrum with EcoRI-digested total wheat DNA, which has a number average molecular weight of 3.2 kilobases and ranges in size from >20 kilobases to <0.1 kilobases (unpublished data). Since the mean insert size compares well with the number average molecular weight of the EcoRI-digested wheat DNA, we conclude that there are no general instabilities of wheat DNA inserts in these experiments. It is possible that a small number of clones containing very simple repeated sequence DNA inserts may not be stable (see below) but in general it is apparent that stable inserts of wheat DNA can be obtained with acceptable efficiency (relative to E. coli controls) in plasmid pAC184 with E. coli host strain HB101.

It is pertinent to mention here some cloning experiments done with pea DNA (R.E. Cuellar and W.F. Thompson, personal communication) that bear on the use of this cloning system. They found that the combination of pAC184 vector DNA and strain HB101 host also provides optimal conditions for obtaining stable pea DNA clones because the size distribution of inserts recovered is the same as that of the pea DNA used in the ligation reaction. The size of pea DNA insertions recovered in plasmid pAC184 with a different E. coli host strain (ES89), however, was much shorter than expected, suggesting that the E. coli host may have some bearing on the recovery of a complete size class distribution of plant DNA inserts.

Our observation that vector pAC184 and strain HB101 provide a suitable combination for stable cloning of cereal DNA is also borne out in cloning of a specific, relatively long DNA sequence, the ribosomal RNA gene units (rDNA) from wheat and barley (10). These units are 9 to 10 kilobases long, as detailed below, and stable clones containing full length inserts were cloned after an enrichment step for these particular sequences. Sixteen of 18 rDNA inserts were full length repeating units, again indicating no marked instability of clones. The rDNA clones were recovered at the frequency expected on the basis of the proportion of the rDNA in the starting DNA.

Both shoot DNA and DNA isolated from dry embryos were used in the cloning of rDNA. This was done in order to test whether shoot DNA, which is more prone to nuclease damage during isolation (18), might provide a lower cloning efficiency. However, the insertion frequency was similar for both types of DNA and the proportion of rDNA clones containing full length repeat units (11 of 13 for barley

shoot DNA vs. 5 of 5 for wheat embryo DNA) suggested that the
source of the DNA did not have any marked effect on its suitability
for cloning.

Experiments using the plasmid pBR322 (19) also provided stable
clones of cereal DNA in E. coli strain HB101. Repeated sequence
units of rye (Secale cereale) DNA and the 5S RNA gene repeating
units of wheat have been cloned in pBR322 (see below). The clones
are stable and insertion frequencies were 5 to 15% of the
transformants recovered.


## CLONES OF A CEREAL POLYPYRIMIDINE SATELLITE DNA

Satellite DNAs have been detected during ultracentrifugation of
cereal DNA in cesium buoyant density gradients (2-4). One particular
satellite DNA, isolated from both wheat and barley in $Ag^+$-$CsSO_4$
gradients, has been the subject of detailed study (20). Renaturation
kinetics indicated that the basic repeating unit of the satellite DNA
was about 5 to 10 nucleotides long and analysis of cRNA transcripts
and restriction endonuclease MboII digests of the satellite further
suggested that it is composed of long, tandem arrays of a simple 12
base-pair repeating unit, $(CTT)_n(CTC)_m$, (i.e., one strand is a poly-
pyrimidine sequence).

In order directly to sequence examples of this satellite DNA and
to examine the stability of this simple repeated sequence when it is
cloned in a bacterial plasmid, recombinant pBR322 plasmids were
prepared containing the barley sequence inserted at the HindIII site
in the tetracycline-resistance promoter. The barley (Hordeum vulgare
var. Sultan) satellite DNA was isolated in an $Ag^+$-$CsSO_4$ buoyant
density gradient and sheared gently by drawing through a fine glass
capillary so that the weight average molecular size was still >5
kilobases. It was prepared for cloning by treatment with E. coli
polymerase I to ensure the presence of flush double-stranded termini
before ligation with synthetic HindIII linker oligonucleotides. The
satellite DNA containing ligated linkers was separated from self-
ligated and unligated linker DNA by velocity sedimentation in a
sucrose gradient, digested with HindIII to reveal the "sticky ends"
and then ligated to HindIII-digested pBR322 DNA. Plasmid DNA was
prepared from six clones containing inserts that hybridized barley
satellite DNA when tested by the colony hybridization technique (21).

We were interested to determine the size of the inserts in
these satellite DNA clones, particularly in view of the observation
that stable inserts greater than 1.5 kilobases could not be obtained
when two simple sequence Drosophila melanogaster satellite DNAs of 5
and 7 base-pair repeating unit length were cloned in a number of
plasmid-bacterial host strain combinations (11). These data,
together with the observation that poly-AT cloning tails tend to
decrease in size (11,22), have been used to suggest that inserts of
very simple repeated DNA sequences may be unstable when cloned in
bacterial plasmids.

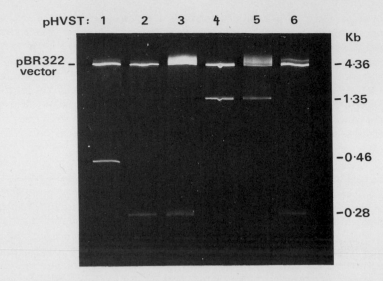

Figure 2.   Insert sizes of barley satellite DNA cloned in plasmid
pBR322 with HindIII linkers.  HindIII-digested plasmid DNA from
recombinant clones was fractionated by electrophoresis on 6%
acrylamide gel.  The vector DNA bands (plasmid pBR322) are indicated
along with the sizes of the inserts in the recombinant plasmids
designated pHVST 1-6.


     Consistent with the observation for the simple sequence D.
melanogaster satellite DNA, we found that the cereal satellite DNA
clones contained inserts that were at most 1.4 kilobases long (Figure
2), even though the DNA used in the ligation reaction was >5 kilobases.  All inserts were susceptible to extensive digestion with
MboII, indicating that they did contain the simple sequence satellite
DNA.  We can only conclude that deletions may readily occur in clones
of this very simple repeated sequence so that the transformants we
recovered after screening are segregational products of deletion
events in the inserted sequence.
     We are currently sequencing the inserted DNA in some of these
clones to determine if the predicted single repeating unit is correct
and to provide a basis for studying the evolution of this very simple
repeated sequence in cereal species.  The first sequenced example was
poly (CTT).


     REPEATED DNA SEQUENCES FROM THE TELOMERIC HETEROCHROMATIN OF RYE

     The telomeres of S. cereale (rye) chromosomes are composed of
constitutive heterochromatin, which accounts for 12 to 18% of S.
cereale DNA (23).  Most of the very highly repeated DNA sequences in
this species are located in this telomeric heterochromatin (6).  We

have characterized repeats accounting for more than 50% of the DNA in
the heterochromatin of S. cereale and have compared the telomeric repeated sequence composition of this species with that in Secale silvestre, the species having the lowest DNA content in the genus (8).

We based our approach on the observation that the portion of rye
DNA spared by most hexanucleotide recognition site restriction endonucleases was greatly enriched for repeated sequences located at the
telomeres of rye. HaeIII digestion of this spared DNA fraction
showed 12 major bands derived from several different repeat families
on gel electrophoresis. We purified these bands from preparative
gels and used them as hybridization probes to screen clone banks.

We prepared clone banks of HaeIII fragments for the analysis of
simple repeats for several reasons. First, HaeIII digestion yielded
a number average fragment size close to the theoretical value (Table
1); second, HaeIII left a higher proportion of the total DNA in bands
than other enzymes, and third, it did not leave a large fraction of
the DNA in fragments too large for plasmid cloning experiments. To
clone HaeIII fragments, total rye chromosomal DNA was digested with
HaeIII, ligated to the chemically synthesized linker sequence
5'CCAAGCTTGG3', which contains the recognition site for HindIII, and
the ligated molecules digested with HindIII in preparation for
ligation to the cloning vector pBR322. Transformants containing
inserted rye DNA were hybridized with repeated sequence probes
prepared as described above. All the probes used were complementary
to various clones and all the positive clones could be reduced to
five classes. That is, several of the probes hybridized to the same
clones suggesting that different DNA fragments in the HaeIII digest
of the spared DNA were of differing lengths or were different
sequence conformations of the same basic sequence.

Figure 3 shows the physical maps of these repeats determined by
hybridizing probes prepared from the cloned examples of the repeats
back to genomic DNA digested with various enzymes. These results
show that the five classes defined six different repeats, three
belonging to the same family. Repeat "a", illustrated in Figure 3a,
is a simple tandem repeat with a length periodicity of approximately
120 base pairs defined by the nuclease HaeIII. Repeat "b",
illustrated in Figure 3b, has a repeat length of 480 base pairs
defined by the endonuclease MboII. This repeat is apparently more
complex in structure than the repeat illustrated in Figure 3a.
HaeIII digestion shows that the 480 base-pair repeat unit contains a
subrepeating unit interspersed with an unrelated sequence (double
thickness line in Figure 3). Sequence analysis on cloned DNA of the
480 base-pair MboII fragments does not reveal obvious subrepetition
within the interspersed sequence. The repeats illustrated in Figure
3c (a 610 base-pair repeat), Figure 3d2 (a 356 base-pair repeat) and
Figure 3d3 (a 630 base-pair repeat) all have complex repeat forms
similar to that of the 480 base-pair repeat. The repeat of Figure
3d1 has a simple repeat unit of approximately 120 base pairs. It is
important to note that the subrepeating units of the complex repeats
illustrated in Figure 3d2 and 3d3 are the same or closely related to

the sequence of the simple repeat illustrated in Figure 3d1.  In situ
hybridization analysis with the cloned examples of these repeats
shows that they are all located predominantly within the telomeric
heterochromatin of S. cereale chromosomes.

     We have shown that there is an interesting correlation between
the presence of these repeats in various Secale species and the
structure of the repeats (8).  S. silvestre, the lowest C-value
member of the Secale, has less than 9% of its total DNA in telomeric
heterochromatin and of the repeats illustrated in Figure 3, it only
contains detectable amounts of the two simple repeats (3a and 3d1),
as judged by DNA hybridization analysis.  S. cereale, on the other
hand, is a high C-value representative of the genus and maintains as
much as 18% of its total DNA content in telomeric heterochromatin.
This species contains both the simple repeats and the more complex
repeats (Figure 3b, 3c, 3d2 and 3d3).  Further, these four complex
repeats account for most if not all of the difference between S.
silvestre and S. cereale DNA amount in the telomeric heterochromatin.
The above mentioned correlation has led to the speculation that each
of the S. cereale-specific repeats has evolved by insertion of DNA
elements into an array of simple repeats followed by a massive
amplification of a portion of the array containing the inserted
sequence.  In each case, the amplification event produces a complex
tandem array consisting of a subrepeating unit interspersed with an
unrelated nonsubrepeating sequence.


          MODES OF EVOLUTION OF REPEATED SEQUENCES IN RYE: A PICTURE
          DEDUCED FROM STRUCTURAL ANALYSIS OF CLONED COMPLEX REPEATS


     Figure 3a illustrates the structure of a simple repeat for S.
cereale.  As discussed in the previous section, this repeat is also
present in the related genus Triticum.  In its simple form, this
repeat is a 120 base-pair tandem repeat and is indistinguishable in
structure in the two Secale species and in Triticum aestivum (wheat).
We have found complex forms of this repeat in all three species (9).
The complex forms were revealed by hybridizing probes prepared from a
cloned example of the 120 base-pair repeat to total genomic DNA of
the three species digested with various hexanucleotide recognition
enzymes and fractionated by gel electrophoresis.  Figure 4 shows the
hybridization pattern to (a) HaeIII, (b) BamHI, (c) BglII, (d)
HindIII and (e) EcoRI digests of S. cereale DNA.  For HaeIII digests
the pattern of hybridization is a series of bands representing a
progression of integral multimers based on the 120 base-pair unit,
with the monomer unit being in the highest stoichiometry.  We assume,
and this can be tested directly by sequencing of cloned examples,
that the multimers arise as a consequence of the loss of restriction
targets by random mutation.  The hexanucleotide enzymes (4b-4e) give
a different hybridization picture.  Most of the DNA in these digests
that is complementary to the probe is high molecular weight.  BamHI
(4b) and EcoRI (4e) digests yield a series of hybridized bands

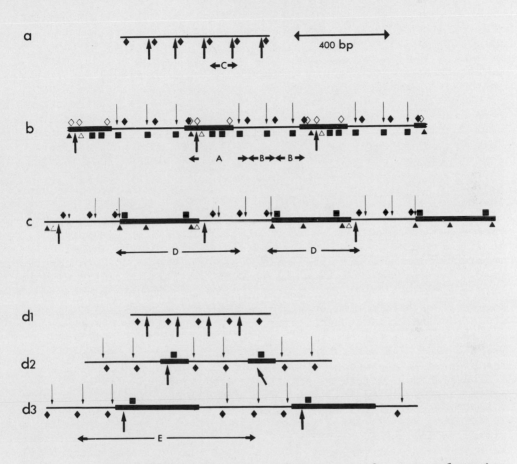

Figure 3.   Physical maps for repeated sequences from rye telomeric
heterochromatin.   ◆ Restriction endonuclease HaeIII sites, ◇
MboI, ▲ MboII, △ TaqI, ■ AluI.  Vertical arrows below the
repeats represent the length of the repeating unit.  Light arrows
above repeats represent lengths of the subrepeating units.  Double
thickness lines represent sequences that interrupt subrepeating
units.  The sections of the repeating units labelled A-E are the
cloned portions of the repeats.

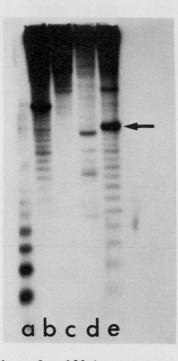

abcde

Figure 4.  Hybridization of a 120 base-pair repeat to rye DNA
digested with various restriction endonucleases.  Rye DNA digested
with (a) HaeIII, (b) BamHI, (c) BglII, (d) HindIII, (e) EcoRI was
fractionated by electrophoresis on 1% agarose gels.  DNA was
transferred to a nitrocellulose filter (31), hybridized by cRNA from
a cloned example of the repeat illustrated in Figure 3a.
Hybridization was in 2 x SSC at 65°C and was revealed by
autoradiography of the washed filter.

that have the same size interval as the progression of fragments in
HaeIII digests.  We assume that these bands arise as a consequence of
recognition target sequences created by random mutation.  In addition
to hybridizing to the series of 120 base-pair interval bands, both
BamHI and EcoRI digests also contain marked hybridization to bands of
much higher stoichiometry.  Further, BglII and HindIII (Figure 4c,4d)
digests that do not produce an integral series of hybridized bands,
do produce major bands of hybridization.
        In order to clone one of these complex repeats, we have taken
advantage of the fact that one of the highest stoichiometry bands
(arrow, Figure 4) in the EcoRI digests was 2.2 kilobases and was
found to comigrate with a band of strong fluorescence in EcoRI
digests of total rye DNA.  DNA of this size was prepared from an
EcoRI digest of total rye DNA and ligated to the plasmid vector

pAC184. Clones containing sequences complementary to the 120 base-pair repeat were selected. Among these were clones of the 2.2 kilo-base repeat. Structural and hybridization analysis showed that this 2.2 kilobase repeat was not a simple tandem array of 120 base-pair units (9). Rather, this repeat contained a dimer of the 120 base-pair repeat interspersed with unrepeated sequences. We have called the portion of the 2.2 kilobase DNA fragment that does not hybridize the 120 base-pair probe "spacer DNA". Probes prepared from pure spa-cer DNA were found to hybridize many repeated and low copy number se-quences. Several repeats, other than the cloned 2.2 kilobase repeat hybridized by the 120 base-pair fragment, are also hybridized by spacer DNA. To explain the results obtained with this clone together with information obtained with interspecies DNA hybridizations (5), we have suggested (9) that cereals contain ancestral repeat elements that have been combined in many permutations of linear order. Throughout evolution, different permutations have been amplified and the 2.2 kilobase repeat represents one such amplified permutation.

## CLONING REPEATING UNITS OF GENES FOR CEREAL RIBOSOMAL RNAs

We have cloned the repeating units for the 18S + 25S ribosomal RNA genes from wheat and barley (10) and the 5S RNA gene repeat units from wheat (W.L. Gerlach and T.A. Dyer, manuscript in preparation). Sequences for each of these gene repeat units were enriched from total DNA with CsCl buoyant density centrifugation in the presence of actinomycin D. The higher average GC content of the ribosomal RNA and 5S RNA gene repeats in plants causes them to bind proportionally more actinomycin D than the bulk of the DNA and they can then be separated in the CsCl density gradient on the basis of this increased binding (24,25). With this procedure, we obtained between 30- and 100-fold enrichment for wheat and barley rDNA and the wheat 5S RNA genes in a number of experiments; hence, the genes that we wished to clone became a workable proportion of the DNA fractions chosen for the experiments.

Experiments on the cloning efficiency of total wheat DNA after centrifugation in actinomycin D/CsCl showed that actinomycin D did not affect the cloning ability of plant DNA, when compared with wheat DNA isolated by our standard procedures (10).

## 18S + 25S rRNA Genes

When native wheat DNA is digested with EcoRI, it shows one size class of ribosomal gene repeating unit 9 kilobases in length, due to the presence of one EcoRI target located within the structural gene for the 25S rRNA (26). Wheat embryo DNA which had been enriched for the ribosomal genes was used in this cloning experiment. Five out of 274 inserts of this DNA into the EcoRI site in plasmid pAC184 hybrid-ized radioactive 18S + 25S rRNA when tested by the colony hybridiza-

tion assay (21).  Isolation and further testing of these inserts
revealed that all five contained full length ribosomal gene repeating
units of 9 kilobases in length.  Analysis of the clones by combined
EcoRI and BamHI restriction endonuclease digestion showed that two
different size classes of repeat differing by about 150 base pairs
had been cloned and a reexamination of the native wheat rDNA used in
the cloning experiment showed that these two size classes were,
indeed, representative of naturally occurring size classes.

The barley shoot DNA, enriched for ribosomal genes, was also
digested with EcoRI and cloned in plasmid pAC184.  Thirteen out of
366 recombinant clones contained sequences complementary to 18S + 25S
rRNA.  When digested with EcoRI, barley DNA yields two size classes
of ribosomal gene repeat units, 9.0 and 9.9 kilobases, and 11 of the
13 clones contained full length inserts of one or the other of these
size classes.  Two of the 13 did not contain full length repeat
units; digestion with a range of endonucleases suggested that one had
a deletion within the rDNA repeat unit while the other had deleted
part of both the rDNA and vector DNA.  Overall, these cloning
experiments showed that the cereal ribosomal gene repeating units
could readily be cloned as stable, full length repeating unit inserts
in plasmid pAC184.

Further experiments on characterizing the distribution of
various restriction enzyme sites in the cloned repeating units and
comparing the digestion patterns with those obtained from native rDNA
revealed a number of features of the organization of the repeating
units.

Tetranucleotide restriction enzymes, which have targets containing the dinucleotide ..CpG.. and which are modified by methylation of the cytosine (e.g., HpaII ..CĊGG.. , HhaI ..GĊGC..), were
found to digest the cloned ribosomal repeats extensively but not the
repeating units in native DNA.  On the other hand, the restriction
enzyme HaeIII (recognition sequence ..GGCC..), which is also modified
by cytosine methylation, digests both the cloned and native repeating
units.  Since the cloned DNA in E. coli does not contain methylated
cytosine, these results were interpreted to mean that the sequence
..CpG.. almost invariably has a methylated cytosine in the natural
ribosomal repeat DNA.  This is similar to somatic Xenopus rDNA where
about 99% of ..CpG.. dinucleotides contain methylated cytosine (27).
The presence of methyl-cytosine in an environment other than ..CpG..
can be inferred from BamHI (GGATĊC) digestion of barley DNA.  Three
of the four BamHI sites in natural barley rDNA are partially resistant to digestion.  Quantitative analysis of digestion products
suggests that the degree of methylation, assumed to be the cause of
the protection from digestion, may be up to 40% at these specific
sites.

Mapping of hexanucleotide restriction enzyme sites (SalGI,
BamHI, BglII) relative to the EcoRI site in the 25S gene indicates
that the length variation within and between wheat and barley rDNA
repeating units can be accounted for by variation in the spacer
region.  The similar distribution of restriction endonuclease sites

within the structural genes suggests that their sequences have been
very highly conserved.

A cRNA made from a cloned repeating unit of wheat rDNA showed
the same hybridization pattern as radioactive 18S + 25S rRNA when
assayed by two procedures: (a) in situ hybridization to wheat meta-
phase chromosomes and (b) hybridization to EcoRI and BamHI digests of
total wheat DNA after transfer to nitrocellulose filters.  This
suggests that sequences present in the ribosomal spacer are not
present in high copy number in other genomic locations.


                              5S RNA Genes


The genes for cytoplasmic ribosomal 5S RNA in wheat are arranged
in two size classes of repeating units of length about 500 base pairs
(26).  These units differ in length by about 80 base pairs and each
contains one BamHI site.  Inspection of the sequences reported for
cereal 5S RNAs (28,29) suggests that the BamHI site is within the
structural gene, 30 base pairs from the 5' terminus.

The DNA fraction enriched for 5S RNA genes by actinomycin D-CsCl
centrifugation was preferentially enriched for the shorter of the two
repeat unit size classes.  This DNA was used in cloning experiments
in which sequences complementary to 5S RNA were recovered as BamHI
fragment inserts in the tetracycline resistance gene of plasmid
pBR322.  Five unit length class inserts were obtained; four contained
the shorter repeating unit while one contained the longer repeating
unit.  Although this number is small, it reflected the preferential
enrichment of the actinomycin D-CsCl step for the shorter repeat
unit.  Two other inserts that contain sequences complementary to 5S
RNA did not have lengths corresponding to unit sizes or multimers of
the 5S gene repeats.  One of these contains a sequence that only
hybridizes 5S RNA under conditions of relaxed stringency and may
therefore contain a genomic sequence related to but not the same as
the 5S RNA sequence.  The other insert hybridizes 5S RNA under stringent conditions and as yet its structure has not been investigated
any further.

The cloned repeating units have been sequenced with a rapid DNA
sequencing technique (30).  The complete sequence of one of the
shorter repeats has been determined; it contains 413 base pairs and
the cloning site was, indeed, the BamHI site within two adjacent
structural genes.  Sequencing of two of the other shorter repeat
units cloned shows a high degree of sequence similarity (only about
2% divergence) between different repeat units.

The longer repeat class has almost completely been sequenced;
only a region within the spacer containing approximately 40 of the
510 base pairs remains to be characterized.  A comparison of the sequences of the longer and shorter repeat units (outlined in Figure 5)
reveals some interesting structural properties.  First, the structural gene sequences show some sequence heterogeneity (about 4%).
However, there is a marked difference between the two genes: the

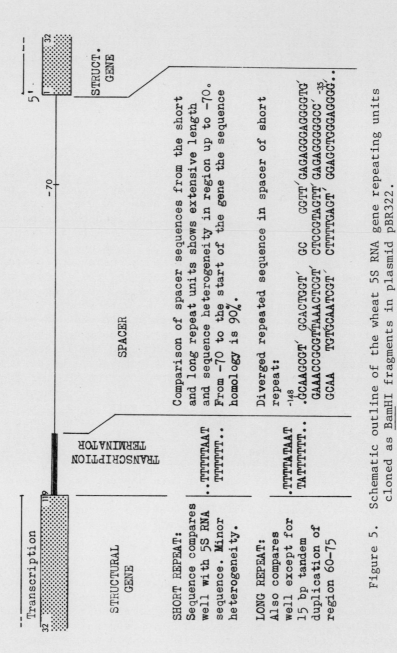

Figure 5.   Schematic outline of the wheat 5S RNA gene repeating units
cloned as BamHI fragments in plasmid pBR322.

structural gene cloned from the longer repeat class contains a tandem
duplication of the 15 base pairs between positions 60 and 75 within
the gene.  Clones of the longer repeat size class are presently being
isolated to determine whether all genes in this repeat size class
contain the duplication.  Second, the units cloned all have tracts of
14 (shorter repeat) or 17 (longer repeat) AT base pairs immediately
following the structural genes.  These are the transcription terminator sequences.  Third, the spacer sequences have a number of
features that will not be described here in detail.  Briefly, however, a comparison of the spacers from the two repeat size classes
shows extensive length and sequence divergency from the terminator of
one structural gene to a position 70 base pairs from the transcriptional start point of the neighboring gene.  From this point to the
start of the structural gene, the two size classes of repeat have
maintained fairly similar nucleotide sequences, suggesting a
functional importance of this sequence preceding the gene.  It is
also interesting to view a part of the shorter repeat unit that has a
diverged repeated sequence of structure ..aaabaaabaaab.. where "a"
and "b" represent different oligonucleotide sequences.  This
structure of a repeated sequence with units consisting of three
subrepeats followed by an unrelated sequence is similar on a small
scale to that deduced for some of the more complex rye repeated
sequences described above.

## CONCLUSIONS

We have cloned several examples of highly repeated sequences and
repeated genes from cereals.  Detailed structural analysis of these
cloned sequences has enabled us to investigate the types of and basis
for structural heterogeneity in repeated sequences.  The importance
and potential of molecular cloning experiments in describing the
repeated sequence DNA component of plant chromosomes is illustrated
by these few examples.

Acknowledgments:  We wish to thank R.B. Flavell, T.A. Dyer, M.
O'Dell, J.Jones and R.D. Thompson for their various contributions to
the experiments reported.  J.R. Bedbrook was the recipient of an EMBO
Long Term Fellowship.  Part of the work was carried out while
W.L. Gerlach was the recipient of a CSIRO Postdoctoral Studentship.
Recombinant DNA experiments were carried out under Category II
containment conditions as defined by GMAG.

## REFERENCES

1   Flavell, R.B., Bennett, M.D., Smith, J.B. and Smith, D.B. (1974)
    Biochem. Genet. 12, 257-269.
2   Huguet, T. and Jouanin, L. (1972) Biochem. Biophys. Res. Commun.
    46, 1169-1174.

3   Ranjekar, P.K., Palotta, D. and Lafontaine, J.G. (1976) Biochim.
    Biophys. Acta 425, 30-40.
4   Gerlach, W.L., Appels, R., Dennis, E.S. and Peacock, W.J. (1978)
    Proc. 5th Int. Wheat Genetics Symp. 1, 81-91.
5   Rimpau, J., Smith, D.B. and Flavell, R.B. (1978) J. Mol. Biol.
    123, 327-359.
6   Appels, R., Driscoll, C. and Peacock, W.J. (1978) Chromosoma 70,
    67-89.
7   Bedbrook, J.R., Gerlach, W.L., Thompson, R.D., Jones, J. and
    Flavell, R.B. (1980) in Emergent Techniques for Crop Improve-
    ment (Rubenstein, I.R., ed.), University of Minnesota Press
    (in press).
8   Bedbrook, J.R., Jones, J., O'Dell, M., Thompson, R.D. and
    Flavell, R.B. (1979) Cell (in press).
9   Bedbrook, J.R., O'Dell, M. and Flavell, R.B. (1979) (submitted
    for publication).
10  Gerlach, W.L. and Bedbrook, J.R. (1979) Nucl. Acids Res. (in
    press).
11  Brutlag, D., Fry, K., Nelson, T. and Hung, P. (1977) Cell 10,
    509-510.
12  Bennett, M.D. and Smith, J.B. (1976) Phil. Trans. Royal Soc.
    (London) B. Biological Sciences 274, 227-274.
13  Hamer, D.H. and Thomas, C.A. (1975) Chromosoma 49, 243-267.
14  Thomas, H.A. and Sherratt, S.H.A. (1956) Biochem. J. 62, 1-4.
15  Spencer, J.H. and Chargaff, E. (1963) Biochim. Biophys. Acta 68,
    18-27.
16  Mann, M.B. and Smith, H.O. (1977) Nucl. Acids Res. 4, 4211-4221.
17  Chang, A.C.Y. and Cohen, S.N. (1978) J. Bacteriol. 134,
    1141-1156.
18  Stern, H. (1968) Methods Enzymol. 12B, 100-112.
19  Bolivar, F., Rodriguez, R.L., Green, P.J., Betlach, M.C.,
    Heyneker, H.L., Boyer, H.W., Crosa, J.H. and Falkow, S. (1977)
    Gene 2, 95-113.
20  Dennis, E.S., Gerlach, W.L. and Peacock, W.J. (1979) Heredity
    (in press).
21  Grunstein, M. and Hogness, D.S. (1975) Proc. Nat. Acad. Sci.
    U.S.A. 72, 3961-3965.
22  Maniatis, T., Sim, G.K., Efstratiadis, A. and Kafatos, F.C.
    (1976) Cell 8, 163-182.
23  Bennett, M.D., Gustafson, J.P. and Smith, J.B. (1977) Chromosoma
    61, 149-176.
24  Hemleben, V., Grierson, D. and Dertmann, H. (1977) Plant Sci.
    Lett. 9, 129-135.
25  Hemleben, V. and Grierson, D. (1978) Chromosoma 65, 353-358.
26  Appels, R., Gerlach, W.L., Dennis, E., Swift, H. and Peacock,
    W.J. (1979) Chromosoma (in press).
27  Bird, A.P. and Southern, E.M. (1978) J. Mol. Biol. 118, 27-47.
28  Barber, C. and Nichols, J.L. (1978) Can. J. Biochem. 56,
    357-364.
29  Payne,P.I. and Dyer, T.A. (1976) Eur. J. Biochem. 71, 33-38.

30    Maxam, A.M. and Gilbert, W. (1977) Proc. Nat. Acad. Sci. U.S.A.
      74, 560-564.
31    Southern, E.M. (1975) J. Mol. Biol. 98, 503-517.

Appendix: DNA Preparation.  The technique which we routinely
use to isolate high molecular weight DNA from cereal embryos is as
follows.  Embryos excised from dry seeds are homogenized in HB buffer
(0.3 M sucrose, 50 mM Tris-HCl pH 8.0, 5 mM $MgCl_2$) and filtered
through three layers of Miracloth.  Triton X-100 is added to 1% and
the eluate centrifuged (5', 500 g).  The pellet is resuspended in 40%
w/w metrizamide (Nyegaard and Co., Oslo) in HB, layered onto a 40% to
50% metrizamide step gradient, centrifuged (20', 32,000 g) and nuclei
are isolated from the 40% to 50% interface.  Nuclei are diluted with
HB, pelleted by centrifugation (5', 2000 g), resuspended in RB buffer
(50 mM Tris-HCl pH 8.0, 20 mM EDTA) and lysed by treatment with
one-eighth volume of 10% Sarkosyl.  The lysate is incubated with
one-eighth volume Calbiochem B grade pronase (5 mg/ml, previously
autodigested 2 hr, 37°C) at 60°C for 5 min and then 37°C for at least
4 hr.  DNA is purified by centrifugation in CsCl ($\rho$ = 1.56 $gm/cm^3$,
40,000 rpm, MSE 8 x 14 ml rotor, 2 days) containing about 1 mg/ml
ethidium bromide.  Ethidium bromide is extracted with isoamyl alcohol
saturated with 5 mM Tris-HCl pH 8.0, 0.25 mM EDTA, and the DNA
dialyzed into 5 mM Tris-HCl pH 8.0, 0.25 mM EDTA for storage.

# THE USE OF RECOMBINANT DNA METHODOLOGY IN APPROACHES TO CROP IMPROVEMENT:  THE CASE OF ZEIN

Benjamin Burr

Biology Department, Brookhaven National Laboratory

Upton, New York  11973

## INTRODUCTION

This chapter is intended to provide a prospective as to what can reasonably be expected when attempts are made to apply recombinant DNA technology to a plant breeding problem.  Most of my remarks will be confined to the problems of protein quality in maize and will bear only indirectly on other problems of crop improvement that may be amenable to current molecular techniques.  It is evident that in this area, recombinant DNA techniques are only useful when employed in conjunction with standard breeding practices and it is the latter that will receive the most attention here.  In this light, I would like to provide an example of how recombinant technology was used as a feasibility test to see if a crop improvement problem was approachable through currently available methods.

## MAIZE AS AN ORGANISM FOR BASIC RESEARCH WITH AN EYE TOWARD APPLICATION

Experiments with crop plants can lead to direct application of findings from basic research.  Thus, the discovery that hybrids of inbred lines expressed heterosis (1) led to the widespread use of hybrid corn.  Geneticists experimenting with cytoplasmic male sterility and nuclear genes that restored fertility also recognized that male sterile plants could replace the process of mechanically detasseling the female parent in the production of hybrid seed (2). In still another example, the shrunken-2 endosperm marker was found to accumulate high levels of sucrose (2a) and has been incorporated

21

into commercial lines for the production of "super sweet" corn.   It
has since been found that shrunken-2 is a mutation in a structural
gene for ADP glucose pyrophosphorylase (2b) and constitutes a nearly
complete block in the conversion of sugar to starch in the endosperm.
On the other hand, methods used for hybrid corn production, high
yield requirements and standards set by industrial processes for
grain quality provide constraints that must be met before laboratory
advances can be utilized in the field.   The list of seemingly
ingenious techniques or discoveries that have had little practical
application is a long one.   Consideration of two examples might be
illuminating.

     1)   Maize geneticists have learned how to select for spon-
taneously produced haploid plants (3).   These plants are sterile but
their chromosome number can be doubled to produce a perfectly
homozygous and fertile plant.   This shortcut to inbreeding would seem
to be ideally suited for the production of parental lines used in
making hybrid seed, but this is not the case.   Plant breeders employ
several generations of self-pollination to obtain their inbred lines.
During this time, they also select parental lines with desirable
qualities of disease resistance, high yield, ability to combine as
parents with other inbreds to produce superior offspring, good plant
form and strength, ability to grow in a dense planting, response to
fertilizers, pesticides and herbicides and adaptation to local
environmental conditions and stresses.   Thus, on the way toward
homozygosity, the plant breeder selects plants that will be good
parents.

     2)   Opaque-2 is a single recessive gene that changes the balance
of proteins in corn endosperm, doubling the quantities of available
lysine and tryptophan (4), the two most limiting essential amino
acids in the kernel.   This allows corn to be fed as a complete
protein diet to swine and humans (5,6).   The mutation, whose
seemingly beneficial properties have been known since 1963, has not
found general use despite intensive efforts to incorporate it into
adapted inbred lines (7).   An undesirable side effect of the mutation
is a softer kernel, which increases susceptibil y to fungal and
insect attack, resulting in a reduction in storage qualities and
poorer milling characteristics.   There is also a 10% decrease in
yield.   These factors have combined to limit widespread use of the
mutation in the United States, where it could serve as animal feed,
and in the more humid South and Central American regions, where
storage is a problem despite the obvious advantages of using the
mutation for human nutrition.   Efforts are underway to find modifiers
that will limit the deleterious side effect of the mutation (8).

     Nevertheless, maize is an attractive organism for the molecular
biologist interested in basic research.   Maize is among the
genetically best known higher plants with some 175 loci mapped on ten
chromosomes (9).   The haploid genome size is 4.5 x $10^9$ base pairs
(10), comparable to that of most mammals.   Male and female flowers
are formed separately so controlled pollinations can easily and
rapidly be made providing hundreds of synchronously developing

kernels from a single operation.  Unlike all other cereals, maize
seeds are not enclosed by bracts, which greatly facilitates screening
for kernel characters.  Since the kernel is large and composed of
specialized tissues, mutants affecting defined biochemical pathways
can be morphologically detected.  Among the best known are the
mutations affecting anthocyanin production in the aleurone, the
outermost layer of the endosperm, and other mutations conditioning
starch biosynthesis in the rest of the endosperm.  These mutations
are mostly of a conditional or dispensable nature.  As in most higher
organisms, different genes perform the same function in different
tissues so that mutations affecting starch synthesis in the endosperm, for instance, exert no effect on the vegetative parts of the
plant.  Apart from electrophoretic variants of isozymes, a dozen
enzyme lesions are presently associated with mutant phenotypes.  Two
loci, waxy, affecting starch granule-bound ADP glucosyl transferase,
and alcohol dehydrogenase-1 are expressed in haploid pollen grains
permitting the detection of rare recombinants between heteroalleles
that can be used in intracistronic mapping (11,12).


                              THE ZEIN SYSTEM

     Our work with the major storage protein of the maize endosperm
began as a result of our interest in the specialized synthesis of
this family of proteins.  The protein is very hydrophobic--normally
being extracted from corn meal with 70% aqueous ethanol--and makes up
about 50% of all the seed protein of corn.  It is contained apparently in noncrystalline form in single membrane-bound organelles
called protein bodies (13).  We found that these organelles had polysomes on their outer surfaces (14).  When polysomes eluted from
isolated protein bodies are placed in an amino acid incorporating
system, only zein polypeptides are made.  Isolation of the protein
bodies thus constitutes an important first step in the purification
of zein mRNAs (15).  When the mRNA is purified away from ribosomal
RNA, a single diffuse band can be seen on denaturing gels.  Zein
mRNAs are estimated to be 1.1 to 1.2 kilobases in length with an
average poly(A) terminus of 110 nucleotides.  The product of
translation of the mRNA in the wheat germ system is 1 to 2000 daltons
larger than the mature protein isolated from the corn seed.  The
additional sequences present in the unprocessed translation products
are located at the amino termini, as are most proteins made on
membrane-bound polysomes, and secreted through a membrane.
Consistent with this hypothesis, zein mRNA translated in Xenopus
oocytes is processed to the same size as native zein and has the same
amino terminal sequences (L. Silver, M. Elzinga, F.A. Burr and B.
Burr, unpublished data).
     When zein polypeptides are separated by charge, a great deal of
heterogeneity is observed (16).  Since zein is sparingly soluble in
8M urea and is heavily amidated (17), some of this heterogeneity
could be explained by aggregation and partial deamidation.  Con-

versely, when zein is separated on the basis of molecular length on
SDS polyacrylamide gels, only two bands are observed:  a heavy chain
of 22,500 daltons and a light chain of 19,000 daltons (14).  Further
evidence of limited heterogeneity is the N-terminal amino acid sequences obtained for the first 36 residues of the light chain and
for the first 23 residues of the heavy chain (L. Silver, M. Elzinga,
F.A. Burr and B. Burr, unpublished data).  Furthermore, cyanogen
bromide cleavage of the heavy chain also produces a discrete
pattern (15).

    If zein is the product of only two genes, its genetic manipu-
lation should be possible and could have important consequences,
since zein constitutes half of the protein in the kernel.  Since
there is less than 0.1 mole of lysine or tryptophan per mole zein
polypeptide (17), it would be desirable to change genetically the
amino acid composition of zein.  As previously mentioned, these two
amino acids essential to the nonruminant animal diet are of limited
quantity in the corn kernel.  If two residues of lysine and tryptophan could be added to each polypeptide without affecting their rate
of accumulation, the amount of these amino acids in the kernel would
double.  Should there be only two zein structural genes, this manipulation could be attempted in a fairly direct manner; larger numbers
of genes would compound the problem.

    It is not certain what effect such substitutions would have on
the conformation of the molecule; any gross change in structure might
be expected to influence packaging and, consequently, the rate of
accumulation.  The most drastic change, of course, would come from
the addition of positively charged lysine chains.  However, there may
be sufficient polar groups in the protein to neutralize these
additional charges.

    In principle, there are conventional methods for making amino
acid substitutions in zein.  Zein imparts the hard, glassy texture
to the outer portion of the corn endosperm.  Mutations that limit
zein accumulation cause the outer portion to have a chalky, opaque
texture.  There are four nonallelic mutations that decrease zein in
the kernel but these do not appear to be structural gene mutations
(7).  Several other mutations also condition this phenotype but do
not affect zein accumulation.  Should there be only a few zein
structural genes, some mutations at these loci (e.g., frameshift
or nonsense mutations) would be expected also to express the opaque
phenotype.  Opaque kernels are easily distinguished from wild-type on
a segregating ear and rare revertants in a large population could be
recovered and examined for new amino acid substitutions at the mutational sites.  Since multiple substitutions are required for each
cistron, a number of these would have to be recombined.  Such recombinants would have to be determined by examining kernels that had
recombined close outside markers.  The number and chromosomal
location of the zein structural genes must be known to utilize the
approach outlined above.

    There is limited knowledge about the genetics of zein and
homologous storage proteins in other cereals.  Righetti et al. (18)

separated zein on one-dimensional isoelectric focusing gels.  As
mentioned previously, there are difficulties with interpreting these
patterns.  Nevertheless, consistent differences in banding patterns
were reported between inbred strains that are codominant in hybrids.
Of 25 or so bands observed by one-dimensional isoelectric focusing, a
group of three and two other individual bands segregated independent-
ly in an F2 generation (19).  Electrophoretic and molecular heterogeneity is also a characteristic of the alcohol-soluble storage proteins of other cereal grains.  In barley, the genes for the two major
groups of hordein are at two loci separated by 10 to 16 map units on
chromosome 5 (20,21).

Mutants are easily obtained in maize by treating the pollen with
mutagen while it is suspended in mineral oil (22).  The pollen is
subsequently used to pollinate a normal ear and the kernels produced
are planted and self-pollinated.  Recessive mutations affecting
kernel characters show up as one-fourth mutant kernels segregating on
self-pollinated ears.  We examined a large population of M2 ears
derived from an ethyl methanesulfonate treatment conducted by
Dr. M.G. Neuffer at the University of Missouri.  Ears segregating
opaque kernels were selected and zein prepared from mutant kernels
and examined on SDS polyacrylamide gels.  Of approximately 300
different samples, three mutants produced a slightly altered pattern
in which a portion of the band (either upper or lower) appeared to be
missing.  One of these is associated with embryo lethality but the
other two are being mapped.  Should further tests with two-
dimensional gels indicate that these are indeed structural mutations
associated with an opaque phenotype, the mutant can then be used to
select for revertants.

## THE USE OF ZEIN cDNA CLONES

The ability to purify the zein mRNAs to apparent homogeneity may
provide a faster way to localize and quantitate the zein structural
genes.  Double-stranded complementary DNA (cDNA) was prepared from
purified zein mRNA and inserted by the A-T tailing method into
bacterial plasmid pMB9 (B. Burr, F.A. Burr, T. St. John, M. Thomas
and R.W. Davis, unpublished data).  Twenty clones were chosen to
represent a range of insert sizes after examination of rapid lysates.
These were characterized for their ability to bind zein mRNA, the
size of the insert and the presence of internal restriction
endonuclease sites.  Eighteen of the twenty clones bound translatable
zein mRNA as assayed by the procedure of Noyes and Stark (23).  The
eighteen positive clones were categorized either as those binding
heavy chain mRNA or those binding light chain RNA.  Inserts including
the A-T tails ranged from 60 to 1195 nucleotides.  The longer ones
were shown to be faithful copies because their hybrids with mRNA were
protected from S1 nuclease digestion when analyzed by the method of
Berk and Sharp (24).  From analyses with eight restriction endonucleases, the cDNA inserts could be grouped, by the presence of

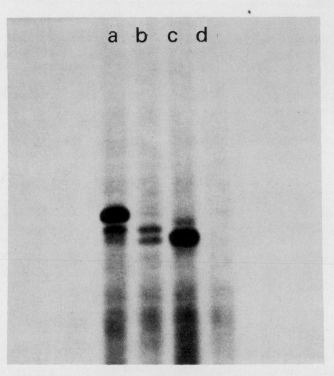

Figure 1.   Recognition of zein cDNA clones by hybridization-
translation.   Zein cDNA clones, in a separate experiment from that
described in the text, were prepared from total endosperm RNA fractionated by size on DMSO-sucrose gradients.   The double-stranded cDNA
was blunt-end ligated with synthetic EcoRI linkers and inserted into
the chloramphenicol resistance gene on plasmid pBR325 (31).   Denatured plasmid DNA was bound to diazotized aminobenxyloxymethyl paper
and hybridized with unfractionated endosperm RNA in 50% formamide at
37° for 24 hrs (32).   Specifically-hybridized RNA was eluted and
translated in a wheat germ cell-free system in the presence of
$^{35}$S-methionine.   (a) Zein heavy chain mRNA clone; (b) and (c) zein
light chain mRNA clones; (d) a clone that hybridizes no translatable
RNA under these conditions.


restriction sites, into three heavy chain types and four to five
light chain types.   This is a minimum estimate of the diversity since
only a few enzymes were used.   It must be conceded, however, that the
apparent nucleic acid heterogeneity among light or heavy chain clones
should be confirmed by hybridizing labeled, cloned DNA back to
restriction-cut maize DNA to show that clones, differentiated by
restriction endonuclease sites, bind to different genomic sequences.
Once the level of cross reaction between genomic sequences is known
from blotting experiments, the clones will be used as probes to

attempt to quantitate the number of structural genes.  The probes can
also be used to localize these sequences of chromosomes either by
quantitative hybridization to lines trisomic or monosomic for each of
the ten maize chromosomes or possibly by in situ hybridization.

The lesson from cDNA cloning experiments is that there are a
minimum of seven mRNA sequences.  This would indicate that there are
at least an equal number of structural genes, which makes the breeding plan previously outlined a very difficult, if not an impossible,
undertaking.  Finding an average four substitutions for each locus
and recombining them would be a very large task.  Furthermore, if
they are unlinked or only loosely linked, transferring them to local-
ly adapted inbreds would present an additional task that commercial
breeders are not prepared or equipped to handle.  There is an outside
chance, however, that a survey of existing corn strains will reveal
some strains that express relatively few zein genes so that fewer
genes could be manipulated with greater effect.

One way to alter the eventual translation product would be to
make changes at the nucleic acid level in cloned zein DNA.  This
certainly can be done (25).  However, the means for inserting the
altered DNA into the genome and selecting for cells expressing the
new sequence are not easy.  Probably the best means of selecting a
recombinant involves the unselected introduction of deletions at a
specific locus by recombination of homologous sequences linked to a
selectable marker on a recombinant plasmid that will not replicate
autonomously in the host cell (26).  This is an important concept
because zein is a developmentally regulated gene product.  It is
synthesized in nondividing endosperm cells and thus is an unlikely
target for direct selection.  However, a major problem remains: while
maize plants can be differentiated from organized tissue in culture
(27) and progress has been made in regenerating plants by continued
selection from cultures derived from a variety of tissues (28), there
is only one report of proliferation from single cells (29).  The
techniques for plating maize cells on a selective medium and obtaining regenerated plants from proliferating colonies are being elabo-
rated but are not yet available.

The zein system presents one of the most accessible approaches
to the study of gene organization in higher plants.  With the number
of apparently full length cDNA clones we have obtained, it will be
interesting to see how some of these closely related sequences have
diverged and to learn how they compare with other eukaryotic messages.  For example, zein mRNAs are very competitive relative to
other endosperm mRNAs in in vitro translation.  It is anticipated
that a sequence with strong complementarity to ribosomal RNA will be
observed in the 5' noncoding region (30).  Among the zein sequences
we expect to find conserved are the signal sequence and a possible
recognition sequence in a noncoding region responsible for directing
the message to the specific membrane site where it is translated.
Finally, the cDNA clones can be used as probes to isolate the zein
coding sequences from the maize chromosomal DNA.

## CONCLUSIONS

Earlier data from our protein studies had suggested that zein was composed of a few polypeptides, perhaps as few as two.  However, analyses of the cDNA clones obtained recently indicate that there is some sequence heterogeneity at the nucleotide level.  Taken together, these results suggest that zein is actually a large family of structurally related proteins and corroborate the heterogeneity observed by isoelectric focusing.  The evidence so far implicates a minimum of ten to twenty genes.  This result means that it will be very difficult to accumulate sufficient mutational changes in the zein genes significantly to affect overall amino acid composition of the grain. Furthermore, lack of genetic linkage may complicate the transfer of altered genotypes to adapted inbred lines.  Although this is a discouraging result, it represents a practical use of recombinant technology in attempting to resolve an agronomic problem.

Acknowledgments:  I would like to acknowledge my colleague in this research, Dr. F.A. Burr, and the technical collaboration of G. Ballantyne and M. Neuberger.  The research upon which this review is based was performed at Brookhaven National Laboratory under the auspices of the U.S. Department of Energy and was supported in part by grant GM-24057 from the National Institutes of Health.

## REFERENCES

1    Jones, D.R. (1918) Bot. Gaz 65, 324-333.
2    Duvick, D.N. (1965) Advan. Genet. 13, 1-56.
2a   Laughnan, J.R. (1953) Genetics 38, 485-499.
2b   Hannah, L.C. and Nelson, O.E. (1976) Biochem. Genetics 14, 547-560.
3    Chase, S.S. (1969) Bot. Rev. 35, 117-167.
4    Nelson, O.E. (1966) Fed. Proc. 25, 1676-1678.
5    Maner, J.H. (1975) in High Quality Protein Maize (Bauman, L.F., ed.), pp. 58-82, Dowden, Hutchinson and Ross, Inc., Stroudsberg, PA.
6    Pradilla, A.G., Harpstead, D.D., Sarria, D., Linares, F.A. and Francis, C.A. (1975) in High Quality Protein Maize (Bauman, L.F., ed.), pp. 27-37, Dowden, Hutchinson and Ross, Inc., Stroudsberg, PA.
7    Nelson, O.E. (1979) in International Symposium on Seed Protein Improvement in Cereals and Grain Legumes, International Atomic Energy Agency, Vienna (in press).
8    Vasel, S.K. (1979) in International Symposium on Seed Protein Improvement in Cereals and Grain Legumes, International Atomic Energy Agency, Vienna (in press).
9    Maize Genetics Cooperative Newsletter, Vol. 51, 124-125, 1977.
10   Phillips, R.L.,. Weber, D.F., Kleese, R.A. and Wang, S.S. (1974) Genetics 77, 285-297.

11   Nelson, O.E. (1958) Science 130, 794-795.
12   Freeling, M. (1976) Genetics 83, 701-717.
13   Duvick, D.N. (1961) Cereal Chem. 38, 374-385.
14   Burr, B. and Burr, F.A. (1976) Proc. Nat. Acad. Sci. U.S.A. 73, 515-519.
15   Burr, B., Burr, F.A., Rubenstein, I. and Simon, M.N. (1978) Proc. Nat. Acad. Sci. U.S.A. 75, 696-700.
16   Mosse, J. (1966) Fed. Proc. 25, 1663-1669.
17   Wall, J.S. (1964) in Proteins and Their Reactions, Symposium on Foods (Schultz, H.W. and Anglemeir, A.F., eds.), pp. 315-341, Avi Publishing Co., Westport, CT.
18   Righetti, P.G., Gianazza, E., Viotti, A. and Soave, C. (1977) Planta 136, 115-123.
19   Soave, C., Suman, N., Viotti, A. and Salamini, F. (1978) Theor. Appl. Genet. 52, 263-267.
20   Oram, R.N., Doll, H. and Koie, B. (1975) Hereditas 80, 53-58.
21   Miflin, B.J., Matthews, J.A., Burgess, S.R. and Shewry, P.R. (1979) in Genome Organization and Expression in Plants (Leaver, C., ed.), Plenum Press, New York, NY (in press).
22   Neuffer, M.G. and Coe, E.H. (1978) Maydica 23, 21-28.
23   Noyes, B.E. and Stark, G.R. (1975) Cell 5, 301-310.
24   Berk, A.J. and Sharp, P.A. (1977) Cell 12, 721-732.
25   Weissmann, C., Nagata, S., Taniguchi, T., Weber, H. and Meyer, F. (1979) in Genetic Engineering (Setlow, J.K. and Hollaender, A., eds.), Vol. 1, pp. 133-150, Plenum Press, New York, NY.
26   Davis, R.W., Struhl, K. and St. John, T. (1979) in Eukaryotic Gene Regulation (Axel, R. and Maniatas, T., eds.), Academic Press, New York and London (in press).
27   Green, C.E. and Phillips, R.L. (1975) Crop Sci. 15, 417-421.
28   Rice, T.B., Reid, R.K. and Gordon, P.N. (1978) in Propagation of Higher Plants Through Tissue Culture (Hughs, K.W., Henke, R. and Constantine, M., eds.), pp. 262-277, Technical Information Center, U.S. Dept. of Energy, Washington, D.C.
29   Potrykis, I., Harms, C.T., Loiz, H. and Thomas, E. (1977) Mol. Gen. Genet. 156, 347-350.
30   Hagenbüchle, O., Santer, M., Steitz, J.A. and Mans, R. (1978) Cell 13, 551-563.
31   Bolivar, F. (1978) Gene 4, 121-136.
32   Smith, D.F., Searle, P.F. and Williams, J.G. (1979) Nucl. Acids Res. 6, 487-506.

# PRODUCTION OF MONOCLONAL ANTIBODIES

Sau-Ping Kwan, Dale E. Yelton and Matthew D. Scharff

Department of Cell Biology
Albert Einstein College of Medicine
1300 Morris Park Avenue
Bronx, New York 10461

## INTRODUCTION

Antibodies have been used extensively in many areas of biology including molecular genetics and biochemistry. Serology has made it possible, for example, to recognize inactive mutant gene products (CRMs), to identify changes in bacterial surfaces and to confirm the presence of specific proteins among the products of cell free translation systems (1,2). More recently, antibodies have been used to identify rare bacterial clones that have been transformed with cloned genes (3,4). Despite the usefulness of immunologic techniques, bacterial geneticists and biochemists are often uncomfortable using antibodies because generating antisera is still more of an art than a science. Immunization frequently results in large amounts of antibody against macromolecules that are barely detectable in the preparation used to immunize the animal. The quality and quantity of antibody varies from one animal to the next and even from one bleeding to another. Results obtained with one antisera may not be reproducible with another. Even highly specific antisera contain antibodies of widely different affinities with different subspecificities and cross reactivities. Such antisera also contain antibodies of different classes and subclasses that vary in their ability to agglutinate and precipitate the antigen or to fix complement.

While these and other related problems have not prevented the effective use of antibodies, immunologists have long sought to improve the reliability of serological tests by generating homogeneous antibodies. With a few antigens, especially bacterial polysaccharides, such attempts have been successful (5). Nevertheless, it has not been possible routinely to generate homogeneous serological

31

reagents against most antigens. In 1975, however, Köhler and
Milstein revolutionized serology by describing a technique for
routinely obtaining monoclonal antibodies (6,7). They found they
could fuse cultured mouse myeloma cells with spleen cells from
immunized animals and that some of the resulting hybrids synthesized
antibody that reacted with the immunizing antigen. Many subsequent
studies have confirmed the applicability of this approach to a wide
variety of antigens (8-10). Such hybrids grow continuously in culture, continue to produce antibody, can be frozen and recovered from
the freezer, and can be injected into the peritoneal cavity of syngeneic mice where they proliferate and induce an ascites containing
large amounts of specific antibody.

The hybridoma technology has many benefits. Since a single
antibody-forming spleen cell fuses with the mouse myeloma cell to
form the hybrid, the antibody produced by the hybrid is a single
antibody molecule and, by definition, monospecific. Since many
hybrids arising from each spleen are screened (see below) and only a
few selected, it is possible to immunize with impure antigens and
obtain specific antibodies. The best hybrids produce antibody levels
as high as 100 µg/ml of tissue culture fluid and 10 mg/ml of ascites
fluid. The continuity of the cell lines insures a perpetual and
virtually unlimited supply of a homogeneous, well-characterized
chemical reagent that can be readily exchanged with other
investigators.

The technique used by many laboratories for producing monoclonal
antibodies is illustrated in Figure 1. BALB/c mice are immunized
with the antigen of interest and then boosted. Two to four days
later the spleen is removed and a single cell suspension of spleen
cells is mixed with cultured, drug-marked mouse myeloma cells in the
presence of polyethylene glycol. The cells are then distributed into
the wells of microculture plates in medium that will selectively kill
the unfused mouse myeloma cells. The spleen cells do not grow well
in culture so only the hybrids between myeloma and spleen cells
survive. Cells are plated in the microculture dishes at a concentration that will give rise to approximately one hybrid clone in
every three wells. Two to three weeks later, as the clones become
grossly visible, the medium is screened for antibody. Clones producing antibody are transferred to larger dishes and, if they continue to produce antibody, are recloned. The subclones are screened
again for antibody production and a few positive clones are grown to
mass culture so they can be both frozen and injected into mice to
form ascites fluid. Each of these steps in illustrated in Figure 1
and will be described in more detail in the sections that follow.


IMMUNIZATION


Monoclonal antibodies have been generated against a variety of
proteins, glycoproteins, glycolipids, nucleic acids and chemically
defined groups (haptens) linked to protein carriers. Unfortunately,

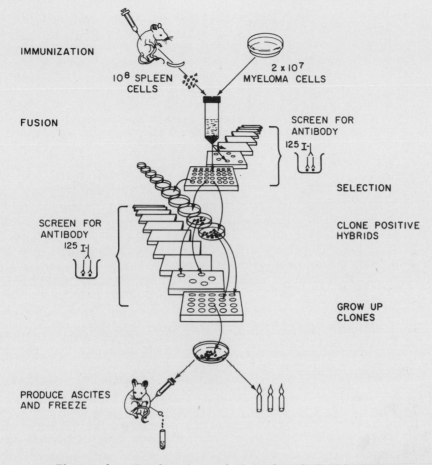

IMMUNIZATION

$10^8$ SPLEEN
CELLS

$2 \times 10^7$
MYELOMA CELLS

FUSION

SCREEN FOR
ANTIBODY
$125$ I-|

SELECTION

SCREEN FOR
ANTIBODY
$125$ I-|

CLONE POSITIVE
HYBRIDS

GROW UP
CLONES

PRODUCE ASCITES
AND FREEZE

Figure 1.   Production of Monoclonal Antibodies

just as with traditional immunization, the likelihood of success is
related to the number of antibody-forming cells in the spleen at the
time of fusion.  This in turn depends on the immunogenicity of the
antigen used.  In general, antigens that induce high titers of serum
antibody will generate many antibody-producing hybrids.  While it is
possible to generate antibody-producing hybridomas from moderately
immunogenic antigens, this is a major undertaking requiring the
screening of very large numbers of hybrids from many mice.  With
current technology, it is generally not possible to produce hybridomas against antigens that induce barely detectable levels of serum
antibody.

    A few simple calculations illustrate the problem.  Each mouse
spleen contains on the average $10^8$ nucleated cells.  The very best
immunogens, such as sheep red blood cells (SRBC), stimulate the rapid
replication and differentiation of antibody-forming cells so that at
the peak of the secondary response each spleen may contain as many as

$10^6$ cells producing antibody against SRBC. Since only one hybrid is usually generated from each $2 \times 10^5$ spleen cells, one would expect to obtain approximately 500 hybrids per spleen and five of these should produce specific antibody. In practice, fusion enriches 4- to 6-fold for antibody-forming hybrids so that 20 to 30 anti-SRBC hybridomas can usually be generated from an optimally immunized mouse. In contrast, a much weaker immunogen that generates serum antibody levels that could be detected with a sensitive assay at 1:100 dilution may stimulate only $10^4$ antibody-forming cells per spleen at the peak of the response. In this case, even with the enrichment that occurs during fusion, it will be necessary to fuse 3 to 5 spleens and screen 1500 to 2500 hybrids to find one hybrid producing reasonable levels of antibody. Since assays for antibody vary enormously in sensitivity, it is difficult to present rules as to when it is possible to obtain antibody-forming hybrids. If a sensitive assay such as enzyme or viral neutralization is being used, we would not attempt a fusion unless we have animals whose serum titers had risen about 100-fold, three days after the boost.

A number of different routes of immunization and adjuvants have been used. Antibody-forming hybridomas have been generated from animals receiving only a single injection of antigen and from animals receiving multiple injections of antigen in complete Freund's adjuvant. In general, it is best to select the simplest immunization regime that will result in a strong antibody response for the antigen being used. It is useful to immunize many (10 to more) mice since one or two will often respond better than the others. Usually, the primary immunization consists of a few hundred micrograms to a few milligrams of antigen administered intraperitoneally in complete Freund's adjuvant. It is important to remember that Freund's adjuvant must be mixed with antigen so as to form a water-in-oil emulsion (11). This can be verified by dropping a small amount of the emulsion on water. If it holds together, the water-in-oil emulsion has been achieved. The animals are then rested for 4 to 6 weeks or until the level of circulating antibody has dropped appreciably. After they have been bled and their antibody levels titered, the mice are injected intravenously with between 100 µg and 5 mg of antigen. The intravenous route is preferred because large amounts of antigen reach the spleen. However, if the antigen will cause emboli or is toxic intravenously, then it is usually injected intraperitoneally in saline. Two to three days later the animals are bled and those animals with the greatest rise in titer are chosen for fusion. It is not unusual for one or two animals to have 10-fold (or more) higher titers than the rest. Fusion is carried out 2 to 4 days after the boost. The need to fuse within 4 days after the boost is one of the few facts on which there is general agreement.

If complete Freund's adjuvant cannot be used, other adjuvants such as mixtures of antigen bound to alum and pertussis vaccine have also been effective. Viruses and some other antigens can be injected in small amounts. The optimum dose for different antigens can vary over a wide range. It is worth repeating that if reasonable levels

of antibody cannot be generated, it will be very difficult to
generate antibody-forming hybridomas.

All of the discussion so far has assumed that BALB/c mice will
be used as a source of spleen cells.  This strain is usually selected
because the mouse myeloma cells currently available are of BALB/c
origin.  The hybrids can therefore be injected back into BALB/c mice
to generate ascites.  However, antibody-producing hybrids have also
been generated by fusing lymphocytes from other strains and rat
spleen cells to both cultured rat and mouse myeloma cells (12,13).

It is much more difficult to obtain antibody-producing hybrids
by fusing mouse myelomas to rabbit spleen cells because the rabbit
chromosomes are rapidly lost and few hybrids continue to secrete
rabbit immunoglobulin.  A similar problem exists when mouse myeloma
cells are fused to human cells except that the loss of human chromosomes is less complete and some mouse x human hybrids do continue to
secrete human immunoglobulin (14,15).  There are few reports of other
species combinations.

In summary, a variety of immunization schemes have been used
successfully but at this time it is not clear whether any one regime
is better than the others.  What is clear is that the better the
antibody response, the better the chance of a successful fusion.
However, since animals having similar antibody responses often give
very different numbers of antibody-forming hybrids, other as yet
unknown factors must also be important.  It should ultimately be
possible to use weak immunogens but this will require techniques to
enrich for the spleen cell that participates in the fusion, increased
fusion frequencies, or both.


FUSION


The technique we use for fusing myeloma cells has been reported
previously (12,16).  Since then, some changes have been made in the
procedure and our current protocol is described here.

A number of mouse myeloma, and one rat myeloma, cell lines have
been shown to be effective as the transformed partner in spleen
fusions.  Table 1 includes a list of the most commonly used cell
lines.  All of these lines lack the enzyme hypoxanthine phosphoribosyl transferase (HPRT).  They are resistant to killing with
6-thioguanine (6-TG) and 8-azaguanine but are killed in medium containing hypoxanthine, aminopterin and thymidine (HAT) (see below).
Cell lines 45.6TG1.7 and $P_3$-X63-Ag8 produce their own heavy and light
chains and hybrids formed with these lines produce immunoglobulins
containing polypeptides from both of the parental molecules as well
as molecules identical to those made by each of the parental cells
(7,17).  The percentage of mixed molecules varies from hybrid to
hybrid.  Such mixed molecules complicate the interpretation of subsequent serological experiments because they have a greatly decreased
affinity for the antigen and could conceivably even have a different
specificity from parental immunoglobulins.  In order to deal with

Table 1

Myeloma Lines Commonly Used in Producing Hybridomas

| Cell line | Immunoglobulin Chains Expressed (H,L) | Ref. |
|---|---|---|
| P$_3$-X63-Ag8 | $\gamma_1$,k | 6 |
| P$_3$-NS1/1-Ag4-1(NS-1) | k (intracellular only) | 7 |
| SP2/0-Ag14 | none | 19 |
| 45.6TG1.7 | $\gamma_{2b}$,k | 18 |
| 4T00.1.L1 | k | 18 |
| X63Ag8.653 | none | 20 |

this problem, variant mouse myeloma cell lines have been generated that either make only the light chain of the parental myeloma (NS-1, 4T00.1.L1) or have lost the ability to synthesize both the heavy and light myeloma polypeptide chains (17-20). It is worth noting that although the NS-1 myeloma does not secrete the myeloma light chain, it is synthesized and, following fusion, hybrid molecules containing the NS-1 light chain are secreted (6). It is obviously preferable to use myeloma cell lines such as Sp2/0-Ag14 (19) and X63Ag8.653 (20) that do not produce their own heavy and light chains.

Prior to fusion, the mouse myeloma cell line is grown in medium containing 5 $\mu$g/ml 6-TG for a few days in order to kill any 6-TG sensitive HPRT$^+$ revertants since these will grow in HAT medium and appear to be hybrids. The drug is washed out and the cells are split daily and grown exponentially (between 2 x $10^5$ cells/ml and 8 x $10^5$ cells/ml) in the absence of drug for a few days before being fused.

Two to four days after boosting the animals, the two or three animals with the best antibody response are sacrificed and the spleens are sterilely removed and placed in a petri dish containing 5 to 10 ml of serum-free growth medium (we use Dulbecco's Modified Eagle's Medium (DME) (Gibco H-21)) supplemented with penicillin, streptomycin, nonessential amino acids and 10% NCTC 109 (Microbio-logical Associates). A single cell suspension is obtained by teasing the spleen with two scalpels or putting it through a wire mesh. The large debris is allowed to settle to the bottom of a 50 ml conical tube (Beckman #CT1620) for 5 min and the supernatant containing single cells is transferred to another tube. The cells are then collected by centrifugation. The nucleated spleen cells are resuspended in 10 ml of serum-free medium and a small sample is diluted in saline containing 0.1% trypan blue and the viable nucle-ated cells are counted in a hemocytometer. Usually between 0.8 and 1.2 x $10^8$ viable nucleated cells are recovered from each mouse spleen and 2 to 4 x $10^8$ cells from each rat spleen.

Simultaneously with the preparation of the spleen cells, the myeloma cells are washed twice in cold serum-free medium and counted. Spleen and myeloma cells are mixed in a ratio between 10:1 and 1:1. Within these limits, we have not been able to detect an effect of the ratio of cells on the yield of hybrids, but have found it convenient to use a ratio of spleen:myeloma of 5:1. The mixture of cells is warmed to 37°C and then centrifuged at room temperature. After aspirating the medium, 0.5 to 1.0 ml of 30% polyethylene glycol (PEG) diluted in warm (37°) DME is added to the pellet. The pellet is quickly suspended with a Vortex and immediately centrifuged for 3 min at 1500 RPM at room temperature. The cells are left in the pellet until they have been exposed to PEG for a total of 8 min. The PEG treatment is ended by slowly adding 10 ml of room temperature DME to the pellet and then gently resuspending the pellet with a pasteur pipette. Some investigators feel that the cells in the PEG are hypertonic and must be slowly diluted over a period of minutes (21). The cells are centrifuged once more and resuspended in DME plus all of the supplements including 10 to 20% FCS and HAT.

The conditions described above have been arrived at empirically using a particular lot of PEG 1000. We know that the concentration and time of exposure to PEG are crucial and have described a simple method for estimating the optimum concentration and time (16). However, it is necessary to confirm this estimate by carrying out a preliminary fusion using a fixed time and different concentrations, usually between 25 and 50%. Different molecular weight PEG has been used by different investigators (21). Sharon and Morrison (personal communication) have suggested that the incubation with PEG should be done at pH 8 and previous reports on fusions with other cell types also reported that higher pHs promoted fusion (22). The pH is easily adjusted by Vortexing the mixture of PEG and DME (driving off $CO_2$) until the desired pH is achieved. Kennett has suggested a slightly different medium (23) and some investigators use aggammaglobulinemic horse serum.

As already discussed, there is a need to increase the fusion frequency. So far, most workers have found PEG far superior to other fusagens, although some workers use Sendai virus (24). It is possible that new more effective transformed cell lines will be described. The fusion frequency is higher when certain sorts of cell fractionations are used but this probably does not result in an absolute increase in antibody-forming hybrids per spleen (25). Finally, fusions are sometimes unsuccessful and the yield of hybrids varies from fusion to fusion, suggesting that there is still much to be learned about the technology.


SELECTION OF HYBRIDS

Since only one out of roughly $2 \times 10^5$ spleen cells forms a viable hybrid with a myeloma cell, it is necessary to place the fused cell mixture in a medium that will select for the growth of the

rare hybrid and eliminate the unfused myeloma and spleen cells; the
HAT selection system of Littlefield is commonly used (26). Mouse
myeloma cells lacking the enzyme hypoxanthine phosphoribosyl transferase (HPRT) serve as the transformed parent. Following fusion, the
cells are grown in selective medium containing hypoxanthine (100 µM),
aminopterin (10 µM) and thymidine (30 µM). The HPRT$^-$ myeloma cells
die because they cannot utilize the exogenous hypoxanthine to synthesize purines and the aminopterin has blocked their endogenous
synthesis of purines and pyrimidines. Hybrids between spleen and
myeloma cells contain the spleen HPRT, utilize the exogenous hypoxanthine and thymidine and continue to grow. The unfused spleen
cells are not killed by the selective medium but they do not grow
well in culture and the few persistent nondividing spleen cells are
rapidly outgrown by the hybrids.

After the fusion, the cells are resuspended in HAT medium at a
density of from 2.5 to 5 x 10$^5$ myeloma cells/ml (plus 1 to 10 times
as many spleen cells) and plated at roughly 0.1 ml/well in 96-well
microculture plates (Linbro #76-003-05). The culture plates are
placed in a heavily humidified atmosphere containing 10% $CO_2$. We
have found that the hybrids do better at this high concentration of
$CO_2$ and require a stable environment in which the pH of the medium
does not fluctuate often for the first few weeks. The wells are fed
with another 50 µl of HAT medium every 5 to 7 days. The myeloma
cells die and disintegrate within the first week, but the spleen
cells can persist two or three weeks. Some spleen cells attach to
the well, appear to phagocytose cell debris and may even proliferate
slowly for a time. Hybrids first appear approximately a week after
fusion as small clones of cells with exactly the same morphology as
the myeloma parent. They grow at various rates and, depending on the
parent myeloma, are grossly visible and ready to be assayed for
antibody production within two to six weeks after fusion. The plates
can be easily scored for hybrids using a mirror (Dynatech #1-200-16).
It is difficult to distinguish between persistent spleen cells and
hybrids as some wells may contain what appear to be clones of slowly
growing cells. If these are spleen cells, they initially grow as a
monolayer whereas hybrids are closely packed and pile up on each
other and are identical in appearance to myeloma clones that arise in
limiting dilution cloning.

Many laboratories prefer to plate out the cells in larger
volumes (1 or 2 ml). This has the advantage of requiring fewer
assays in the initial screen for antibody-forming hybrids. However,
each well generally contains multiple hybrids making it essential to
clone very early to avoid overgrowth by irrelevant hybrids. Furthermore, given the differential growth rates of the hybrids, this
system may favor the fast-growing hybrids and miss slower-growing,
antibody-forming hybrids in the initial screen (depending on the
sensitivity of the assay) (see below).

Antibody-forming hybrids are plated into 24-well tissue culture
plates (Linbro #76-033-05) (Figure 1) and then to petri dishes.
While in the 24-well dishes, they are fed with complete growth medium

supplemented with hypoxanthine (100 μM) and thymidine (30 μM) (HT medium) through several divisions to dilute out the aminopterin. Once well-established, the cell lines are maintained in DME supplemented with penicillin, streptomycin and 10 to 15% FCS or aggamma-globulinemic horse serum.

We have found that some batches of serum support the growth of hybrids better than others. We screen a number of batches by limiting dilution cloning (see below) of the parental myeloma cell and use the lots which give the highest (50% or better) cloning efficiency.

## SCREENING FOR POSITIVE HYBRIDS

Because a fusion can generate 100 to 500 hybrids growing at different rates, it is essential to have a rapid and sensitive method of identifying antibody-producing hybrids. It is time-consuming and virtually impossible to continue to propagate all the hybrids beyond the wells in which they are initially plated. The screening assay should therefore require less than 200 μl of culture medium and less than two or three days to perform. A number of convenient assays exist. We shall describe a few that are widely used and have proven reliable.

The type of assay employed depends on the nature of the target antigen. For cell-surface antigens or soluble antigens that are easily attached to a solid support, one of the several binding assays is probably best. Impure antigens and those available only in small amounts may require precipitation or neutralization of an activity of the antigen by the monoclonal antibody. These assays are technically more difficult because a monoclonal antibody can usually bind to only one site on a nonpolymeric antigen. If this binding is not sufficient to precipitate or neutralize the antigen, an indirect reagent such as an anti-mouse (rat) immunoglobulin antiserum may be added to precipitate the hybridoma antibody and, with it, the antigen.

Assays that exploit biological activities of immunoglobulins, especially complement fixation, are generally not suitable unless indirect reagents are added. Many subclasses of mouse Ig, for example, do not fix complement. Those subclasses of IgG that do fix complement, do so much less efficiently that does IgM.

A binding assay is depicted in Figure 1. The well contains either target cells or a soluble antigen that has been adsorbed to the surface (27). Supernatants from hybrid-containing wells are incubated to allow the antibody to bind to the antigen (37° for 1 hr or 4° overnight). The target well is thoroughly washed to remove unbound Ig and $^{125}$I-labelled antibodies, specific for mouse Ig, are added and incubated. To insure that all classes of immunoglobulin are being detected, it is often easier to prepare anti-light chain reagents, rather than anti-Ig against all heavy chain subclasses. In the mouse, 98% of all serum antibodies contain kappa light chains, so $^{125}$I-anti-kappa chain should detect 98% of all monoclonals. After thorough washing, the wells are counted and those significantly above background are presumed to indicate the presence of the desired monoclonal in the supernatant. Methods for preparing and iodinating the anti-immunoglobulin reagents have been reported in detail elsewhere (28). If target cells contain receptors for the carboxy-terminal end of the immunoglobulin molecule, it may be necessary to cleave the molecule into an Fab fragment that contains only the antigen-binding aminoterminal portion.

Reagents coupled to either alkaline phosphatase or horse radish peroxidase (ELISA assays) (29) provide an alternative to $^{125}$I-antibodies. After the last wash, the enzyme's substrate is added. If the enzyme is present, the substrate is hydrolyzed yielding a colored product whose presence is detected visually. The reaction can be quantitated using a spectrophotometer. The advantage of the ELISA assay is in avoiding $^{125}$I, which requires a special license and has a short half-life. Horse radish peroxidase conjugates are available commercially (Cappel).

Protein A from Staphylococcus aureus is known to bind many subclasses of mouse immunoglobulin and some investigators have used $^{125}$I-linked or enzyme-conjugated protein A in binding assays (30,31). Because protein A does not bind to mouse IgM and binds especially well to rabbit IgG, many of these assays employ a rabbit anti-mouse Ig serum followed by protein A. This sandwich technique detects all Ig classes and is more sensitive than the direct assay. In addition, $^{125}$I protein A is commercially available (Amersham).

Many protein antigens are easily coupled to red blood cells (RBC) with glutaraldehyde. Protein (1 to 10 mg) is incubated with 10 ml of 10% RBC in phosphate-buffered saline containing 0.25% glutaraldehyde at 25° for 30 min to 1 hr. Cells are washed free of the glutaraldehyde with cold PBS and may be stored at 4° for several months with preservatives to inhibit bacterial growth. These target cells may be used in the binding assays mentioned above and are very convenient for hemagglutination assays. In hemagglutination assays for monoclonal antibodies it is again preferable to add anti-mouse Ig (usually at a 1 to 100 dilution or greater) to facilitate the cross-linking of red blood cells (RBC) by monoclonals that are poor agglutinins. Antigen-coupled RBC (prepared sterilely) have been used to screen for positive hybridomas by adding the RBC directly to wells containing the hybrids (32). In positive wells, the RBC are seen to agglutinate and/or form rosettes with the hybrid cells.

Finally, use of the fluorescence-activated cell sorter (FACS) has been reported in screening and cloning hybridomas (33). Sharon, Morrison and Kabat have reported a method for initially plating the fused cells in soft agarose over fibroblast feeders and screening individual clones by replica immunoadsorption (34).

In cases where the screening assays are extremely sensitive (such as the $^{125}$I binding assay) it is sometimes possible to detect antibody secreted by the persistent spleen cells. In such cases, a very large percentage of wells may initially appear positive. Repeating the screen at a higher dilution of supernatant will

eliminate negative wells with small amounts of polyclonal antibody.
This will also eliminate hybridomas producing either small amounts of
antibody or antibodies with very low affinities.  When multiple
hybrids are available, only the ones with the highest titers should
be preserved.

## CLONING CELLS

It was observed with the earliest hybridomas that the cultures
sometimes stopped producing antibody (35).  This instability is no
doubt due in part to the loss of heavy or light chain synthesis
associated with chromosome loss and/or overgrowth by contaminating
clones or variants that arise during the propagation of the hybridomas.  To minimize overgrowth by contaminating hybrids and non-
producing variants it is important to clone the hybridomas early.
Even after cloning, some hybrids are very unstable while others need
to be recloned only occasionally.  If many hybrids making the desired
homogeneous antibody can be generated, it is easier to discard
unstable hybrids than to try to maintain them.  When a newly-
generated hybridoma has a unique specificity and must be preserved,
it is possible to maintain it by repeated cloning or enriching for
antigen-binding cells by any of a variety of methods (36).

Two methods may be used to isolate individual clones of cells.
Either method can fail to produce true clones if two or more cells
adhere to each other during plating, so the procedure should be
repeated at least twice to insure clonality.  The first method is
cloning by limiting dilution.  The hybrid is diluted to a very low
density (1 to 10 cells/ml) and plated at 0.1 ml/well in 96-well
microculture plates.  Because of the suboptimal cell density, the
cloning efficiency may be low by this method (<10%) and each
well in which cells grow can be assumed to contain a single clone.
If the cloning efficiency of certain hybrids is prohibitively low, it
can often be raised by adding primary cells (normal spleen cells,
thymocytes, fibroblasts) as feeders that the hybrid will eventually
outgrow.  When the clone is large enough (covering 10 to 50% of the
well's surface), the medium can be tested for antibody.

The second method is cloning in soft agarose and has been
reported in detail previously (37).  The cells are plated over a
layer of fibroblasts in medium containing approximately 0.25%
agarose.  The agarose, previously solubilized by autoclaving, gels
during a brief incubation at 4°C, immobilizing the cells and their
progeny.  After approximately 10 days, the clones are large enough to
be picked with a micropipet and placed in 50 μl of medium in a microculture plate.  The clones are fed as necessary and can usually be
tested for antibody production within a week after they are picked.
The method has a generally higher cloning efficiency due to the
fibroblast feeders.  Furthermore, antisera to mouse immunoglobulin
can be added to the medium such that a precipitate forms around those
clones secreting immunoglobulin (37).  This allows direct visual-

ization of even the rare antibody-secreting cell in cultures largely
overgrown by nonproducers and greatly reduces the work of picking
and screening.

## FREEZING CELLS

After generating a sufficient quantity of the monoclonal reagent
from a hybrid and as protection against tissue culture mishaps, the
cells can be frozen. Exponentially growing cells are collected by
centrifugation and resuspended at a density of at least $10^7$ cells/ml
in cold (0°) freezing medium containing 10% dimethylsulfoxide (DMSO)
and 90% fetal calf serum. DMSO is toxic to cells so they must be
kept cold after being resuspended in the freezing medium. Samples of
from 0.2 to 1.0 ml are placed in sterile glass vials that can be
sealed with a propane torch. The vials should be stored at a temperature below -80°C.

To put the cells back into culture, a vial is quickly thawed and
the cells washed once in 10 ml of medium at 4°. The cells are resuspended initially in approximately 1 ml of medium and placed in an
incubator. A few hours later, the number of viable cells can be
determined and the volume adjusted to bring the cells to a density
of $5 \times 10^5$/ml.

## PRODUCING ASCITES

Ascites fluid from tumor-bearing animals usually contains from
100 to 1000 times as much antibody as spent culture medium and thus
is the best source of high titer reagents. Syngeneic mice receive
two intraperitoneal injections at a seven day interval of 0.2 ml of
Pristane (38) and are injected intraperitoneally one day later with
$10^7$ cells/mouse. One to two weeks later, the mice develop an ascites
that may be tapped by inserting a #18 gauge needle into the abdominal
cavity. The fluid is collected into a tube containing an anti-
coagulant, then centrifuged to remove cells, sampled and stored at
-20°C. Each mouse can usually be tapped several times, at one or two
day intervals, yielding between 2 and 10 ml of ascites (containing
from 1 to 10 mg/ml of monoclonal antibody).

All of the mouse myeloma lines at present commonly used in
fusions originated as tumors in BALB/c mice. If the donor of the
primary lymphocytes is of another strain, and especially one with a
different H-2 haplotype, ascites should be produced in Fls of the
parental strains to avoid tumor rejection.

Ascites can also be obtained from xenogeneic hybridomas using
immunosuppressed animals or nude mice. We have used irradiated (600
Rads) mice to obtain ascites fluid from such hybrids. The irradiation is done the day after the second Pristane injection and cells
are injected the day after irradiation. Because the animals die

sooner, it is advantageous to inject more cells (2 to 3 x $10^7$/mouse) to generate ascites more quickly.

## PURIFICATION

Monoclonal antibodies can be purified from ascites fluid or tissue culture medium by antigen affinity chromatography with the same techniques used for heterogeneous antibodies.  However, large amounts of antigen often are not readily available.  The monoclonal antibody is the only mouse immunoglobulin in tissue culture medium, and because it is made by so many cells in the animal, the major mouse immunoglobulin in ascites fluid.  Significant purification can be obtained either by using protein A columns (Pharmacia), which bind most classes of immunoglobulin, or by ion exchange chromato- graphy (39).  This is not possible with antisera because they are so heavily contaminated with the many other serum antibodies produced by the animal.

## CONCLUSION

The production of monoclonal antibodies, while feasible for most antigens, is not a trivial undertaking.  It is a rather expensive and time-consuming technology because of the amount of cell culture required; it can take as long as four to six months to generate a stable, cloned hybridoma, assuming all goes well and the first fusions generate the desired hybrids.

Considering the difficulties in generating hybridomas, in many cases it may be more advantageous to obtain antisera in the tradi- tional way.  This might be true, for example, if only a limited amount of serum is needed and the antigen is easily purified.  In many instances, the heterogeneity of antisera makes them easier to use.  Most rabbit antisera readily precipitate and agglutinate antigens and fix complement.  A given monoclonal may lack any of these properties, but it may be possible to generate defined mixtures of different monoclonals that will provide a defined reagent with benefits of heterogeneous antisera.  If a monoclonal cross reacts in a way that interferes with its use, this cross reactivity cannot be adsorbed out, so other monoclonals that do not cross react will have to be generated.

In spite of these problems, it seems inevitable that most serology will ultimately be carried out with monoclonal antibodies. This is partly aesthetic and partly due to the real benefits. Impure antigens can be used to obtain pure antibodies.  Once a stable hybridoma producing large amounts of a desired antibody is generated and frozen away, the investigator can depend on an indefinite supply and unlimited amount of that particular antibody.  Monoclonal anti- bodies can be selected for properties that are particularly useful in a specific procedure.  If one wishes to construct an affinity column

to purify antigen, for example, antibodies with either very high or
very low affinity are less useful than antibodies with intermediate
activity.  It is possible to select among a collection of monoclonals
for the most appropriate one.  In addition, rare specificities can be
cloned out and amplified using the hybridoma technique.  This should
make it possible to obtain large amounts of antibody against minor
antigenic specificities.  Monoclonal antibodies have already proved
very useful in studying viral antigens, tumor and differentiation
antigens and a number of immunologic questions (8), and will
certainly be equally useful in a wide variety of genetic studies.

Acknowledgments:  S.-P. Kwan is a fellow of the Cancer Research
Institute, Inc.  D.E. Yelton is a medical scientist trainee supported
by grant 5TS-GM1674 from the National Institute of General Medical
Sciences.  Development of some of the technology reported here was
supported by grants from the National Institutes of Health (AI-10702
and AI-5231) and the National Science Foundation (PCM75-13609).

## REFERENCES

1    Broome, S. and Gilbert, W. (1978) Proc. Nat. Acad. Sci. U.S.A.
     75, 2746-2749.
2    Erlich, H.A., Cohen, S.N. and McDevitt, H.P. (1978) Cell 13,
     681-689.
3    Skalka, A. and Shapiro, L. (1976) Gene 1, 65-79.
4    Sanzey, B., Mercereau, O., Ternynck, T. and Kourilsky, P. (1976)
     Proc. Nat. Acad. Sci. U.S.A. 73, 3394-3397.
5    Krause, R.M. (1970) Adv. Immunol. 12, 1-56.
6    Köhler, G. and Milstein, C. (1975) Nature 256, 495-497.
7    Köhler, G. and Milstein, C. (1976) Eur. J. Immunol. 6, 511-519.
8    Current Topics in Microbiology and Immunology: Lymphocyte Hybri-
     domas (1978) (Melchers, F., Potter, M. and Warner, N., eds.),
     Vol. 81, Springer-Verlag, Berlin, Heidelberg and New York.
9    Monoclonal Antibodies (Kennett, R. and McKearn, T., eds.),
     Plenum Press, New York, NY (in press).
10   Immunol. Rev. (in press).
11   Horwitz, M.S. and Scharff, M.D. (1969) in Fundamental Techniques
     in Virology (Habel, K. and Salzman, N.P., eds.), pp. 253-262,
     Academic Press, New York and London.
12   Galfre, G., Howe, S.C. and Milstein, C. (1977) Nature 266,
     550-552.
13   Galfre, G., Milstein, C. and Wright, B. (1979) Nature 277,
     131-133.
14   Schwaber, J. (1975) Exp. Cell Res. 93, 343-354.
15   Levey, R. and Dilley, J. (1978) Proc. Nat. Acad. Sci. U.S.A. 75,
     2411-2415.
16   Gefter, M.L., Margulies, D.H. and Scharff, M.D. (1977) Somat.
     Cell. Genet. 2, 231-236.

17    Margulies, D.H., Cieplinski, W., Dharmgrongartama, B., Gefter,
      M.L., Morrison, S.L., Kelly, T. and Scharff, M.D. (1977) Cold
      Spring Harbor Symp. Quant. Biol. 41, 781–791.

18    Margulies, D.H., Kuehl, W.M. and Scharff, M.D. (1976) Cell 8,
      405–415.

19    Schulman, M., Wilde, C.D. and Kohler, G. (1978) Nature 276,
      269–270.

20    Kearney, J.F., Radbruch, A., Liesegang, B. and Rajewsky, K.
      (1979) J. Immunol. 123, 1548–1550.

21    Herzenberg, L.A., Herzenberg, L.A. and Milstein, C. (1978) in
      Handbook of Experimental Immunology (Weir, D.M., ed.), Vol. 2,
      pp. 251–257, Blackwell Scientific Publications, Oxford, London,
      Edinburgh and Melbourne.

22    Croce, C.M., Koprowski, H. and Eagle, H. (1972) Proc. Nat. Acad.
      Sci. U.S.A. 69, 1953–1956.

23    Kennett, R.H., Denis, K.A., Tung, A.S. and Klineman, N.R. (1978)
      Curr. Topics Microbiol. Immunol. 81, 77–91.

24    Köhler, G. (1976) in Immunological Methods (Lefkovits, I. and
      Pernis, B., eds.), pp. 391–396, Academic Press, New York, San
      Francisco and London.

25    Andersson, J. and Melchers, F. (1978) Curr. Topics Microbiol.
      Immunol. 81, 130–139.

26    Littlefield, J.W. (1964) Science 145, 709–710.

27    Gearhart, P.J., Sigal, N.H. and Klineman, N.R. (1975) J. Exp.
      Med. 141, 56–71.

28    Jensenius, J.C. and Williams, A.F. (1974) Eur. J. Immunol. 4,
      91–97.

29    King, T.P. and Kochoumian, L. (1979) J. Immunol. Methods 28,
      201–210.

30    Brown, J.P., Klitzman, J.M. and Hellstrom, I. (1977) J. Immunol.
      Methods 15, 57–66.

31    Engvall, E. (1978) Scand. J. Immunol. 8(7), 23–31.

32    Claflin, L., and Williams, K. (1978) Curr. Topics Microbiol.
      Immunol. 81, 107–109.

33    Parks, D.R., Bryan, V.M., Oi,V.T. and Herzenberg, L.A. (1979)
      Proc. Nat. Acad. Sci. U.S.A. 76, 1962–1966.

34    Sharon, J., Morrison, S. and Kabat, E.A. (1979) Proc. Nat. Acad.
      Sci. U.S.A. 76, 1420–1424.

35    Williams, A.F., Galfre, G. and Milstein, C. (1977) Cell 12,
      663–673.

36    Current Topics in Microbiology and Immunology: Techniques for
      Separation and Selection of Antigen Specific Lymphocytes (1978)
      (Haas, W. and Von Boehmer, H., eds.), Vol. 84, Springer-Verlag,
      Berlin, Heidelberg and New York.

37    Coffino, P., Baumal, R., Laskov, R. and Scharff, M.D. (1972) J.
      Cell. Physiol. 79, 429–440.

38    Potter, M., Pumphrey, J.G. and Walters, J.W. (1972) J. Nat.
      Cancer Inst. 49, 305–308.

39    Fahey, J.L. and Terry, E.W. (1978) in Handbook of Experimental
      Immunology (Weir, D.M., ed.) Vol. 1, Chapter 8, Blackwell
      Scientific Publications, Oxford, London, Edinburgh and
      Melbourne.

# MEASUREMENT OF MESSENGER RNA CONCENTRATION

S.J. Flint

Department of Biochemical Sciences
Princeton University
Princeton, New Jersey  08540

## INTRODUCTION

Messenger RNA (mRNA) species, the work horses of gene expression, have been the subject of much interest since the original demonstrations of their existence in phage-infected E. coli some two decades ago (1-3). In the intervening period, we have learned a great deal about the basic properties and mode of functioning of mRNA but our knowledge is far from complete, particularly when it comes to eukaryotic mRNA synthesis and its regulation.

In bacteria, the products of transcription of the DNA genome serve directly as mRNA. The majority of prokaryotic mRNA species are short-lived, so that regulation of gene expression is achieved primarily at the level of transcription. In eukaryotic cells, on the other hand, all mRNA species are subject to some form of post-transcriptional modification and most appear to derive from larger precursors, the primary products of transcription. Post-transcriptional processing events include the addition of poly(A) to the 3'-termini of RNA sequences destined to function as mRNA, capping of their 5'-termini, methylation, both of the capping residues and at internal sites, and specific cleavage and ligation processes (splicing). The majority, if not all, of these maturation steps take place in the eukaryotic cell nucleus and the resulting mature mRNA molecules must be transported to cytoplasmic polyribosomes where they are translated. As a result, transcription and translation are separated topographically and, therefore, temporally in the eukaryotic cell; regulatory processes might therefore intervene at any one of the steps leading to production of mature mRNA in addition to transcription. Eukaryotic cell mRNA populations also differ from those of prokaryotes in being significantly more stable and, given the increased size of the eukaryotic genomes, of greater sequence complexity.

47

        In view of these differences, it is not too surprising that
analysis of eukaryotic mRNA synthesis and its regulation has progressed more slowly than study of corresponding events in prokaryotes.  Despite the difficulties inherent to the study of eukaryotic
RNA sequences (and their precursors), a variety of techniques that
permit quantitative analysis have been developed over the last 10 or
so years.  These methods, which may be applied to the study of individual RNA species or to populations of different RNA molecules, have
led to some success in the elucidation of a variety of problems including: 1) determination of the fraction of the genome encoding both
mRNA sequences and their precursors; 2) analysis of the nature of
informational DNA sequences complementary to populations of mRNA and
nuclear RNA and the ways in which different kinds of DNA sequences
are arranged relative to one another in the genome; 3) determination
of the concentration of different kinds of RNA molecules within a
given population and the frequency of representation and sequence
complexity of different sub-fractions within RNA populations; 4)
measurement of the concentration within an RNA population of RNA
molecules transcribed from a specific gene, and 5) the description of
alterations of the concentration of a specific RNA in a population in
response to various stimuli, including those of differentiation and
development, neoplastic transformation and hormone treatment.
        Although incomplete, this list illustrates two kinds of ques-
tions addressed by quantitative analysis of eukaryotic cell RNA populations: those concerned with the quantitation and description of
the organization within the genome of DNA sequences complementary to
mRNA and those concerned with regulation of gene expression and the
molecular mechanisms by which it is mediated.
        In all but a few rare instances, the quantitative methods em-
ployed rely upon hybridization between the RNA preparation(s) of
interest and DNA sequences complementary to it.  In most eukaryotic
cells, however, any specific kind of mRNA (and its precursor(s)) are
present at only a few (to a few hundred) copies per cell among many
thousands of other kinds of RNA molecules.  Moreover, because as much
as 95% of the DNA sequences constituting a eukaryotic genome may not
be complementary to RNA, the DNA sequences of any given gene represent a very small fraction of the total DNA.  The haploid genome of
man, for example, comprises about $1.8 \times 10^{12}$ daltons of DNA (4)
whereas an average gene would be expected to be of the order of $10^6$.
In many cases, therefore, total RNA populations, rather than individual RNA species, have been analyzed with such DNA probes as
complementary DNA (cDNA), made in vitro, or genomic DNA.  These
experiments can provide much general information about the structure
and expression of the eukaryotic genome but clearly are not ideal for
the study of subtle regulatory processes, a purpose more likely to be
achieved through investigation of the expression of individual genes.
Until quite recently, this goal could be attained only by the use of
certain specialized systems in which very large amounts of one, or a
small number, of specific gene product(s) are made.  In these cases,
it is possible to obtain specific probes, usually in the form of DNA

made from a purified mRNA template; but such systems are the exception rather than the rule.  Thus, a major hindrance to the study of
expression of individual genes, and therefore of the mechanisms by
which their expression might be regulated, has been the problem of
obtaining specific probes.  With the advent of molecular cloning
techniques, this stumbling block is, in principle, completely removed,  as it should be possible to obtain large amounts of any DNA sequence of interest.  It is anticipated that the study of specific
genes will become a more common pursuit in the near future, adding
greatly to our understanding of gene expression and its regulation in
eukaryotic cells.  This article describes the methods that have been
developed for the quantitation of mRNA, concentrating on examples
drawn from the study of eukaryotic cell RNA sequences.  As mentioned
previously, hybridization between DNA and RNA forms the basis of most
of these methods.  The various hybridization protocols, and their
potential application with cloned DNA, are considered here.

## GENERAL FEATURES OF HYBRIDIZATION EXPERIMENTS

    The basic experimental manipulations are very similar in all
hybridization protocols; the RNA and DNA to be reacted are mixed and
incubated under optimal annealing conditions (see below).  If the DNA
component of the reaction is not single-stranded, then it is denatured just prior to addition of the RNA.  At time zero and at suitable intervals thereafter, samples are withdrawn from the hybridization mixture and assayed for the amount of nucleic acid that has
entered DNA:RNA hybrid.  The precise nature of the assay naturally
depends upon whether the DNA or RNA component of the hybridization
reaction is labeled.  When unlabeled RNA is hybridized with labeled
DNA, the fraction of input DNA entering hybrid may be determined by
either measurement of the amount of DNA that becomes resistant to S1
nuclease or chromatography on hydroxylapatite.  In the former case,
the samples are adjusted to conditions of salt, $Zn^{+2}$ concentration
and pH optimal for the nuclease (5,6) and the amount of resistant DNA
is determined by precipitation with trichloracetic acid after S1
digestion.  Comparison with samples of the reaction mixture that have
been precipitated without S1 nuclease digestion then yields an
estimate of the fraction of input DNA that is resistant—that is, has
formed a DNA:RNA hybrid.  In the latter assay, the sample is passed
through a column of hydroxylapatite of suitable size at 60°C (7).
The column is then washed extensively with 0.12 M to 0.14 M phosphate
buffer, pH 6.8, and single-stranded DNA is eluted.  Double-stranded
DNA and DNA:RNA hybrids may then be eluted with phosphate buffer of
0.40 to 0.45 M.  By measuring the amount of radioactivity recovered
in the two fractions, it is possible to determine both the total
radioactivity in the sample and the fraction that behaves as a
DNA:RNA hybrid.  Under these conditions of chromatography, duplex DNA
with a physical length of less than 40 to 50 base pairs does not bind
efficiently to hydroxylapatite in 0.125 to 0.14 M phosphate buffer

(8). It is therefore important that the labeled DNA probe exceeds
this length. On the other hand, the labeled DNA should not be too
large; any DNA:RNA hybrid in which the hybrid region exceeds 40 to 50
base pairs will register as duplex under these conditions, even when
the hybrid carries long segments of unhybridized DNA. In contrast,
the S1 nuclease assay will register only that fraction of the DNA
probe that is actually hydrogen-bonded to complementary nucleic
acid. The use of labeled DNA that is too long can therefore lead to
significant over-estimation of the fraction of the DNA entering the
DNA:RNA hybrid when the hydroxylapatite assay is used.

When the hybridization mixture contains labeled RNA and unlabeled DNA, the fraction of input RNA that has formed a hybrid is
measured on the basis of its resistance to RNase in concentrations of
salt sufficient to stabilize the hybrids. The hybridization mixture
is digested with RNase A, and often with RNase T1 and T2, and the
amount of labeled RNA surviving digestion is determined by trichloracetic acid precipitation or chromatography on Sephadex columns.
Labeled RNA binds quantitatively to hydroxylapatite in 0.12 M phosphate buffer at 60°C. However, in 0.2 M phosphate buffer, containing
8 M urea and 1% SDS at 40°C, no single-stranded RNA will bind,
although both DNA:RNA duplexes and RNA:DNA hybrids bind efficiently
(9). Thus, this modified hydroxylapatite assay can be used to monitor hybridization with labeled RNA.

In some protocols, one of the two components of the reaction may
be attached to some form of solid support or the reaction may be performed under nonaqueous conditions. Nevertheless, the basic experimental manipulations follow the outline presented in the preceding
paragraphs. Moreover, certain parameters that determine the rate of
annealing are common to all methods of hybridization. In addition to
the concentration and nucleotide sequence complexity of the reacting
nucleic acid species, these parameters include the following.

(a) Salt Concentration (10-12). The rate of the annealing
reaction increases significantly with increasing NaCl concentration
up to 0.4 to 0.5 M. Above this value, the effects of further increases in salt concentration are much less dramatic. Most hybridization reactions are therefore performed in the presence of high
salt concentrations, but to facilitate comparisons between different
sets of data, kinetic parameters are usually expressed as the values
they would exhibit in 0.12 M phosphate buffer (12).

(b) Temperature (10,13,14). The rate of the reaction exhibits
a bell-shaped curve depicting dependence on incubation temperature
with a fairly broad optimum, 20 to 30° below the melting temperature
(Tm) of the DNA. The Tm is itself influenced by the G+C content of
the nucleic acid, but these effects are relatively small and usually
ignored.

(c) Viscosity (11,15,16). As any hybridization reaction
depends upon collision between complementary nucleic acid segments,
it is not surprising that the rate is influenced by the viscosity of
the solution. When the hybridization is performed at Tm - 25°C, the
rate is, in fact, inversely proportional to viscosity (11). Many of

the methods discussed in subsequent sections aim to achieve conditions of vast DNA or RNA excess: that is, require the use of very
high concentrations of DNA or RNA.  Thus, viscosity, is likely to
become of some importance, but it is a parameter for which little
allowance is made in most experiments.

       d)  Physical Length of Reacting Nucleic Acid Segments (11).  In
addition to the nucleic acid sequence complexity (10,11,17,18), the
rate of DNA:DNA reassociation or DNA:RNA hybridization is strongly
influenced by the physical length of the nucleic acid segments, the
second-order rate constant for renaturation of a DNA preparation
being proportional to the square root of the single-stranded length
(11).  It is therefore of some importance that nucleic acids are of a
standard length in all reactions to be compared.  In most DNA:RNA
hybridization experiments, DNA is degraded by shearing or controlled
alkaline hydrolysis to segments a few hundred nucleotides in length
before addition to the hybridization mixutre.  However, little
attention is generally paid to the physical state of the RNA.  As it
has been reported that reaction rates are retarded when long driver
nucleic acid is hybridized to a short (labeled) tracer (19), error
may be introduced when the rates of hybridization of different RNA
preparations of different length to the same DNA probe are compared.

      In summary, it is clear that a number of parameters can
influence the rate of hybridization reactions.  It is therefore
crucial that these remain constant in a comparison of the kinetic
parameters of different reactions.


## RNA EXCESS HYBRIDIZATIONS

      Many of the quantitative analyses of unfractionated RNA
populations of eukaryotic cells, or of specific RNA species,
performed in the past relied upon hybridization of excess, unlabeled
driver RNA to a tracer, labeled DNA.  This is because it has generally proved more practical to obtain sufficiently large quantities
of RNA than of an appropriate DNA.  As the RNA is not labeled, it is
the steady-state concentration of a specific kind of RNA or of a
family of RNA sequences that is measured.  Until the advent of
molecular cloning techniques, the labeled DNA available for use in
RNA excess hybridization experiments was of two types: genomic DNA
labeled in vivo or, more recently in vitro by nick translation, and
complementary DNA (cDNA) copies of a specific mRNA or an unfractionated mRNA population.


## cDNA Probes

      The discovery of retroviral reverse transcriptases (20,21)
provided a novel means of synthesizing labeled DNA.  When reverse
transcriptase is supplied with a poly(A)-containing mRNA template, an
oligo(dT) primer and the four deoxyribonucleoside triphosphates, the

enzyme will synthesize DNA copies of the RNA template (22-24). In the presence of labeled substrates, it is possible to synthesize cDNA of significantly higher specific activity than can be achieved by in vitro labeling. This is an important advantage, making it far easier to attain the necessary RNA excess to drive hybridization reactions to completion. Such high concentrations of driver RNA are expecially necessary when investigating total mRNA populations which may be of very high complexity, with individual components represented at low frequency, or in the study of specific mRNA species that are rarely expressed.

Under these conditions, a single-stranded cDNA probe reacting with a vast excess of RNA, the simplest reaction for a homogeneous DNA with its complementary RNA will display pseudo first-order kinetics and may therefore be described by the reaction

$$\frac{dC}{dt} = -kR_oC \qquad\qquad\qquad [1]$$

where C = the concentration of DNA remaining single-stranded at time, t,

$R_o$ = the initial and constant concentration of complementary RNA, and

k = the rate constant of the pseudo first-order reaction.

Upon integration and putting $C = C_o$ when $t = 0$, equation [1] becomes

$$\ln \frac{C_o}{C} = kR_ot \qquad\qquad\qquad [2]$$

or

$$\frac{C_o}{C} = e^{kR_ot} \quad . \qquad\qquad\qquad [3]$$

When 50% of the final amount of input cDNA that can enter hybrid is in DNA:RNA hybrid, $C = C_o/2$ and $R_ot$ is defined as $R_ot_{1/2}$. Thus, from equation [2]

$$\ln \frac{C_o}{C_o/2} = kR_ot_{1/2} \qquad\qquad\qquad [4]$$

$$k = \ln2/R_ot_{1/2} = 0.691/R_ot_{1/2} \quad . \qquad\qquad\qquad [5]$$

Ideally, the rate of the reaction is inversely proportional to the $R_ot_{1/2}$ value expressed in moles nucleotide $l^{-1}$ x seconds, which depends directly on the sequence complexity of the RNA (18,26). This has been demonstrated experimentally by an analysis of the hybridization kinetics of α-globin mRNA, a mixture of rabbit α- and β-globin mRNA and encephalomyocarditis virus RNA with their respective cDNAs (27). In all cases, the reactions displayed the predicted first-order kinetics and the measured $R_ot_{1/2}$ values of the three RNA preparations were proportional to their sequence complexities.

Usually, less than 100% of the labeled cDNA enters hybrid at the termination of the reaction (examples are discussed in subsequent paragraphs). It is therefore necessary to make a correction to the equations given above for the proportion of cDNA that is reactive. It is also convenient to introduce the term D, the concentration of cDNA in hybrid at time t, the parameter actually measured by the assays described previously. Thus,

$$D = C_0 - C$$

$$\frac{D}{C_0} = \frac{C_0 - C}{C_0}$$

$$= 1 - \frac{C}{C_0}$$

from equation [3]

$$\frac{D}{C_0} = 1 - e^{-kR_0 t} \tag{6}$$

If P represents the proportion of cDNA that can react, then,

$$\frac{D}{C_0} = P - e^{-kR_0 t}$$

from equation [5]

$$\frac{D}{C_0} = P - e^{-0.691 R_0 t / R_0 t_{1/2}} \tag{7}$$

Two kinds of information may be obtained directly from an analysis of the kinetics of hybridization of excess RNA with a labeled cDNA. The sequence complexity of an unknown RNA population may be determined by comparing the hybridization of the unknown RNA with its cDNA to that of an RNA of known complexity with cDNA copied from it. Alternatively, the concentration of an RNA of known complexity in different RNA populations may be determined by comparison of the rates of hybridization ($R_0 t_{1/2}$ values) of the cDNA made from the mRNA with both the pure mRNA itself and with unfractionated RNA populations of interest. Numerous examples of both kinds of application exist in the literature. Initial experiments performed with cDNA probes were of the former type and were directed towards characterization of complex mRNA populations present in eukaryotic cells (26–33). In experiments of this kind, cDNA made against the total mRNA population of the cell of interest, usually defined as the total cytoplasmic or polysomal population of poly(A)-containing RNA molecules, is hybridized to the mRNA. Typical reactions, between chicken liver cDNA and its mRNA template and between HeLa cell cDNA and its mRNA template, are depicted in Figure 1 as plots of the fraction of the labeled cDNA hybridized against $\log R_0 t$. In both cases

A. HeLa cell cDNA and its mRNA template (26). B. Chicken liver cDNA and its mRNA template (31).

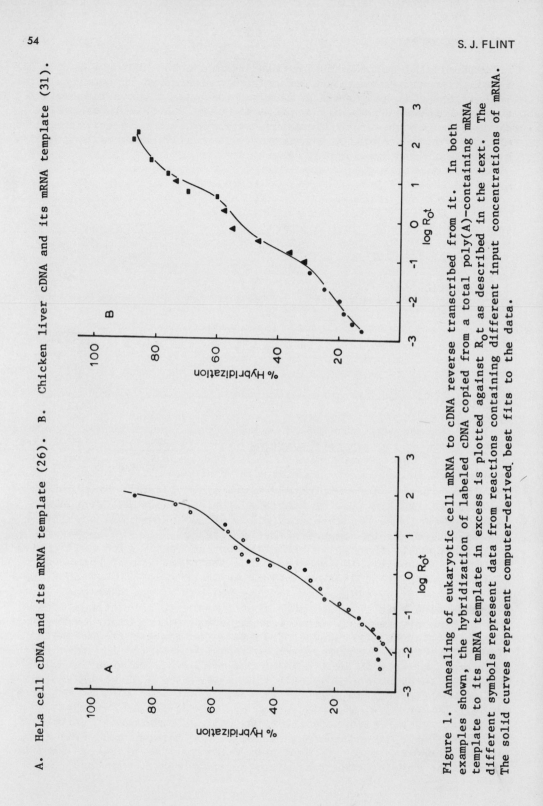

Figure 1. Annealing of eukaryotic cell mRNA to cDNA reverse transcribed from it. In both examples shown, the hybridization of labeled cDNA copied from a total poly(A)-containing mRNA template to its mRNA template in excess is plotted against $R_{o}t$ as described in the text. The different symbols represent data from reactions containing different input concentrations of mRNA. The solid curves represent computer-derived best fits to the data.

illustrated in Figure 1 the hybridization reaction extends over 4 to 5 orders of magnitude of $R_0t$ values, reaches a plateau when less than 100% of the cDNA has hybridized (about 87% and 85% in the examples shown in Figures 1A and 1B, respectively) and comprises several kinetic components. In the experiment shown in Figure 1A, three kinetic components of the reaction can be distinguished quite readily by visual inspection. However, detailed, and nonsubjective, analysis of this kind of reaction is best achieved by computer analysis, as might be inferred from the sort of result shown in Figure 1B.

In this situation, when the RNA is not a homogenous population, $D/C_0$ (equation [6]) represents the sum of a number of individual reactions, $D1/C_0 + D2/C_0 + D3/C_0$, each with its one terminal value of the proportion of cDNA entering hybrid, given by P1, P2, P3...Pn.

Then, from equation [7]

$$D/C_0 = Dn/C_0 \\ = Pn - e^{-0.691 Pn\, R_0t/R_0t_{1/2n}} \qquad . \qquad [8]$$

A computer program is then used to find the values of P and $R_0t_{1/2}$ for different numbers of components of the reaction that minimize the use of squares of deviations of the observed data points ($D/C_0$) from the calculated curve. These values, then, define each kinetic component of the reaction, that is each class of RNA sequences within the total population, and can be used to calculate both complexity and, with additional information, abundance of each RNA class. In Table 1, for example, are given the values of Pn and $R_0t_{1/2n}$ obtained from such a computer analysis of the hybridization curve shown in Figure 1B, which comprises three kinetic components. As shown in Table 1, these values are used to calculate the $R_0t_{1/2}$ value, which would be displayed by each class of RNA if it were pure and comprised all the mRNA represented by the value $R_0$. This $R_0t_{1/2}$ value may also be corrected for the poly(A) sequences of the mRNA, as these do not contribute to the hybridization reaction. The poly(A) content of an mRNA can, for example, be determined by titrating the mRNA preparation with labeled poly(U) (34,35). Bishop and his colleagues estimated that the HeLa cell mRNA preparations of Figure 1B contained about 8.5% by weight poly(A) and they applied a 10% correction to obtain the values given in column 4 of Table 1 (26). In order to interpret these data further, it is necessary to compare corrected $R_0t_{1/2}$ values with the $R_0t_{1/2}$ of a standard RNA of known complexity, annealing to its cDNA under identical conditions of hybridization. In the example given in Table 1, the number average molecular weight of the HeLa cell mRNA was estimated by sucrose gradient centrifugation to be $6.4 \times 10^5$ (26). The $R_0t_{1/2}$ value of purified $\alpha$-globin mRNA hybridizing with its cDNA was measured by Bishop and his colleagues to be $3 \times 10^{-4}$ moles liter$^{-1}$sec. The molecular weight of $\alpha$-globin mRNA was taken as $2 \times 10^5$, so that the $R_0t_{1/2}$ of a pure mRNA of molecular weight $6 \times 10^5$ was calculated to be $9 \times 10^{-4}$ moles liter$^{-1}$sec. The corrected $R_0t_{1/2}$ values shown in column 4 of Table 1 are then divided by this estimate of the $R_0t_{1/2}$ value of a pure

Table 1

Annealing of HeLa Cell mRNA and cDNA

| Trans- ition | % cDNA hybridized $Pn$[a] | $R_o t_{1/2}$ Observed | $R_o t_{1/2}$ Corrected[b] | Number of[c] differing sequences | Copies/cell[d] of each sequence |
|---|---|---|---|---|---|
| 1 | 0.22 | $5 \times 10^{-2}$ | $1.5 \times 10^{-2}$ | 17 | 8,090 |
| 2 | 0.28 | $9 \times 10^{-1}$ | $3.35 \times 10^{-1}$ | 370 | 470 |
| 3 | 0.55 | 45 | 29.9 | 33,000 | 9 |

[a]Obtained by computer analysis of the hybridization data shown in Figure 1B.

[b]Calculated from $R_o t_{1/2n}$ observed assuming that the RNA of each class was pure, i.e. contributed all the RNA represented by the value $R_o$. This value is also corrected for the presence of nonhybridizing poly(A) sequences in the mRNA.

[c]Calculated from $R_o t_{1/2n}$ corrected by dividing the $R_o t_{1/2}$ value, $9 \times 10^{-4}$ moles liter$^{-1}$ sec of a pure mRNA of $6 \times 10^5$ (the number average molecular weight of the HeLa cell mRNA preparation) estimated by comparison with an $R_o t_{1/2}$ value of $3 \times 10^{-4}$ moles liter$^{-1}$ sec exhibited by rabbit $\alpha$-globin mRNA, $2 \times 10^5$ (26).

[d]Calculated on the basis that $10^6$ HeLa cells contain 0.625 µgm of poly(A)-containing mRNA (26). One HeLa cell contains $6 \times 10^{23} \times 6.25 \times 10^{-7}/6 \times 10^5 \times 10^6$ copies of mRNA molecules, each of $6 \times 10^5$, or $6.25 \times 10^5$ copies of mRNA. The fraction of these present in each RNA class is given by the P value. Thus each different class 1 mRNA sequence is present at a concentration of $6.25 \times 10^5 \times 0.22/17 = 8090$.

mRNA species of the same average size of the HeLa cell mRNA population to yield the number of different sequences, each of $6 \times 10^5$, in each of the three RNA classes, shown in column 5, Table 1. The frequency with which each member of each class is represented, in copies per cell, can also be calculated from the total amount of mRNA present in the cell, Avogadro's number and the proportion of the total mRNA represented in each class, as outlined in Table 1.

The data summarized in Table 1 reveal that three abundant classes of mRNA can be distinguished in HeLa cells, although the vast majority of the different mRNA species are represented only rarely in the mRNA population. This pattern is quite typical of eukaryotic cell mRNA populations and has been described for a variety of cell

types including chicken oviduct and liver (172), mouse tissues
(173,174), rat liver (33), Drosophila melanogaster cells in culture
(28) and yeast (32).  Some comparisons between different mRNA populations, such as those present in different tissues of the same
organism, those present at different developmental stages and those
present in a certain cell type in different states (growing compared
to resting, transformed compared to nontransformed) have also been
made (28-31,36,37).  These studies reveal that while the great
majority of mRNA sequences appear common in any comparison of mRNA
sequences, some specific differences can be detected, particularly
when cDNA probes are fractionated into subpopulations of different
abundance by hybridization with mRNA to a specific $R_o t_{1/2}$ value
followed by fractionation of hybridized and unhybridized cDNA by
chromatography on hydroxylapatite or S1 nuclease digestion.  In this
way, Hastie and Bishop (30) were able to investigate the expression
of mouse kidney abundant, middle frequency and low frequency mRNA
sequences in other mouse tissues.  These authors concluded that the
majority of kidney mRNA sequences were also present in liver and
brain, although sequences that are abundant in one tissue may be
present in lower frequency classes in other tissues.  While permitting a general description of eukaryotic cell mRNA populations, it is
clear that this approach is not well-suited to detailed investiga-
tions of regulatory events that result in only small changes in the
overall RNA populations.

     More tailored to this end is the use, as probe, of cDNA specific
for one type of mRNA.  Such cDNA probes can be synthesized when it is
possible to obtain a purified mRNA; that is, when one kind of mRNA is
present in a certain cell type in sufficient amounts that it can be
purified by physiochemical procedures.  The method most commonly
employed has been fractionation on the basis of size in sucrose
gradients or polyacrylamide gels, under denaturing or nondenaturing
conditions, following isolation of the cytoplasmic or polysomal,
poly(A)-containing mRNA fraction by chromatography on oligo(dT)-
cellulose.  In this way, $\alpha$- and $\beta$-globin mRNAs (38-40), immunoglobulin mRNAs (41-43) and ovalbumin mRNA (44-46) from appropriate
sources, have been significantly purified as has histone mRNA from
poly(A)-lacking RNA fractions (47-49).  Some purification, 5- to
10-fold, of a specific mRNA that is of average size and does not
constitute a major fraction of the mRNA, can also be achieved by
immunoprecipitation of polysomes containing the mRNA and nascent
polypeptides specified by it (50-54).  An mRNA that is not especially
abundant may also be purified by hybridization.  An mRNA population,
enriched for the mRNA of interest, is transcribed into cDNA in the
usual way.  This cDNA, and the mRNA from which it was copied, are
then subjected to a limited hybridization such that only the most
abundant cDNA and mRNA sequences enter hybrid.  Separation of the
cDNA:RNA hybrid from unhybridized RNA and release of mRNA from the
hybrid then provide a preparation both of the most abundant mRNA
sequences in the population and cDNA complementary to it (see, for
example, ref. 55).  During any of these purification procedures, the

mRNA of interest is frequently monitored by translation in an in
vitro protein synthesizing system.  As we shall see, it is necessary
to obtain an estimate of the degree of purity of the final mRNA
preparation.  This may be made from the products of in vitro translation (see below) or from a determination of the fraction of the
mRNA having the characteristics of the specific mRNA, for example, by
size fractionation of a labeled mRNA preparation.  Once purified, the
mRNA is copied by reverse transcriptase into cDNA which is then
hybridized both to the template, "pure" mRNA and to other RNA preparations of interest under conditions of RNA excess.  The rate of
hybrid formation is assayed as described previously and plotted
against the product of the initial (and constant) RNA concentration,
$R_0$ and time of incubation, t, to give hybridization curves of the
sort depicted in Figure 2A.  As the complexity of the RNA under study
is constant in this kind of experiment, the rates of the two reactions being compared are determined solely by the concentrations of
RNA sequences complementary to the cDNA in the two hybridization
mixtures, all other parameters being equal.  Thus, as the rate
constant is inversely proportional to $R_0 t_{1/2}$ (equation [5]), the
concentration of mRNA sequences in the unknown sample may be calculated from the relation

$$\frac{R_0 t_{1/2} \text{ (mRNA)}}{R_0 t_{1/2} \text{(unknown)}} = \frac{\text{\% mRNA sequences in unknown}}{\text{\%mRNA sequences in the "pure"}} \cdot \qquad [9]$$
$$\text{mRNA preparation}$$

The $R_0 t_{1/2}$ values for the reactions of cRNA with the "pure"
mRNA and unknown RNA samples may be estimated directly from curves of
the sort shown in Figure 2A as the $R_0 t$ value (in moles liter$^{-1}$ sec)
at which half the maximum amount of cRNA that can hybridize has
annealed with RNA.  However, these estimates can be more conveniently and accurately made from double-reciprocal plots of hybridization against $R_0 t$, of the sort shown in Figure 2B.  These are based
on empirical equations of the form

$$\frac{1}{D} = \frac{h}{R_0 t} + \frac{1}{Ds} \qquad\qquad\qquad [10]$$

where
        D  =  fraction DNA hybridized at time t,
        Ds =  fraction DNA hybridized at infinite time,
        $R_0$ =  initial and constant RNA concentration,
        h  =  a constant,
which can be used to describe excess hybridization reactions (56,57).
In an ideal case, in which all the probe enters hybrid at the
termination of the reaction, such a plot of the hybridization data
should yield a straight line intersecting the ordinate at 0.01 (100%
hybridization).  In the example shown in Figure 2B, intersection is
at a value of 1/D of 0.012 so that a value of 85% is taken as the
fraction of cDNA hybridized in the presence of saturating amounts of

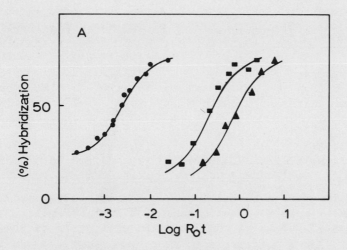

(A) The data are plotted as the extent of hybridization against log $R_{0}t$.

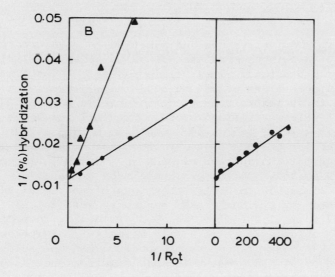

(B) Double-reciprocal plots of (A) as discussed in the text.

Figure 2.  Hybridization of ovalbumin cDNA with various hen oviduct RNA fractions (58).  The hybridization of cDNA reverse transcribed from purified ovalbumin mRNA to this template (●-●-●), polysomal RNA (▲-▲-▲) and nuclear RNA (■-■-■) in both panels.

Table 2

Hybridization of Ovalbumin cDNA to Hen Oviduct RNA Fractions

| RNA | Observed $R_o t_{1/2}$ mole liter$^{-1}$ sec | %[a] oval-bumin mRNA | µg RNA/mg DNA Total RNA | Ovalbumin[b] mRNA | Copies[c] ovalbumin mRNA/mg DNA |
|---|---|---|---|---|---|
| Polyri-bosomal | $1.58 \times 10^{-1}$ | 1.20–1.50 | 116 | 13.9–17.4 | $13.9$–$17.4 \times 10^{12}$ |
| Nuclear | $5.58 \times 10^{-1}$ | 0.33–0.42 | 229 | 0.28–0.36 | $0.25$–$0.36 \times 10^{12}$ |

[a]Calculated from the observed $R_o t_{1/2}$ value of ovalbumin cDNA hybridizing with "pure" ovalbumin mRNA of $2.5 \times 10^{-3}$ mole liter$^{-1}$ sec and an estimated purity of the ovalbumin mRNA of 75 to 95%.

[b]Values in column 3 multiplied by those in column 4.

[c]Calculated assuming a molecular weight of $6 \times 10^5$ (44,46) for ovalbumin mRNA.

complementary RNA. $R_o t_{1/2}$ values can then be read directly from such curves, in this case as the $R_o t$ value at which 42.5% of the input DNA is hybridized. Thus, it is possible to obtain by extrapolation accurate values of Ds and hence $R_o t_{1/2}$ even when insufficient RNA is available to achieve saturation of the cDNA probe.

In the experiment illustrated in Figure 2 and Table 2, $R_o t_{1/2}$ values of $2.5 \times 10^{-3}$, $1.58 \times 10^{-1}$ and $5.58 \times 10^{-1}$ moles liter$^{-1}$ sec were measured for the reactions of the "pure" ovalbumin mRNA, hen oviduct polysomal RNA and hen oviduct nuclear RNA, respectively, with ovalbumin cDNA (58). As the ovalbumin mRNA preparations were estimated to be 75 to 95% pure, the fractions of polysomal and nuclear RNA contributed by ovalbumin mRNA sequences were calculated from equation [9] to be 1.2 to 1.5% and 0.33 to 0.42%, respectively. As illustrated in Table 2, it is a trivial matter to express these data in terms of RNA mass or numbers of specific RNA molecules, provided that the amount of RNA present in each fraction under study (with allowance made for recoveries during experimental manipulation) and the molecular weight of the specific mRNA are known. If the amount of RNA or DNA present in one cell (and in this case the fraction of cells in the oviduct that synthesize ovalbumin) can also be estimated, RNA concentrations can then be expressed as copies of specific RNA per cell.

Given the amounts of useful information that can be gleaned from this relatively simple experimental approach, it is not surprising to discover that it has been applied widely to the study of individual eukaryotic genes and their regulation. Some examples include: the hormonal induction of mRNAs specifying egg white proteins in hen oviduct (58-62); the regulation of globin gene expression in erythro-blastic leukemia cells (63-65), and after transformation of chick embryo fibroblasts (66) and in human thalassemias (67,68); the syn-thesis of α-crystallin mRNA during lens induction in chick embryo (69); synthesis of albumin mRNA in rat liver (55); synthesis of protamine mRNA during development of trout testis (70); the ampli-fication of dihydrofolate reductase genes in methotrexate-resistant cells in culture (71); the synthesis of human placental lactogen mRNA during pregnancy (72).

However, interpretation of all experiments with cDNA probes must be made with some caution. One problem is the failure of all the probe (100% of the input cDNA) to enter the hybrid in the presence of saturating amounts of RNA. It is usual to observe 70 to 90% satur-ation, even when cDNA is hybridized to the mRNA template from which it was transcribed. As the nonhybridizing sequences must be comple-mentary to the mRNA, it would seem that a small fraction of the cDNA sequences exist in such a physical state that they either cannot enter hybrid or that the DNA:RNA hybrids formed do not register as hybrid in the assay used (see above). Provided this problem is recognized, it is a simple matter to correct for it. Much more serious are problems introduced by an unequal representation of sequences comprising the mRNA template in a cDNA product. Early attempts at reverse transcription generally employed one labeled deoxyribonucleoside triphosphate substrate at a concentration suf-ficiently lower than the other three to be rate-limiting. Under these conditions, a cDNA product of high specific activity can be synthesized, but a significant fraction, if not the majority, of the cDNA made is shorter than the mRNA template (22-24). As the reverse transcription of an mRNA is absolutely dependent on the presence of a primer, in these reactions oligo(dT), this observation implies that short cDNA chains are copies of only 3' sequences of the mRNA and that copies of the 5' end are present only in the less abundant, longer products. Consistent with this interpretation is the presence of relatively more oligo(dT) in the total cDNA product reverse trans-cribed from a globin mRNA template than in cDNA that is a full-length copy of the template (73) and all that is known of the mechanisms by which the natural templates, retroviral RNA genomes, are transcribed (see, for example, ref. 74). Thus, there is little doubt that the cDNA made in vitro in the presence of limiting deoxyribonucleoside triphosphate concentrations does not comprise an equimolar mixture of the template mRNA sequences; rather, sequences complementary to the 3' end of the mRNA will be represented preferentially. The extent of over-representation of such sequences cannot be predicted, nor can it be measured and corrected for conveniently. This problem is there-fore best avoided by the use of cDNA preparations in which all the sequences of the mRNA template are represented uniformly, namely cDNA that is a full-length copy of the template. It is now apparent that this goal can be achieved through the use of equimolar concentrations of all four deoxyribonucleoside triphosphates in the reverse trans-

cription reaction (74-76). The amount of full-length cDNA made in
such reactions apparently can also be increased by the use of limiting concentrations of $Mg^{+2}$ (77) and high concentrations of reverse
transcriptase (78) in the reaction as well as by the addition of
0.04 M sodium pyrophosphate (79). Full-length cDNA may be fractionated from shorter products by electrophoresis, preferably under
denaturing conditions, eluted from the gel and degraded to lengths of
300 to 400 nucleotides before use in hybridization experiments.

## Double-Stranded DNA Probes

An alternative source of labeled DNA is genomic DNA. The DNA of
the majority of eukaryotic organisms includes highly-repetitive,
middle-repetitive and nonrepetitive (unique or single-copy) DNA sequences (18). The complexity of the latter fraction varies from
about 20 times that of E. coli DNA in Drosophila to upwards of 500
times the E. coli value in mammals (80). Hybridization experiments
in DNA excess (see next section) established that the majority of
eukaryotic cell mRNA sequences are transcribed from the nonrepetitive
DNA fraction. This fraction, denoted here as single-copy DNA
(scDNA), as well as total genomic DNA, has been used as probe in RNA
excess hybridization experiments. The scDNA fraction can be isolated
by reannealing denatured, total genomic DNA of the organism under
study to $C_ot$ values at which most of the highly- and middle-repetitive sequences have reannealed, but the majority of scDNA sequences
have not. The single-stranded scDNA is then separated from reassociated, double-stranded DNA by hydroxylapatite chromatography and is
usually subjected to at least one more cycle of reassociation and
chromatography. Single-copy DNA prepared in this fashion is generally free of repetitive DNA sequences (see, for example, ref. 81). In
early experiments, genomic DNA was labeled, in vivo, but nick-transla-
tion of the purified DNA in vitro (82) now permits higher specific
activities to be attained.

In the first application of this method, Galau et al. (81)
hybridized sea urchin [3]H-scDNA to different preparations of sea
urchin gastrula polysomal mRNA in vast RNA excess. All mRNA preparations behaved in the same fashion and saturation of the scDNA
probe was achieved soon after a calculated $R_ot$ value of 200, with no
significant further reaction by an $R_ot$ value of 1200. In RNA excess,
the hybridization reaction can, in principle, again be described
(25,170) by the equation

$$\frac{C}{C_o} = e^{-kR_ot} \quad . \tag{11}$$

As demonstrated by Galau et al. (81), hybridization to labeled
genomic DNA does indeed display the predicted first-order kinetics,
provided that the RNA concentration is sufficiently high.

The sequence complexity of the mRNA reacting with scDNA may be estimated very simply from the observed saturation data. In the experiments of Galau et al. (81), for example, 1.02% of the input $^3$H-scDNA entered hybrid at saturation with polysomal mRNA. However, when the same labeled scDNA preparation was reannealed with an excess of total sea urchin DNA, only 75% entered the hybrid, presumably because the remainder of the labeled DNA was rendered too small by the purification procedures to which it was subjected to form detectable DNA:RNA hybrids. Thus, the 1% of input $^3$H-scDNA probe observed to hybridize with mRNA actually represents 1.35% of the input DNA or 2.7% of the total single-copy DNA sequences of the genome. The sea urchin genome comprises $8.12 \times 10^8$ base pairs, 75% of which are found in the single-copy fraction. Thus, the complexity of the mRNA is $8.12 \times 10^8 \times 0.75 \times 0.27$ which is $1.7 \times 10^7$ nucleo-tides or, assuming an average value of 1200 nucleotides (not includ-ing poly(A)) for an animal cell mRNA, some 14,000 different mRNA sequences (81). As this DNA:RNA hybridization reaction does exhibit first-order kinetics, the proportion of the driver RNA hybridizing to the DNA probe can be calculated from the measured rate of the reac-tion by comparing the observed $R_o t_{1/2}$ value with that of a standard reaction with an RNA of known complexity, as described in the pre-vious section for cDNA probes. Galau et al (81), for example, esti-mated that only 8% of the total mRNA hybridized to the scDNA probe, suggesting that the vast mass of the polysomal mRNA must consist of RNA sequences repeated a large number of times and having a complex-ity of only 5 to 10% that of the mRNA molecules hybridizing to scDNA (81). Clearly, then, the data collected from hybridization of an excess of RNA with a double-stranded scDNA probe can provide esti-mates of both the complexity of complementary RNA sequences and the proportion of the total mRNA population of different classes of eukaryotic cell mRNA: that is, the same kinds of information as obtained by hybridization with labeled cDNA (see Table 1). Simi-larly, complex DNAs of this kind have also been used to compare pop-ulations of unique RNA sequences, such as those found in the poly(A)-containing and poly(A)-lacking fractions of the same cell (83), in polysomal and nuclear fractions (84) and in the mRNA extracted from different cell types of the same organism (85).

Until very recently, labeled double-stranded DNA probes of a specific nature had not been employed in RNA excess hybridization experiments, simply because such discrete DNAs were not available. With the advent of molecular cloning, however, isolation of such specific DNA segments is no longer an impediment and some studies of this kind have appeared in the literature. A particularly inter-esting example is that reported by Tsai et al. (86) in which the expression of mRNA-coding sequences of the ovalbumin gene was com-pared to the expression of intervening sequences. In the chicken genome, the DNA sequences encoding ovalbumin are separated into eight nonadjacent segments by seven sets of noncoding intervening sequences (87-91). The intervening sequences appear to be trans-

cribed, although they are not present in mature ovalbumin mRNA (62).
Single-stranded DNA complementary to mRNA was prepared by hybridization of denatured, nick-translated, cloned DNA fragments to excess,
purified ovalbumin mRNA.  S1 nuclease digestion and alkaline hydrolysis were then used to separate the hybridized DNA from DNA remaining single-stranded and RNA, respectively.  In this way, a probe
equivalent to a full-length cDNA copy of purified ovalbumin mRNA was
prepared (62,86).  Two DNA probes for intervening sequences of the
ovalbumin gene were obtained in the form of cloned EcoRI fragments of
genomic DNA, termed $OV_{2.4}$ and $OV_{1.8}$ (92), only 7% and 5%, respectively, of the sequences of which hybridize to purified ovalbumin
mRNA (62).  To examine the expression of coding and intervening
sequences, these probes were hybridized to various RNA fractions
isolated from hen oviduct (86).  With all RNA preparations tested,
essentially 100% of the coding-sequence probes and 45 to 50% of the
intervening DNAs sequence $OV_{2.4}$ and $OV_{1.8}$ DNA probes entered
hybrid in the presence of saturating amounts of RNA, as expected from
the nature of the labeled DNAs if only one strand of genomic, coding
and intervening DNAs are transcribed in vivo.  The fact that only 45
to 50% of labeled, double-stranded DNA enters hybrid at saturation
indicates that the reaction is in such vast RNA excess that there is
no opportunity for DNA:DNA reassociation to occur.  Thus, valid
comparisons between the reactions with single-stranded and
double-stranded DNA probes can be made.  Some of the results obtained
are summarized in Table 3.  These are based on the principles
discussed previously for RNA excess hybridizations and permit
calculations of the absolute concentration, in copies/cell, of RNA
sequences complementary to coding and intervening sequences of the
ovalbumin gene.  From the data summarized in Table 3, for example, it
is clear that although intervening sequences of the ovalbumin gene
are transcribed and present in nuclear RNA, they are essentially
restricted to that organelle and do not reach the polyribosomes.
Thus, this kind of quantitative experiment with discrete DNA probes
can provide much valuable information about the expression  and
regulation of eukaryotic genes.  Undoubtedly, this approach will gain
in popularity as additional, specific cloned DNA segments become
available.


## Comparison of RNA excess Hybridization with cDNA and scDNA Probes

Under ideal circumstances, measurements of mRNA complexity and
concentration made for a given mRNA preparation with both labeled
cDNA and scDNA should yield similar results.  In a few systems,
comparable estimates have, in fact, been obtained by the two methods
(31,93), but in the majority of cases, considerable discrepancies
between the two sets of data have been observed; some examples are
listed in Table 4.  The magnitude of the discrepancies, two-fold or
greater, is such that they are unlikely to reflect only experimental

Table 3

Hybridization of Oviduct RNA to Structural and Intervening
Sequences of the Ovalbumin Gene

| Probe | RNA | $R_0t_{1/2}$ observed[a] mole liter$^{-1}$sec | % ovalbumin RNA in total RNA | RNA:DNA | Molecules[b] RNA/tubular gland cell |
|---|---|---|---|---|---|
| Structural | Total | $5.4 \times 10^{-1}$ | $5.6 \times 10^{-1}$ | 3.5 | 58,000 |
| gene | Nuclear | $9.2 \times 10^{-1}$ | $3.3 \times 10^{-1}$ | 0.25 | 2,500 |
| sequences | Polysomal | $5.5 \times 10^{-1}$ | $5.5 \times 10^{-1}$ | 2.9 | 47,000 |
| Intervening | Total | $1.4 \times 10^{-2}$ | $2.1 \times 10^{-3}$ | 3.5 | 228 |
| sequences, | Nuclear | $1.0 \times 10^{-1}$ | $3.0 \times 10^{-2}$ | 0.25 | 233 |
| $OV_{2.4}$ | Polysomal | $1.3 \times 10^{3}$ | $2.3 \times 10^{-4}$ | 2.9 | 20 |

[a]Calculated as described in Figure 2 with $R_0t_{1/2}$ values of $3 \times 10^{-3}$ mole liter$^{-1}$ sec for pure ovalbumin mRNA hybridizing to single-stranded complementary DNA and $2.9 \times 10^{-3}$ mole liter$^{-1}$ sec for $OV_{2.4}$ DNA hybridizing to complementary RNA, calculated from the complexity of $OV_{2.4}$ DNA, 1900 nucleotide pairs compared to that of ovalbumin coding sequence probe, 1930 nucleotides (62).

[b]Calculated from the relation

$$\text{molecules RNA}_{ov} = \text{Fraction} \times \frac{\text{RNA}}{\text{RNA}_{ov}} \times \frac{\text{RNA}}{\text{DNA}} \times 2.6 \times 10^{-12} \times \frac{6.02 \times 10^{23}}{\text{m.w. RNA}}$$
$$(\text{g DNA/cell})$$

using molecular weights of the structural and $OV_{2.4}$ RNA sequences of 1930 and 1900 nucleotides, respectively. Also taken into account in the values presented in column 6 is the fact that only 80% of the total oviduct cells are tubular gland cells producing ovalbumin.

Table 4

Comparison of cDNA and scDNA Hybridization Data

| Organism | Number of unique RNA sequences | | Refs. |
|---|---|---|---|
| | scDNA Hybridization | cDNA Hybridization | |
| Yeast | 4,000 | 3,000 | 32 |
| Mouse L cells | 26,000 | 10,000 | 109-111 |
| Rat Liver | 31,000 | 23,000 | 33 |
| Parsley leaves | 25,000 | 9,200 | 102 |
| Tobacco leaves | 27,000 | 12,000 | 100,108 |

error.  As discussed in an earlier section, cDNA hybridization may
suffer from problems introduced by unequal representation of different sequences of the template mRNA in the cDNA product.  On the
other hand, it has been suggested recently (94) that the genomic
scDNA fraction isolated by limited reassociation of total genomic DNA
may not reflect accurately the scDNA sequence content of the cell.
This idea stems from the observation that in many eukaryotic genomes,
unique DNA sequences are intimately interspersed with repetitive DNA
sequences (for example, refs. 95,96) and the demonstration that some
unique DNA sequences do not behave as unique during reassociation,
even when the DNA is fragmented to relatively short lengths (about
400 nucleotides) because they are linked to short, repetitive DNA
sequences (97).  Thus, it is likely that scDNA probes, especially
those prepared from the DNA of higher plants, which contain very
large amounts of repetitive sequences interspersed throughout the
majority of unique DNA sequences (98-102), do not, in fact, contain
all the unique or single-copy sequences of the genome.  It has
therefore been suggested (94) that the scDNA fraction isolated by
limited reassociation is likely to be enriched for the longer unique
sequences of the genome, including those complementary to mRNA,
resulting in over-estimation of numbers of mRNA sequences when such
DNA is hybridized to mRNA.

Another potential source of over-estimation of the fraction of
an scDNA probe complementary to mRNA arises from the presence of
intervening sequences in the DNA probe.  As discussed previously for
the ovalbumin gene, such sequences are not expressed as mRNA.  However, they will register as DNA:RNA hybrids in the hydroxylapatite
assay if they are present in a segment of scDNA that also includes a
stretch of coding sequences of sufficient length, a minimum of 50
nucleotides (8), to bind the entire scDNA segment when the coding
sequences have hybridized to mRNA.  This source of error can be
eliminated readily by the use of assays that measure only the
fraction of labeled scDNA that has actually formed a hybrid with
mRNA, such as digestion with S1 nuclease.

Given the problems associated with these RNA excess hybridization methods, it is not surprising that discrepancies of the sort
illustrated in Table 4 exist in the literature.  What is more important, perhaps, is the fact that the majority of the difficulties
noted in this discussion can be obviated by the newer technologies
available, at least in the study of individual mRNA species.  Most
significantly, all problems associated with uncertainties about the
precise nature of the DNA probe can be eliminated by the use of
full-length cDNA preparations or well-characterized specific cloned
DNA segments, or subfragments derived from a cloned DNA.  In addition, in vitro labeling techniques permit labeled DNA of high
specific activity to be prepared so that it becomes far easier to
attain levels of RNA excess at which even the most infrequently-
represented RNA sequences within a population will hybridize.
Indeed, the use of high specific activity probes should also minimize
problems introduced by viscosity effects and nucleic acid degradation

during the hybridization as lower concentrations of input RNA and/or
shorter incubation times are needed to attain saturation of the
lower amounts of DNA that can be used.  Thus, we are now in the
happy position of being able to avoid most of the parameters
affecting RNA excess hybridization reactions that are difficult
to predict quantitatively.

## DNA EXCESS HYBRIDIZATION

### Labeled DNA Probes

The steady state concentration of RNA sequences complementary to
a given DNA can also be measured by hybridization between labeled,
single-stranded DNA and unlabeled RNA under conditions of DNA excess.
The hybridization of varying concentrations of RNA and complementary,
single-stranded DNA may be described by the second-order rate
equation

$$\frac{C_o}{C} = \frac{R_o e^{K_h t}(R_o - C_o) - C_o}{R_o - C_o}$$  [12]

where $R_o$ and $C_o$ are the initial concentrations of RNA and
complementary single-stranded DNA, C is the concentration of DNA
remaining single-stranded at time t and $K_h$ is the second-order rate
constant (107).  Using the approximation $e^x = 1 + x$ for the initial
stage of hybridization,

$$\frac{C_o}{C} = \frac{R_o (1 + K_h t (R_o - C_o) - C_o}{(R_o - C_o)}$$

$$= R_o K_h t \frac{(R_o - C_o) + (R_o - C_o)}{R_o - C_o}$$

$$= 1 + K_h R_o t \ .$$  [13]

Thus, to determine $K_h$, known concentrations of DNA and pure
complementary RNA are allowed to anneal and the observed
hybridization data plotted as $C_o/C$ against $R_o t$ to yield a straight
line with an intercept of 1 on the ordinate and a slope equal to $K_h$.
In a mixture of RNA in which only a fraction, f, is complementary to
the probe DNA, $R_o = fC_r$ where $C_r$ is the total input RNA
concentration,

$$\frac{C_o}{C} = 1 + K_h f C_r t \ .$$  [14]

Thus a plot of $C_o/C$ versus $C_r t$ for the hybridization of the RNA of interest to an appropriate single-stranded DNA probe yields a straight line whose slope equals $K_h f$. As $K_h f$ can be measured empirically from the hybridization of the standard DNA and RNA, f can be calculated readily from the measured slopes of such plots. This procedure assumes only that the physical state of the RNA under test is the same as that of the standard RNA. It is, however, necessary to obtain an estimate of the value of $C_o$, the maximum amount of input DNA capable of hybridizing with the most abundant RNA, examined during the initial phase of the reaction. This value can be determined empirically by allowing the reaction to continue to saturation and assuming the RNA sequences complementary to the DNA constitute a homogenous population or the experiment can be designed such that all, or a known fraction, of the sequences comprising the labeled DNA are complementary to the RNA under assay (107). In an analysis of the expression of SV40 early genes in cells infected by mutant and wild-type virus, for example, Reed and his colleagues (107) used as probe labeled DNA of the E strand of the SV40 genome, 50% of which had been established to be complementary to early viral RNA. Using a standard reaction of the same labeled DNA and cRNA transcribed from the SV40 genome by E. coli RNA polymerase in vitro, which is asymmetric and contains all sequences of the SV40 genome in approximately equimolar proportions (103), these authors obtained an estimate of $K_h$ equal to 2.58 ml/$A_{260}$ hr, where $A_{260}$ is a unit of nucleic acid mass equivalent to 40 $\mu$g of RNA (107). This value could then be used to quantitate the changes in early viral mRNA accumulation observed in cells infected by certain SV40 mutants using the methods outlined in the previous paragraph.

An alternative treatment of hybridization between unlabeled RNA and labeled complementary single-stranded DNA that avoids the necessity of direct measurement of $K_h$ has also been suggested (104). We can write the second-order reaction between varying, relative concentrations of RNA and complementary single-stranded DNA (171,172) in the form

$$\frac{dh}{dt} = K_h \ (P-h) \ (R-h) \qquad\qquad\qquad [15]$$

where  h = molar concentration of DNA:RNA hybrid at time t,
       P = initial DNA concentration,
       R = initial RNA concentration,
      $K_h$ = second-order rate constant.
Upon integration, and with h=0 when t=0, equation [15] gives

$$\frac{1}{(P-R)}\ln\frac{R(P-h)}{P(R-h)} = K_h t \quad . \qquad\qquad\qquad [16]$$

If we define $f_h$ as the fraction of single-stranded DNA in hybrid at time t, $f_h = h/P$ , and Z as the initial ratio of complementary RNA:DNA concentrations, Z = R/P and substitute in equation [16]

$$\frac{1}{(1-Z)}\ln\frac{Z(1-f_h)}{(Z-f_h)} = K_h Pt \quad . \qquad\qquad [17]$$

The value of the term on the right of equation [17] can be calculated
by assuming that the rate constant of this DNA:RNA annealing reaction
is equal to the rate constant of DNA renaturation under identical
conditions. Then $K_h P$ would equal the reciprocal of the time required
for half-renaturation of complementary denatured DNA with initial
concentration 2P, a parameter that is readily measured or calculated
for DNA of known complexity. When this assumption is used to estimate $K_h Pt$, it is found from equation [17] that at high values of this
function, i.e. in the presence of high concentrations of input DNA
and long incubation times, Z is approximately proportional to the
fraction of input labeled single-stranded DNA that enters the
hybrid, $f_h$.

In reconstruction experiments, SV40 cRNA transcribed by E. coli
RNA polymerase in vitro has been hybridized with $^{32}$P-labeled DNA of
the complementary DNA strand of defined restriction endonuclease
fragments of SV40 DNA under conditions of DNA excess (104). As
predicted, $f_h$ was indeed observed to be a linear function of the
RNA:DNA ratio (Z), at least during the initial phase of the hybridization reactions, at values of Z < 0.6 (104). Moreover, theoretical
curves calculated from equation [17] for the range of values of Z
used in the reconstruction experiment described the observed data
points extremely well (104).

Thus, this treatment would appear to be valid, suggesting that
relatively little error is introduced by the major assumption made,
that of equal rates of DNA:DNA reassociation and DNA:RNA annealing.
In fact, direct comparisons between the rate constants of the two
kinds of reaction suggest that they differ by a factor of only 1.25
to 1.30 (105,106) (see below for further discussion). Thus, the
molar ratio of complementary RNA to DNA probe sequences could be
determined from measurement of the fraction of input DNA that enters
hybrid in the presence of increasing concentrations of RNA, when the
input DNA concentration is sufficiently high to drive the reaction to
completion in the incubation time used. In practice, the quantity of
input RNA, x mg, that is required to half-saturate the labeled DNA
under conditions of initial DNA excess is measured and is assumed to
contain one-half the molar amount of DNA, y μg, that can be saturated
by the RNA preparation under study. Thus, one knows immediately that
x mg of the RNA contains y/2μg of RNA complementary to the probe DNA;
that is, the fraction of unlabeled RNA complementary to the labeled
DNA is measured. This value may be expressed as copies of mRNA per
cell by calculations like those outlined in the legend to Table 2.

Although this DNA excess hybridization method has been developed
with the separated strands of viral DNA or purified restriction
endonuclease fragments of viral DNA (104,107), it is applicable
whenever a specific DNA complementary to the RNA of interest can be
isolated, labeled (in vivo or in vitro) and the strands of DNA

separated.  Although DNA excess is required, the quantities of
single-stranded DNA required to meet the conditions described are, in
fact, not excessive when individual RNA species are being studied.
In a study of SV40 gene expression in cells infected by wild-type and
mutant virus, Reed et al. (107), for example, employed only 0.25 ng
of labeled single-stranded viral DNA in each hybridization reaction.
Thus, large quantities of neither DNA probe nor test RNA are consumed
in such experiments.  The only impediment in the application of this
technique with cloned DNA probes might be a failure to achieve separation of the strands of the DNA.

## Unlabeled DNA Probes

The classic use of DNA excess hybridization is that of excess
unlabeled DNA and a labeled RNA preparation, permitting saturation
experiments to be performed and the fraction of input labeled RNA
complementary to the DNA to be measured.  Provided that other parameters, such as the molecular weight (or average molecular weight) of
the RNA and its specific activity, can also be determined, this
experimentally-measured value can be used to calculate complementary
RNA concentrations in terms of RNA mass, or numbers of RNA molecules,
as outlined in previous sections.  This method can be applied to the
study of either steady-state RNA populations or newly-synthesized RNA
molecules, depending on whether the RNA is labeled for relatively
long or short periods of time.

The DNA probes originally employed in DNA excess hybridization
reactions comprised total genomic DNA or genomic DNA fractionated on
the basis of its reiteration frequency by limited reassociation.
Such DNA is fragmented, denatured and hybridized in aqueous solution
to small quantities of the labeled mRNA of interest.  Under conditions of sufficient DNA excess, this kind of hybridization reaction
exhibits the predicted first-order kinetics and the data obtained are
usually presented as a plot of percentage hybridization against $C_o t$,
where $C_o$ is the initial (and constant) DNA concentration.  The interpretation of kinetic parameters measured in such experiments depends,
as with RNA excess hybridization experiments, on comparison with a
standard reaction in which an RNA and its complementary DNA of known
complexity are annealed under identical conditions.  It is, for
example, possible to estimate the reiteration frequency of the DNA
sequences hybridizing to the labeled RNA tracer from the experimentally measured $C_o t_{1/2}$ values of the reactions with the unknown
and standard RNA preparations by calculations analogous to those
described for RNA excess hybridization.

When applied to total eukaryotic cell mRNA, early experiments of
this kind revealed that the vast majority of the mRNA sequences were
transcribed from unique sequence DNA (112-117), although reiteration
of specific mRNA sequences, such as the histone mRNAs, could be
detected readily (118-120).  This approach can be extended to gain
further insights into the nature and expression of mRNA sequences and

their arrangement in the genome through the use of fractionated
genomic DNA probes, such as scDNA and sequences contiguous to repetitive DNA sequences (see, for example, ref. 121) and by similar analysis of nuclear RNA sequences (see, for example, refs. 9,122,126).

All quantitative analyses of DNA excess hybridization data,
however, require determination of the saturation value at the termination of the reaction (so that the value of $C_{o}t_{1/2}$ may be determined). This may be estimated by visual inspection of hybridization
curves and corrected for the observed reannealing of the DNA (9,
123). However, even when most of the driver DNA has reassociated,
the fraction of labeled RNA that enters hybrid is frequently observed
to be less than 75% of the input RNA (11,121,124,125). The reason
behind this failure to drive the tracer RNA completely into hybrid is
not understood. It is noteworthy that under conditions of vast DNA
excess, the rate of DNA:RNA hybridization is significantly retarded
relative to that of DNA reassociation (121,124,125). Obvious
explanations of this latter phenomenon include competition between
homologous DNA and RNA sequences for a complementary DNA sequence or
a decreased reactivity of single-stranded DNA on molecules that
contain duplex DNA structures (124). These can be excluded by the
finding that DNA excess reactions between purified single-stranded
DNA and its complementary RNA are also retarded relative to
hybridization of the same DNA to its complementary DNA strand (125).
This phenomenon must be taken into account when curves are fitted to
the observed hybridization data with second-order or pseudo
first-order rate equations (see, for example, ref. 121). Moreover,
the failure to attain 100% hybridization of the tracer RNA by the
termination of the hybridization suggests that parameters such as the
$C_{o}t_{1/2}$ value of a reaction can best be measured from double-
reciprocal plots of the fraction of RNA in hybrid against DNA concentration, as discussed previously.

Although simple in principle, it is clear from the preceding
paragraphs that DNA excess hybridization is not so straightforward in
practice when the driver DNA can not only reanneal with the tracer
RNA but also renature with complementary DNA sequences during the
reaction. A variety of procedures have therefore been adopted to
prevent DNA:DNA reassociation during the hybridization. The first
relied on immobilization of denatured DNA upon nitrocellulose
filters, such that reassociation of complementary DNA strands is
inhibited (127). Under these conditions, however, the rate of
hybridization is significantly slower than with DNA in solution
(127-129) simply because DNA segments are not free, thus reducing the
chance of collision with complementary RNA sequences. In order to
estimate the amount of mRNA hybridized at saturation, it is therefore
frequently necessary to extrapolate the experimental hybridization
data to infinite DNA concentration (or infinite time). A variety of
artifacts have been associated with filter hybridization. During the
course of the reaction, for example, noncovalently-bound DNA may
become detached from the filter. Any released DNA is then available
to hybridize with labeled RNA in solution above filter-bound DNA,

resulting in an underestimation of the fraction of labeled RNA complementary to filter-bound DNA. Underestimation may also result from loss of DNA:RNA hybrids from filter as hybridization continues. If the DNA probe is completely complementary to the RNA under study, then all sequences of a DNA segment will eventually enter hybrid, leaving no free single-stranded regions through which the DNA can bind to the filter. Thus, hybrids may be lost from the filter as individual DNA chains become saturated (130). Despite these problems, filter hybridization has been very widely employed and even now is used with discrete DNA probes, such as those derived from viral DNAs (for example, ref. 131) or purified by molecular cloning (ref. 132, for example).

Problems associated with the loss of DNA from nitrocellulose filters can be circumvented by covalent coupling of DNA to a solid matrix such as Sepharose or cellulose (133-138). This approach also renders DNA excess hybridization with single-stranded cDNA probes a practicable alternative. The use of single-stranded cDNA in excess, of course, avoids any of the difficulties introduced by concomitant DNA reassociation during the DNA:RNA hybridization, but application of this method has been impeded by the limited availability of sufficient quantities of cDNA to attain saturation. When the cDNA is coupled to a solid support, it can be recovered at the end of the hybridization, making this approach more economical and therefore more attractive. Another advantage of hybridization with the driver DNA covalently attached to a resin like Sepharose or cellulose is the high rates of reaction observed. At high ratios of bound DNA to tracer labeled RNA, the rate of hybrid formation approaches that observed when the DNA driver is in solution (139). Finally, whenever the driver DNA is attached to some kind of support, it becomes easier to reduce background values. The amount of labeled RNA entering hybrid is most frequently assayed as the RNA resistant to RNase digestion in 0.3 M NaCl. Under these conditions, however, any double-stranded RNA will also survive digestion so that a background of RNase-resistant material is observed in the absence of any added DNA, especially when the labeled tracer comprises nuclear RNA. When DNA:RNA hybrids are linked to a solid matrix, such RNase-resistant material can be separated from hybrids simply by extensive washing; when hybridization is carried out in solution, such a facile separation is not possible. The attainment of low backgound is particularly important to the analysis of RNA species that represent a small proportion of the total population. DNA excess experiments with DNA coupled to a solid matrix have, among other applications, been used to study the synthesis of globin RNA (139,140) and to purify and analyze viral mRNA species (134-136,141).

An alternative that has yet to be adopted widely is the use of conditions of solution hybridization that permit the formation of DNA:RNA hybrids but not DNA:DNA reassociation. In the presence of high concentrations of formamide, DNA:RNA hybrids are significantly more stable than the equivalent DNA:DNA duplexes (57,142). It is therefore possible to incubate mixtures of denatured DNA and com-

plementary RNA at a temperature greater than the $T_m$ irreversible for
the DNA, but below that of corresponding DNA:RNA hybrids.  As this
temperature depends on the G+C content of the nucleic acids under
study, it is necessary to determine it empirically for the DNA under
study and the hybridization conditions employed, generally 80%
formamide, 0.04 M NaCl and 0.04 M Pipes buffer, pH 6.4 (142).  This
method has the advantage of convenience and the assurance that all
labeled complementary RNA can be driven into hybrid (see, for
example, ref. 143) but, like the others described, has its drawbacks
too.  These include the following.

(a)  A slow rate of hybridization, approximately one-twelfth that
of an equivalent reaction in aqueous solution (142), necessitating
the use of high DNA concentrations or long incubation times.

(b)  Relatively high backgrounds, presumably because of the
formation of RNase-resistant RNA duplexes that cannot be separated
from DNA:RNA hybrids by any simple procedure.  With labeled nuclear
RNA, for example, 1 to 3% of the input RNA has been observed to be
resistant to digestion with a mixture of RNases A, T1 and T2 after
incubation in the absence of DNA (143).

(c)  Failure to find suitable conditions of incubation if the
DNA includes sequences of widely-differing G+C content.  A temperature at which even the DNA sequences with the highest content of
G+C cannot reanneal may, for example, be too high to permit formation
of DNA:RNA hybrids that have a significantly lower G+C content.
Alternatively, G+C rich DNA sequences may reanneal at a temperature
of incubation that favors formation of hybrids between sequences that
are A+T rich.  However, incubation of the hybridization mixture over
a reverse temperature gradient can be used to ensure that all RNA
sequences have the opportunity to hybridize with their complementary
DNA sequences.

This method, then, is presently most suited to the analysis of
labeled RNA sequences that represent a fairly high proportion of the
total RNA population under study, such as certain classes of viral
RNA (for example, ref. 143).


NONHYBRIDIZATION METHODS


In Vitro Translation

Translation of mRNA in a heterologous, cell-free protein
synthesizing system (144-146) provides a sensitive assay for the
detection of a specific mRNA, particularly when translation systems
with a low endogenous background, such as that from micrococcal
nuclease-treated extracts of rabbit reticulocytes (147), are used.
At RNA concentrations below saturation of the system, the amount of a
given protein made is proportional to the amount of its mRNA that is
added, so that it is possible to estimate the concentration of a
given mRNA present in different mRNA preparations by quantifying the

amount of labeled polypeptide made in response to the addition of
constant amounts of the different RNAs.  The absolute amount of the
mRNA of interest cannot be determined by this method, but some
estimate of the fraction of the total mRNA that specifies the
polypeptide(s) of interest and comparison between different mRNA
populations can be made.  This method is frequently employed to
monitor mRNA purification procedures and to estimate the final degree
of purity of a specific mRNA.  It has also been used to investigate
the response of particular cell types to external stimuli, i.e. to
demonstrate the presence of enhanced amounts of mRNAs specifying the
egg-white proteins ovalbumin, ovamucoid and conalbumin following
estrogen and/or progesterone treatment of hen oviduct (148,149), to
investigate the induction and stability of interferon mRNA (150,151)
and the induction of vitellogenin mRNA in Xenopus liver by estradiol
(152).  When applying this method, it is important to be sure that
the translation system is not saturated with mRNA and that mRNA
preparations are free of potential inhibitors of translation, such as
hemin (153,154) and double-stranded RNA (155,156).  Unlike the other
methods described here, this cannot be applied to mRNA sequences
present in immature precursor molecules.


                    Direct Measurement of Specific RNA Species

        In certain rather special circumstances, it is possible to
obtain an estimate of the concentration of specific mRNA by direct
measurement of RNA concentration or the amount of label in the RNA.
Clearly, this approach hinges upon the ability to separate the mRNA
of interest from all others or to label it exclusively.  Such special
circumstances include the following.
        (a)  Those in which the mRNA of interest comprises a very high
proportion of the total mRNA population and can therefore be purified
with relative ease.  It has, for example, been reported that globin
mRNA constitutes 98% of the poly(A)-containing mRNA fraction in circulating reticulocytes of anemic mice (140).
        (b)  Those in which the mRNA exhibits unusual structural
features that facilitate its purification.  The histone mRNAs, for
example, are of a characteristic, and relatively small, size and,
unlike the great majority of eukaryotic cell mRNAs, lack 3'-terminal
poly(A) sequences (47).
        (c)  Those in which only specific mRNAs are labeled or trans-
lated as occurs following infection with certain animal viruses.  At
appropriate times after infection with adenoviruses (143,160) or
rhabdoviruses (157-159) only viral mRNA sequences are labeled,
whereas in picornavirus-infected cells only viral mRNA species are
found in polyribosomes (161,162).  Similarly, when Drosophila cells
are subjected to heat-shock, a very restricted class of mRNA molecules are translated and can be isolated from polyribosomes
(163-167).

In such cases, the cells can be labeled with RNA precursors under conditions suitable for specific labeling and the appropriate mRNA fraction purified and assayed for the specific mRNAs of interest, e.g. by size fractionation. The amount of label present in each individual species of mRNA can then be determined directly by counting. Because of the difficulty in determining the specific activity of the individual labeled mRNAs of interest, this method is again most suited to the determination of relative concentrations of an mRNA, or mRNAs, under different circumstances. For example, it has been applied to an analysis of the effects of temperature-sensitive mutations in the genome of the rhabdovirus, vesicular stomatitus virus, on the synthesis of individual viral mRNA species (168).

## CONCLUSIONS

Quite a variety of hybridization methods that can permit the estimation of absolute as well as relative concentrations, have been exploited in the measurement of mRNA concentration over the last decade. The examples used to illustrate these various methods have been drawn from the literature describing eukaryotic cell mRNAs, but these techniques can be applied to any other RNA population of the cell, such as total nuclear RNA or HnRNA. The choice of method to be adopted is governed first by whether steady-state RNA or newly-synthesized RNA is the subject of investigation. In the latter case, the experimenter is limited to DNA excess hybridization between pulse-labeled RNA and unlabeled DNA. In the past, experimental design has also often been determined by the availability of reagents, most notably of large quantities of suitable DNA probes. Thus, RNA excess hybridization has in general been employed more frequently. However, many of these barriers can be lifted when the DNA sequences complementary to the RNA of interest can be purified and amplified by molecular cloning. The nature of the RNA sequences to be analyzed may also influence the choice of specific method. RNA sequences that represent only a very small fraction of the total RNA population, for example, may perhaps be best detected using single-stranded immobilized DNA. On the other hand, convenience of obtaining the necessary reagents is not a factor to be ignored. In summary, no hard and fast rules can be drawn to govern the choice of hybridization method, which may in the end be determined as much by preference of the investigator as by the considerations listed here.

As intimated in this article, quantitative analysis of both complex mRNA populations and specific mRNA sequences has provided a great deal of information about the general organization and expression of the genome of a variety of eukaryotic cells and the regulation of expression of a few individual genes. It is therefore only to be expected that studies of this kind will flourish as the quality, variety and sophistication of DNA probes increase through the application of molecular cloning methods and their associated technologies.

REFERENCES

1    Brenner, S., Jacob, F. and Meselson, M. (1961) Nature (London)
     190, 576-580.
2    Volkin, E. and Astrachan, L. (1957) in The Chemical Basis of
     Heredity (McElroy, W.D. and Glass, B., eds.), pp. 686-695, Johns
     Hopkins Press, Baltimore, MD.
3    Hall, B.D. and Spiegelman, S. (1961) Proc. Nat. Acad. Sci.
     U.S.A. 47, 137-161.
4    Britten, R.J. and Davidson, E.H. (1969) Science 165, 349-357.
5    Sutton, D. (1971) Biochim. Biophys. Acta 240, 522-531.
6    Vogt, V.M. (1973) Eur. J. Biochem. 33, 192-200.
7    Kohne, D.E. and Britten, R.J. (1971) Proc. Nucl. Acid Res. 2,
     500-512.
8    Wilson, D.A. and Thomas, C.A. (1973) Biochim. Biophys. Acta 331,
     333-340.
9    Smith, M.J., Hough, B.R., Chamberlin, M.E. and Davidson, E.H.
     (1974) J. Mol. Biol. 85, 103-126.
10   Britten R.J. and Kohne, D.E. (1966) Carnegie Inst. Wash. Year B.
     65, 78-106.
11   Wetmur, J.G. and Davidson, N. (1968) J. Mol. Biol. 31, 349-370.
12   Britten, R.J. and Smith, J. (1969) Carnegie Inst. Wash. Year B.
     68, 378-386.
13   Marmur, J., Rownd, R. and Schildkraut, C.L. (1963) Prog. Nucl.
     Acid Res. 1, 231-300.
14   Nygaard, A.P. and Hall, B.D. (1964) J. Mol. Biol. 9,. 125-142.
15   Subirana, J.A. and Doty, P. (1966) Biopolymers 4, 171-187.
16   Thrower, K.J. and Peacocke, A.R. (1966) Biochim. Biophys. Acta
     119, 652-654.
17   Marmur, J. and Doty, P. (1961) J. Mol. Biol. 3, 585-594.
18   Britten, R.J. and Kohne, D.E. (1968) Science 161, 529-540.
19   Chamberlin, M.E., Galau, G.A., Britten, R.J. and Davidson, E.H.
     (1978) Nucl. Acids Res. 5, 2073-2094.
20   Temin, H. and Mizutani, S. (1970) Nature (London) 226,
     1211-1213.
21   Baltimore, D. (1970) Nature (London) 226, 1209-1211.
22   Ross, J., Aviv, H., Scolnick, E. and Leder, P. (1972) Proc. Nat.
     Acad. Sci. U.S.A. 69, 264-268.
23   Verma, I.M., Temple, G.F., Fan. H. and Baltimore, D. (1972)
     Nature New Biol. 235, 163-167.
24   Kacian, D.L., Spiegelman, S., Bank, A., Terada, M., Metafora,
     S., Dow, L. and Marks, P.A. (1972) Nature New Biol. 235,
     167-169.
25   Bishop, J.O. (1972) Acta Endocrinol. Suppl. 161, 247-276.
26   Bishop, J.O., Morton, J.G., Rosbash, M. and Robertson, M. (1974)
     Nature (London) 250, 199-204.
27   Bishop, J.O., Beckmann, J.S., Campo, M.S., Hastie, N.D.,
     Izquierdo, M. and Perlman, S. (1975) Phil. Trans. Royal Soc. B.
     272, 147-157.
28   Levy, B.W. and McCarthy, B.J. (1975) Biochemistry 14, 2440-2446.

29    Ryfell, G.V. and McCarthy, B.J. (1975) Biochemistry 14,
      1379-1385.
30    Hastie, N.D. and Bishop, J.O. (1976) Cell 9, 761-774.
31    Axel, R., Fiegelson, P. and Schutz, G. (1976) Cell 7, 247-254.
32    Hereford, L.M. and Rosbash, M. (1977) Cell 10, 453-462.
33    Savage, M.J., Sala-Trepat, J.M. and Bonner, J. (1978)
      Biochemistry 17, 462-467.
34    Gillespie, D., Marshall, S. and Gallo, R.P. (1972) Nature New
      Biol. 236, 227-231.
35    Bishop, J.O., Rosbash, M. and Evans, D. (1974) J. Mol. Biol. 85,
      75-86.
36    Williams, J.G. and Penman, S. (1975) Cell 6, 197-206.
37    Williams, J.G., Hoffman, R. and Penman, S. (1977) Cell 11,
      901-907.
38    Lockard, R.E. and Lingrel, J.B. (1969) Biochem. Biophys. Res.
      Commun. 37, 204-209.
39    Gaskill, P. and Kabat, D. (1971) Proc. Nat. Acad. Sci. U.S.A.
      68, 72-75.
40    Gould, H.J. and Hamlyn, P.H. (1973) Proc. Nat. Acad. Sci. U.S.A.
      70, 173-178.
41    Honjo, T., Packman, S., Swan, D., Nau, M. and Leder, P. (1974)
      Proc. Nat. Acad. Sci. U.S.A. 71, 3659-3663.
42    Delovitch, T.L. and Baglioni, C. (1973) Proc. Nat. Acad. Sci.
      U.S.A. 70, 173-178.
43    Tonegawa, S., Bernardini, A., Weinmann, B.J. and Steinberg, C.
      (1974) FEBS Lett. 30, 301-304.
44    Woo, S.L.C., Rosen, J.M., Liakros, C.D., Robberson, D.L., Chow,
      Y.C., Busch, H., Means, A.R. and O'Malley, B.W. (1975) J. Biol.
      Chem. 250, 7027-7039.
45    Rosen, J.M., Woo, S.L.C., Holder, J.W., Means, A.R. and
      O'Malley, B.W. (1975) Biochemistry 14, 69-78.
46    Haines, M.E., Carey, N.H. and Palmiter, R.D. (1974) Eur. J.
      Biochem. 43, 549-560.
47    Adesnik, M. and Darnell, J.E. (1972) J. Mol. Biol. 67, 397-406.
48    Jacobs-Lorena, M., Baglioni, C. and Boren, T.W. (1972) Proc.
      Nat. Acad. Sci. U.S.A. 69, 2095-2099.
49    Grunstein, J., Levy, S., Schedl, P. and Kedes, L. (1973) Cold
      Spring Harbor Symp. Quant. Biol. 38, 717-724.
50    Palacios, R., Sullivan, D., Summers, M.M., Kiely, M.L. and
      Schimke, R.T. (1973) J. Biol. Chem. 248, 540-548.
51    Shapiro, D.J. and Schimke, R.T. (1975) J. Biol. Chem. 250,
      1759-1764.
52    Schnechter, I. (1974) Biochemistry 13, 1875-1885.
53    Taylor, J.M. and Schimke, R.T. (1973) J. Biol. Chem. 248,
      7761-7668.
54    Groner, B., Hynes, N.E., Sippel, A.E., Jeeps, S., Huu, M.C.N.
      and Schutz, G. (1977) J. Biol. Chem. 252, 6666-6674.
55    Strair, R.K., Yap, S.H. and Shafritz, D.A. (1977) Proc. Nat.
      Acad. Sci. U.S.A. 74, 4346-4350.

56   Bishop, J.O., Robertson, F.W., Burns, J.A. and Melli, M. (1969)
     Biochem. J. 115, 361-370.

57   Birnsteil, M.L., Sells, B.H. and Purdon, L.F. (1972) J. Mol.
     Biol. 63, 21-29.

58   Cox, R.F., Haines, M.E. and Emtage, J.S. (1974) Eur. J. Biochem.
     49, 225-236.

59   Harris, S.E., Rosen, J.M., Means, A.R. and O'Malley, B.W. (1975)
     Biochemistry 14, 2072-2081.

60   McKnight, G.S., Pennequin, P. and Schimke, R.T. (1975) J. Biol.
     Chem. 250, 8165-8110.

61   Palmiter, R.D., Moore, P.B., Mulvihill, E.R. and Emtage, S.
     (1976) Cell 8, 557-572.

62   Roop, D.R., Nordstrom, J.L., Tsai, J.Y., Tsai, M.-J. and
     O'Malley, B.W. (1978) Cell 15, 671-685.

63   Ross, J., Gielen, J., Packman, S., Ikawa, Y. and Leder, P.
     (1974) J. Mol. Biol. 87, 697-714.

64   Orkin, S.H. and Swerdlow, P.S. (1977) Proc. Nat. Acad. Sci.
     U.S.A. 74, 2475-2479.

65   Lo, J.C., Aft, R., Ross, J. and Mueller, G.C. (1978) Cell 15,
     447-453.

66   Groudine, M. and Weintraub, H. (1975) Proc. Nat. Acad. Sci.
     U.S.A. 72, 4464-4468.

67   Benz, E.J., Forget, B.H., Hillman, D.G., Cohen-Solal, M.,
     Pritchard, J., Cavallesco, J., Prensky, W. and Housman, D.
     (1978) Cell 14, 299-312.

68   Old, J.M., Proudfoot, N.J., Wood, W.G., Longley, J.L., Clegg,
     J.B. and Weatherall, D.J. (1978) Cell 14, 289-298.

69   Shinohara, T. and Piatigorsky, J. (1976) Proc. Nat. Acad. Sci.
     U.S.A. 73, 2808-2812.

70   Iatrou, K. and Dixon, G.H. (1977) Cell 10, 433-441.

71   Schimke, R.T., Kaufman, R.J., Alt, F.W. and Kellems, R.F. (1978)
     Science 202, 1051-1055.

72   McWilliams, D., Callahan, R.C. and Boime, I. (1977) Proc. Nat.
     Acad. Sci. U.S.A. 74, 1024-1027.

73   Efstratiadis, A., Maniatis, T., Kafatos, F.C., Jeffrey, A. and
     Vournakis, J.N. (1975) Cell 4, 367-378.

74   Haseltine, W.A., Kleid, D.G., Panel, A., Rothenberg, E. and
     Baltimore, D. (1976) J. Mol. Biol. 106, 109-131.

75   Collett, M.S. and Faras, A.J. (1975) J. Virol. 16, 1220-1228.

76   Rothenberg, E. and Baltimore, D. (1976) J. Virol. 17, 168-174.

77   Rothenberg, E. and Baltimore, D. (1977) J. Virol. 21, 168-178.

78   Friedman, E.Y. and Rosbash, M. (1977) Nucl. Acids Res. 4,
     3455-3471.

79   Myers, J.C., Spiegelman, S. and Kacian, D.L. (1977) Proc. Nat.
     Acad. Sci. U.S.A. 74, 2840-2843.

80   Britten, R.J. and Davidson, E.H. (1971) Quart. Rev. Biol. 46,
     111-138.

81   Galau, G.A., Britten, R.J. and Davidson, E.H. (1974) Cell 2,
     9-20.

82   Rigby, P.J., Dieckmann, M., Rhodes, C. and Berg, P.J. (1977) J.
     Mol. Biol. 113, 237-251.
83   Grady, L.J., North, A.B. and Campbell, W.P. (1978) Nucl. Acids
     Res. 5, 697-712.
84   Bantle, J.A. and Hahn, W.E. (1976) Cell 8, 139-150.
85   Galau, G.A., Klein, W.K., Davis, M.M., Wold, B.J., Britten, R.J.
     and Davidson, E.H. (1976) Cell 7, 487-505.
86   Tsai, M.-J., Tsai, S.Y. and O'Malley, B.W. (1979) Science 204,
     314-316.
87   Dugaiczyk, A., Woo, S.L.C., Lai, E.C., Mace, M.L., McReynolds,
     L. and O'Malley, B.W. (1978) Nature (London) 274, 328-333.
88   Dugaiczyk, A., Woo, S.L.C., Colbert, D.A., Lai, E.C., Mace, M.L.
     and O'Malley, B.W. (1979) Proc. Nat. Acad. Sci. U.S.A. 76,
     2253-2257.
89   Garapin, A.C., Cami, B., Roskara, W., Kourilsky, P. LePennec,
     J.P., Perrin, F., Gerlinger, P., Cochet, M. and Chambon, P.
     (1978) Cell 14, 629-639.
90   Breathnach, R., Benoist, C., O'Hare, K., Gannon, F. and Chambon,
     P. (1978) Proc. Nat. Acad. Sci. U.S.A. 75, 4853-4857.
91   Gannon, F., O'Hare, K. Perrin, F., LePennec, J.P., Benoist, C.,
     Cochet, M., Breathnach, R., Royal, A., Garapin, A., Cami, B. and
     Chambon, P. (1979) Nature (London) 278, 428-434.
92   Woo, S.L.C., Dugaiczyk, A., Tsai, M.-J., Lai, E.C., Catteral, F.
     and O'Malley, B.W. (1978) Proc. Nat. Acad. Sci. U.S.A. 75,
     3688-3692.
93   Rozek, C.E., Orr, W.C. and Timberlake, W.E. (1978) Biochemistry
     17, 716-722.
94   Kiper, M. (1979) Nature (London) 278, 279-280.
95   Davidson, E.H., Hough, B.R., Amenson, C.S. and Britten, R.J.
     (1973) J. Mol. Biol. 77, 1-23.
96   Manning, J.E., Schmid, C.W. and Davidson, N. (1975) Cell 5,
     159-172.
97   Graham, D.E., Neufeld, B.R., Davidson, E.H. and Britten, R.J.
     (1974) Cell 1, 127-137.
98   Flavell, R.B., Bennett, M.D., Smith, J.B. and Smith, D.B. (1974)
     Biochem. Genet. 12, 257-269.
99   Flavell, R.B. and Smith, D.B. (1976) Heredity 37, 231-252.
100  Zimmerman, J.L. and Goldberg, R.B. (1977) Chromosoma 59, 227-252.
101  Walbot, V. and Dune, L.S. (1976) J. Mol. Biol. 105, 503-521.
102  Kiper, M. and Herzfeld, F. (1978) Chromosoma 65, 335-351.
103  Westphal, H. (1970) J. Mol. Biol. 50, 407-420.
104  Flint, S.J. and Sharp, P.A. (1976) J. Mol. Biol. 106, 749-771.
105  Hutton, J.R. and Wetmur, J.G. (1973) J. Mol. Biol. 77, 495-500.
106  Galau, G.A., Britten, R.J. and Davidson, E.H. (1977) Proc. Nat.
     Acad. Sci. U.S.A. 74, 1020-1023.
107  Reed, S.I., Stark, G.R. and Alwine, J.C. (1976) Proc. Nat. Acad.
     Sci. U.S.A. 73, 3083-3087.
108  Goldberg, R.B., Hoschek, G. and Kamalay, J. (1978) Cell 14,
     123-131.

109   Kleinman, L., Birnie, G.D., Young, B.D. and Paul, J. (1977)
      Biochemistry 16, 1218-1223.
110   Birnie, G.D., Macphail, E., Young, B.D., Getz, M. and Paul, J.
      (1974) Cell Differentiation 3, 221-230.
111   Mauron, A. and Spohr, G. (1978) Nucl. Acids. Res. 5, 3013-3032.
112   Gelderman, A.H., Rake, A.V. and Britten, R.J. (1971) Proc. Nat.
      Acad. Sci. U.S.A. 68, 172-176.
113   Greenberg, J.R. and Perry, R.P. (1971) J. Cell Biol. 50, 774-787.
114   Goldberg, R.B., Galau, G.A., Britten, R.J. and Davidson, E.H.
      (1973) Proc. Nat. Acad. Sci. U.S.A. 70, 3516-3520.
115   Dina, D., Crippa, M. and Beccari, E. (1973) Nature New Biol. 242,
      101-105.
116   Firtel, R.A., Jacobson, H. and Lodish, H.F. (1972) Nature New
      Biol. 239, 225-228.
117   Melli, N., Whitfield, C., Rao, K.V., Richardson, M. and Bishop,
      J.O. (1971) Nature New Biol. 231, 8-12.
118   Grunstein, M., Schedl, P. and Kedes, L. (1973) in Molecular
      Cytogenetics (Hamkalo, B.A. and Papaconstantinou, J., eds.),
      pp. 115-123, Plenum Press, New York, NY.
119   Weinberg, E.S., Birnsteil, M.L., Purdom, I.F. and Williamson, R.
      (1972) Nature (London) 240, 225-228.
120   Grunstein, M. and Schedl, P. (1976) J. Mol. Biol. 104, 323-349.
121   Davidson, E.H., Hough, B.R., Klein, W.H. and Britten, R.J. (1975)
      Cell 4, 217-238.
122   Holmes, D.S. and Bonner, J. (1974) Biochemistry 13, 841-848.
123   Klein, W.H., Murphy, W., Attardi, G., Britten, R.J. and Davidson,
      E.H. (1974) Proc. Nat. Acad. Sci. U.S.A. 71, 1785-1789.
124   Smith, M.J., Britten, R.J. and Davidson, E.H. (1975) Proc. Nat.
      Acad. Sci. U.S.A. 72, 4805-4809.
125   Galau, G.A., Smith, M.J., Britten, R.J. and Davidson, E.H. (1977)
      Proc. Nat. Acad. Sci. U.S.A. 74, 2306-2310.
126   Hames, B.D. and Perry, R.P. (1977) J. Mol. Biol. 109, 437-453.
127   Gillespie, D. and Spiegelman, S. (1965) J. Mol. Biol. 12,
      829-842.
128   Spiegelman, G.B., Haber, J.E. and Halverson, H.O. (1973)
      Biochemistry 12, 1234-1242.
129   Flavell, R.A., Birfelder, E.J., Sanders, J.P.M. and Bors, P.
      (1974) Eur. J. Biochem. 47, 535-543.
130   Haas, M., Vogt, M. and Dulbecco, R. (1972) Proc. Nat. Acad. Sci.
      U.S.A. 69, 2160-2164.
131   Nevins, J.R. and Darnell, J.E. (1978) Cell 15, 1477-1493.
132   Schibler, U., Marcu, K.B. and Perry, R.P. (1978) Cell 15,
      1495-1509.
133   Gilham, P.T. (1968) Biochemistry 7, 2809-2813.
134   Shih, T.Y. and Martin, M.A. (1974) Biochemistry 13, 3411-3418.
135   Noyes, B.E. and Stark, G.R. (1975) Cell 5, 301-310.
136   Gilboa, E., Prives, C.L. and Aviv, H. (1975) Biochemistry 14,
      4215-4220.
137   Arndt-Jovin, D.J., Jovin, T.M., Bahr, W., Frischauf, A.M. and
      Marquardt, M. (1975) Eur. J. Biochem. 54, 411-418.

138    Poonian, M.S., Schlabach, A.J. and Weissbach, A. (1971)
       Biochemistry 10, 424–427.
139    Levy, S. and Aviv, H. (1976) Biochemistry 15, 1844–1847.
140    Bastos, R.N., Volloch, Z. and Aviv, H. (1977) J. Mol. Biol. 110,
       191–203.
141    Prives, C., Gilboa, E., Revel, M. and Winocur, E. (1977) Proc.
       Nat. Acad. Sci. U.S.A. 74, 457–461.
142    Casey, J. and Davidson, N. (1977) Nucl. Acids Res. 4, 1539–1552.
143    Beltz, G.A. and Flint, S.J. (1979) J. Mol. Biol. 131, 353–374.
144    Roberts, B.E. and Paterson, B.M. (1970) Proc. Nat. Acad. Sci.
       U.S.A. 70, 2330–2334.
145    Hunt, T. and Jackson, R.J. (1974) in Modern Trends in Human
       Leukaemias (Neth, R., Gallo, R.C., Spiegelman, S. and Stohlman,
       F., eds.), pp. 300–307, J.F. Lehmanns Verlag, Munich.
146    Lane, D.C., Marbaix, G. and Gordon, J.B. (1971) J. Mol. Biol. 61,
       73–91.
147    Pelham, H.R.B. and Jackson, R.J. (1976) Eur. J. Biochem. 67,
       247–256.
148    Palmiter, R.D. and Smith, L.T. (1973) Nature New Biol. 246,
       74–76.
149    Rhoads, R.E., McKnight, S.M. and Schimke, R.T. (1973) J. Biol.
       Chem. 248, 2031–2039.
150    Cavalieri, R.L., Havell, E.A., Vilcek, J. and Pestka, S. (1977)
       Proc. Nat. Acad. Sci. U.S.A. 74, 4415–4419.
151    Greene, J.J., Diefenback, C.W. and Tso, P.O.P. (1978) Nature
       (London) 271,81–83.
152    Farmer, J.R., Henshaw, E.C., Berridge, M.V. and Tata, J.R. (1978)
       Nature (London) 273, 401–403.
153    Preston, C.M. (1979) J. Virol. 29, 275–284.
154    Giglioni, B., Gianni, A.M., Comi, P., Ottolenghi, S. and Runnger,
       D. (1973) Nature New Biol. 246, 99–102.
155    Legon, S., Jackson, R.J. and Hunt, T. (1973) Nature New Biol.
       241, 150–152.
156    Ehrenfeld, E. and Hunt, T. (1971) Proc. Nat. Acad. Sci. U.S.A.
       68, 1075–1078.
157    Hunter, T., Hunt, T., Jackson, R.J. and Robertson, H.D. (1975) J.
       Biol. Chem. 250, 409–417.
158    Huang, A.S. and Wagner, R.R. (1965) Proc. Nat. Acad. Sci. U.S.A.
       54, 1579–1584.
159    Wertz, G.W. and Younger, J.S. (1970) J. Virol. 6, 476–484.
160    Weck, P.K. and Wagner, R.R. (1978) J. Virol. 25, 770–780.
161    Lindberg, U., Persson, T. and Philipson, L. (1972) J. Virol. 10,
       909–919.
162    Willems, M. and Penman, S. (1966) Virology 30, 355–367.
163    Ehrenfeld, E. and Lund, H. (1977) Virology 80, 297–308.
164    Tissieres, A., Mitchell, H.K. and Tracy, V.M. (1974) J. Mol.
       Biol. 84, 389–398.
165    Lewis, M., Helmsing, P.J. and Ashburner, M. (1975) Proc. Nat.
       Acad. Sci. U.S.A. 72, 3604–3608.

166    McKenzie, S.L., Henikoff, S. and Meselson, M. (1975) Proc. Nat. Acad. Sci. U.S.A. 72, 1117–1121.
167    Koninkx, J.F.J.G. (1975) Biochem. J. 152, 17–22.
168    Bressmann, H. Levy, B.W. and McCarthy, B.J. (1978) Proc. Nat. Acad. Sci. U.S.A. 75, 759–763.
169    Clinton, G.M., Little, S.P., Hagen, F.S. and Huang, A.S. (1978) Cell 15, 1455–1462.
170    Britten, R.J. (1969) Carnegie Inst. Wash. Year B. 67, 332–335.
171    McKnight,G.S. (1978) Cell 14, 403–413.
172    Monahan, J.J., Harris, S.E. and O'Malley, B.W. (1976) J. Biol. Chem. 251, 3738–3748.
173    Jaquet, M., Affara, N.A., Robert, S., Jacob, H., Jacob, F. and Gros, F. (1978) Biochemistry 17, 69–79.
174    Meyuhas, O. and Perry, R.P. (1979) Cell 16, 139–148.

# DNA CLONING IN MAMMALIAN CELLS WITH SV40 VECTORS

D.H. Hamer

Recombinant DNA Unit
National Institute of Allergy and Infectious Diseases
National Institutes of Health
Bethesda, Maryland    20205

## INTRODUCTION

The past few years have witnessed tremendous advances in the
cloning of animal cell genes in bacteria.  With this technology, it
is possible to purify and amplify essentially any gene for which
either a nucleic acid probe or functional test is available.  Once
the gene has been cloned, it is a straightforward matter to determine
its structure and, ultimately, its entire sequence.  Such studies
provide a crucial framework for understanding the regulation and
expression of the gene, especially when it is possible to compare its
sequence with those of several closely related genes.  However, there
is an important limitation to cloning in bacteria; because the genetic machinery is so different in prokaryotes and eukaryotes, it is
not possible to test directly the function of suspected regulatory or
control regions.  For this purpose, it is necessary to return the
gene, and mutants derived from it, to a eukaryotic cell where their
function can be tested.
The first attempts to apply recombinant DNA technology to the
introduction of DNA segments into animal cells have utilized simian
virus 40 (SV40) as the vector.  This is not surprising in view of the
intense effort that has been devoted to understanding the structure,
replication and transcription of this virus.  Indeed, the SV40 genome
is probably the best characterized of all eukaryotic DNAs.  An additional impetus for using SV40 as a vector is its ability to transform
cells from a wide variety of species.  This raises the possibility of
using SV40 recombinants to introduce foreign genetic information into
cultured cells, or even animals, in a stable fashion.
This chapter describes the various strategies that can be used
to construct SV40 recombinants and to propagate them either as

viruses or as stable genetic elements. Included are new techniques
for forming the recombinant molecules by cloning in E. coli rather
than by exclusively in vitro manipulations. In addition, evidence is
presented for the utilization of eukaryotic regulatory and processing
signals in the SV40 system. Finally, several unique advantages of
SV40 recombinants as genetic and biochemical tools are discussed.
But first, it is necessary to review briefly the key features of the
life cycle and molecular biology of SV40.

THE SV40 GENOME

     The genome of SV40 is a 5200 base-pair, covalently closed DNA
circle. For convenience, the genome is divided into map units with
the 0.00/1.00 point at the unique EcoRI restriction endonuclease site
(Figure 1). The structural and functional organization of the viral
genome have been described in detail in two recent reviews (1,2).
The complete nucleotide sequence of the viral DNA has been indepen-
dently determined by two groups (3,4).
     SV40 can undergo two types of interactions with cultured mamma-
lian cells. In African green monkey kidney cells, which are the permissive host, the virus undergoes a productive cycle. The virus
first attaches to the cell, penetrates it and uncoats. Shortly
thereafter, the viral early gene region, which extends counterclockwise from 0.67 to 0.17 map units, is transcribed. This region,
constituting the A complementation group, encodes the large form (T-
antigen) and small form (t-antigen) of the viral tumor antigen. T-
antigen is involved in both the initiation of viral DNA synthesis and
in transformation and is required for both productive and nonproductive infection. The role of t-antigen is less certain, but it is not
required for productive infection in tissue culture. The next step
in the viral life cycle is the bidirectional replication of the viral
DNA from the unique origin at 0.67 map units. Finally, the viral
late gene region, extending clockwise from 0.67 to 0.17 map units, is
expressed. These sequences code for the major viral coat protein VP1
(B/C complementation group) and the minor structural proteins VP2 and
VP3 (D complementation group. At about three days postinfection, the
cells are killed and a burst of infectious virus is released. SV40
does not shut off the synthesis of host cell macromolecules during
the productive cycle. Even late in infection, normal or increased
amounts of cellular DNA, RNA and proteins are produced.
     In other types of cells, termed nonpermissive or semipermissive,
this cycle is aborted prior to the onset of viral DNA replication.
Instead, part or all of the viral genome becomes stably associated
with the host cell, usually through integration into the chromosomal
DNA. Such cells are said to be transformed and can be recognized by
a variety of parameters related to loss of normal growth characteristics, e.g., ability to form colonies in agar or to cause tumors in
animals. Transformed cells invariably express viral early functions
but generally do not synthesize late gene products. SV40 transforma-

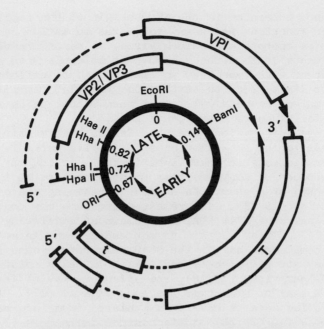

Figure 1.  A map of the SV40 genome.

tion has been observed in several mammalian species including rat,
mouse, hamster and human.

The transcription of SV40 is rather complex.  However, we will
be interested only in the two major species of late mRNA, termed 19S
and 16S.  Both of these mRNAs consist of short 5' leader sequences,
extending clockwise from approximately 0.71 to 0.75 map units, covalently attached to longer body sequences.  The body of the 19S mRNA,
which codes for VP2 and probably VP3, extends clockwise from 0.76 map
units to the polyadenylation site at 0.17 map units.  The 16S mRNA,
which encodes VP1, has a body extending from 0.94 to 0.17 map units.
Both types of mRNAs are produced in large quantities late in the
lytic cycle; 16S mRNA comprises about 10% and 19S mRNA about 1% of
the total cellular mRNA synthesis.  Further details of the structure
and synthesis of the SV40 late mRNAs can be found in references 5-11.


SV40 VECTORS


Productive Infection

Most of the experiments with SV40 hybrids have involved propagation of the recombinant molecules as virus in productively infected
monkey cells.  The only cis-dominant function required for the repli-

cation of such molecules is the SV40 origin of DNA replication. All
of the other functions necessary for virus formation can be supplied
in trans by an appropriate helper virus. A second requirement in
such experiments is that the recombinant molecule have a length 70%
to 100% that of wild-type SV40 so that it can be efficiently encapsidated. This limit is quite strict; for example, the plaque-forming
ability of molecules only 10% longer than wild-type is reduced by
almost 100-fold (Hamer, unpublished data).

Defective Reiteration Mutants. Serial high multiplicity passage
of SV40 leads to the accumulation of defective particles with deletions, rearrangements and host substitutions in the viral DNA (12-
18). The genomes of many of these mutants consist of tandem, head-
to-tail repeats of a short DNA segment containing the origin of viral
DNA replication (19-23). The repeated sequence may contain as little
as 150 base pairs of SV40 DNA, the remainder consisting of host cell
sequences (24). These mutants do not contain any intact SV40 genes
and do not complement any of the SV40 temperature-sensitive mutants.
However, they are efficiently replicated in cells coinfected with
wild-type helper virus (25,26), and oligomers having lengths between
70% and 100% that of wild-type are encapsidated (22). A variety of
such reiteration mutants have been isolated from late passage stocks
either by plaquing (25,26) or biochemical techniques (22).

The reiteration mutant genomes can be cleaved with restriction
endonucleases to generate short DNA fragments suitable as cloning
vectors. The vector fragment containing the SV40 origin of replication is joined to the foreign gene of interest and the recombinant
molecules are used to infect monkey cells together with wild-type DNA
as helper. The resulting virus stock then contains a mixture of
recombinant and wild-type particles. This scheme has been used to
form recombinants carrying bacteriophage λ DNA segments ranging from
500 to 2300 base pairs (27-29). In these experiments, the recombinant genomes carried two or three SV40 replication origins, thus
ensuring their efficient replication.

A theoretical advantage of the reiteration mutant vehicles is
that they might be used to clone DNA fragments with lengths approach-
ing that of wild-type SV40. However, it is not yet certain that such
recombinants, which would contain only one replication origin, would
replicate sufficiently well to compete with the wild-type virus used
as helper. A disadvantage of these vectors is that there is no simple method of selecting for recombinant viruses. Also, the reiteration mutants are rather unstable and this may lead to the accumulation of unexpected recombinants in the virus stock. A final disadvantage of these vectors is that they generally lack the viral RNA
processing sites required for the expression of foreign genes that
lack their own processing sites.

Late Region Deletions. A second type of vector can be prepared
by excising sequences from the late gene region of the wild-type SV40
genome. Fortuitously, there are several unique restriction endonuclease sites in the late region (Figure 1). These facilitate the

preparation of molecules with deletions ranging from 485 base pairs
(cleavage with HpaII plus HaeII) to 2183 base pairs (cleavage with
HpaII plus BamHI). Alternatively, one can cleave at a unique site
in the late region, then partially digest with an enzyme that has
several sites in the genome and isolate the desired vector fragment
by gel electrophoresis. For example, partial digestion with HindIII
at 0.945 map units together with BamHI generates a 1040 base pair
deletion (30).

Such molecules can be used as vehicles for the propagation of
DNA fragments with lengths less than or equal to the deleted region.
The viral and foreign DNAs are joined by ligation of the cohesive
termini produced by some restriction endonucleases (31,32), by using
oligonucleotide linkers (33,34), or by the poly(dA)-poly(dT) tailing
method (35,36). The resulting recombinants retain both the SV40
origin of DNA replication and the entire early or A gene region.
Consequently, they can be propagated as virus by coinfection with a
temperature-sensitive SV40 A gene mutant (tsA) as the complementing
helper. When such mixed infections are carried out at the nonpermissive temperature of 41°, virus is produced only by those cells
infected with both the recombinant, which provides early gene functions, and the helper, which provides late gene functions. SV40 late
region deletion vehicles have been used to clone a bacteriophage λ
DNA fragment (37), an E. coli tRNA gene (38,39), yeast tRNA genes (S.
Goff and P. Berg, personal communication), Drosophila histone sequences (P. Berg, personal communication), a rabbit β-globin cDNA
(30,40), chromosomal mouse $\beta^{MAJ}$-globin (41,42) and α-globin gene
sequences (M. Kaehler, D. Hamer and P. Leder, unpublished data),
mouse immunoglobulin variable region genes (D. Hamer and P. Leder,
unpublished data) and sea urchin histone sequences (64).

An example of the use of a late region deletion vehicle to clone
an E. coli suppressor tRNA gene is shown in Figure 2 (38,39). In
this experiment, the SV40 vector segment extending clockwise from the
EcoRI to the HpaII site was ligated to an 870 base-pair HpaII-EcoRI
fragment of bacteriophage DNA containing the $tRNA^{Tyr}su^{+}III$ gene.
Because HpaII and EcoRI produce different cohesive termini, the two
fragments could be joined in only one orientation, such that the tRNA
structural sequence was on the same strand as the viral late coding
sequences. The resulting circular recombinant molecules were purified by CsCl/ethidium bromide density gradient centrifugation, mixed
with tsA helper DNA and used to infect monkey cells. This generated
a virus stock containing about 30% recombinant and 70% helper
genomes. Analysis of the recombinant molecules by nucleic acid
hybridization, restriction endonuclease cleavage and electron
microscopic visualization of heteroduplexes revealed that the
bacterial sequence had not suffered any substantial alterations
during its sojourn in monkey cells.

There are several advantages to the late region deletion
vehicles. First, their ability to complement tsA mutants allows
easy selection for recombinants. Second, by selecting an appropriate

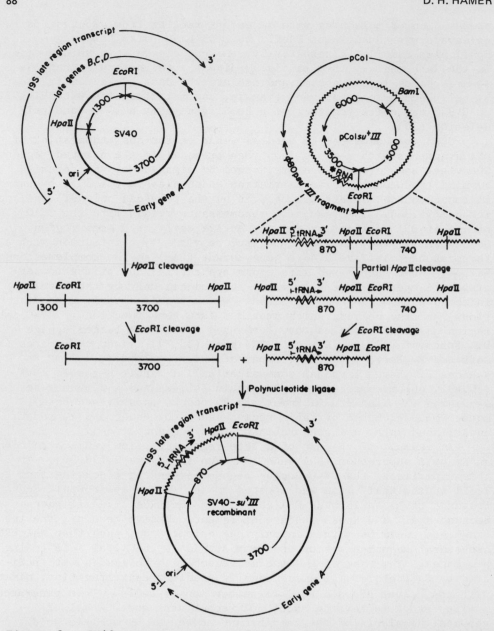

Figure 2. SV40 carrying an E. coli tRNA suppressor gene. The SV40
vector contained the sequences extending from the EcoRI to the HpaII
site. The bacterial DNA fragment, which was obtained from a plasmid
carrying the tRNA suppressor gene, also had EcoRI and HpaII termini.
Ligation of these two DNAs generated recombinant molecules in which
the tRNA structural sequence replaced the viral late region coding
sequences (38).

vector and orientation of the inserted fragment, one can determine whether or not the foreign DNA will be expressed under the control of the SV40 late region promoter and RNA processing sites. Third, they can be used for nonproductive as well as productive infection. Their chief disadvantage is that the site of the inserted fragment is limited to 2500 base pairs.

Early Region Deletions. SV40 vectors can also be prepared by excision of early gene region sequences (B. Howard and P. Berg, personal communication; P. Gruss and G. Khoury, personal communication). For example, cleavage of the viral DNA with TaqI at 0.57 map units and BclI at 0.19 map units generates a 2080 base-pair deletion. Recombinants made with such vectors can be propagated as virus by coinfection with a temperature-sensitive late gene mutant as the complementing helper. In principle, these vectors offer many of the same advantages afforded by the late region deletions. However, the expression of foreign genes cloned in this fashion has not yet been studied in detail.


Nonproductive Infection


Studies with temperature-sensitive mutants, deletion mutants and DNA fragments have established that an intact early gene region and origin of DNA replication are sufficient for transformation by SV40 (1,2). Accordingly, any of the late region deletion vehicles described above can be used for nonproductive as well as productive infection. If the recombinant molecules are sufficiently small to be encapsidated, they can be propagated as virus in permissive cells prior to transformation of the nonpermissive host. Alternatively, the recombinant molecules constructed in vitro can be used directly for transformation.

The former strategy was used to introduce the E. coli suppressor tRNA gene, tRNA$^{Tyr}$su$^+$III, into mammalian cells (43,44). The SV40-tRNA recombinant was initially propagated as a virus, as described in the previous section, then purified from the helper DNA by gel electrophoresis. This preparation was used to infect rat embryo cells, which are nonpermissive for SV40, and transformed clones were selected on the basis of their altered morphology. The DNA from these transformants was analyzed for SV40 and E. coli tRNA sequences by both reassociation kinetics and gel-transfer hybridization analysis of restriction fragments. These experiments showed that the transformants contained about two copies per haploid genome of both the SV40 vector and bacterial tRNA sequences. Surprisingly, most of the recombinant genomes appeared in the form of free episomes rather than integrated into the host chromosomes. The SV40-tRNA recombinant was also used to infect monkey cells in the absence of helper virus. This led to the isolation of clones carrying the recombinant genome as an episome at levels of about 10,000 copies per cell. Unfortunately, attempts to repeat this type of experiment with other recombinant molecules have been unsuccessful (G. Fareed, personal communication).

This general scheme has also been used to introduce sea urchin histone sequences into cultured rat cells (64). In this case, however, the recombinant genomes were found to be integrated into the host chromosomes rather than replicating as episomes. It was also shown that the SV40-sea urchin recombinants could be recovered from the transformed cells after fusion with permissive cells.

DNA transformation with recombinant molecules prepared in vitro has been used to transduce bacteriophage λ sequences into mouse cells (45). In this study, a 22,000 base-pair seqment of λ DNA was joined to the SV40/EcoRI+HpaII vector and used to infect mouse 3T3 cells. Gel-transfer hybridization analysis of the DNA from three transformants showed that each contained both SV40 and λ sequences. By using various restriction endonucleases, evidence was obtained both for linkage of the SV40 and λ sequences and for integration into the chromosomal DNA. However, it was not possible to deduce the mechanism of integration nor the extent of the λ sequences retained.

Despite the preliminary nature of these experiments, the advantages of transformation versus lytic infection are clear. First, the host cells are not killed, so the foreign DNA can be introduced as a stable genetic element. Second, since there is no requirement for encapsidation, it is possible to use large foreign DNA segments. On the other hand, the low copy number of the recombinant genomes in the transformed cells may make detection of the foreign gene products difficult.

## SV40 Vectors Cloned in E. coli

The construction of SV40 transducing viruses can be substantially simplified by performing the initial cloning steps in bacteria rather than mammalian cells. The first step in this scheme is to construct a bacterial plasmid carrying an SV40 vector segment. This plasmid-SV40 recombinant is then used as a vehicle to clone the foreign gene of interest. Finally, the plasmid-SV40-foreign gene hybrid is cleaved with an appropriate restriction endonuclease so as to generate an SV40-foreign DNA recombinant molecule that can be used to infect mammalian cells.

With this idea in mind, we have recently constructed several E. coli plasmid pBR322 derivatives carrying useful SV40 vehicle sequences (D. Hamer, M. Kaehler and P. Leder, unpublished data). These include the complete SV40 genome cleaved at either the EcoRI or BamHI site, the segment extending clockwise from the BamHI to the EcoRI site and the segment extending clockwise from the BamHI to the HpaII site (see Figure 1). In the last of these hybrids, the SV40 BamHI and HpaII sites were converted to EcoRI sites by ligation to synthetic oligonucleotide linkers (33,34) prior to cloning in the unique EcoRI site of pBR322. The resulting recombinant can be linearized by partial digestion with EcoRI and used as a vector to clone any DNA fragment that has either natural or synthetic EcoRI termini. (If the foreign DNA fragment contains an internal EcoRI site, it can be pro-tected by methylation with the EcoRI modification enzyme prior to the linking step.)

The formation of SV40 recombinants by cloning in E. coli rather than mammalian cells has two major advantages: it is quicker and cheaper. This is because bacterial clones (e.g., for colony hybridi-zation) can be grown up in a few hours on inexpensive agar plates. In contrast, an SV40 plaque assay on monkey cells can take as long as two weeks and requires expensive tissue culture media, serum, a $CO_2$ incubator, etc. The bacterial cloning strategy may prove particular-ly useful for constructing a large series of recombinants with dele-tion or point mutations.

## Detection and Purification of Recombinants

Recombinants that retain one or more intact SV40 genes can be detected by their ability to complement an appropriate temperature-sensitive helper virus at the nonpermissive temperature (37-39). Individual clones of these recombinants can be obtained either by a direct DNA plaque infection (26) or by an infectious centers method (25). If the recombinant requires a wild-type helper, individual clones can be screened by restriction endonuclease digestion (27) or by hybridization (28). The latter can be accomplished most effi-ciently by transferring the plaques directly to a nitrocellulose filter and hybridizing with a probe directed against the foreign DNA (46).

## EFFECTS OF THE STRUCTURE OF LATE REGION DELETION VEHICLES ON FOREIGN GENE EXPRESSION

The SV40 late gene region contains signals for RNA initiation, processing and translation. One can determine whether or not these signals will be utilized in a recombinant virus by the appropriate selection of vector and the orientation of inserted fragment. Thus, to study the expression of a complementary DNA (cDNA) sequence, it may be necessary to retain the viral promoter, splice junction and polyadenylation site, but eliminate the viral translational start signals. In contrast, to observe the expression of an intact chromo-somal gene, it may be desirable to remove all of these viral signals.

## Promoter

Foreign DNA fragments inserted in the SV40 late gene region can be transcribed from the viral late promoter to generate hybrid RNAs (38-42). The exact location of the SV40 late region promoter(s) is not known. The capped 5' termini of both the 16S and 19S mRNAs are

quite variable (9-11,47-52) and deletion mutants lacking the major
cap site are viable (53,54).  Consequently, it has not yet been
possible to develop a vector lacking the late region promoter yet
retaining the ability to replicate.  This problem can be overcome
simply by cloning the foreign gene in the anti-sense orientation
relative to the late promoter, i.e., such that the coding or structural sequences of the foreign gene are on the opposite strand as the
SV40 late coding sequences.  In this orientation, the foreign DNA
will be efficiently transcribed only if its own promoter is functional.  Recent experiments with mouse α-globin hybrids (M. Kaehler,
D. Hamer and P. Leder, unpublished data) suggest that anti-sense RNA
initiated from the SV40 promoter does not seriously interfere with
the expression of the mouse gene.

## Splice Junctions

The major late mRNA splice junctions span 0.715-0.760/0.765 map
units for the 19S RNA and 0.760/0.935 map units for the 16S RNA (3-
11).  By a judicious choice of restriction enzymes, these junctions
can be retained or removed from the vector.  Thus, cleavage with
HpaII plus BamHI removes both junctions, HaeII plus BamHI removes the
16S but not the 19S junction and EcoRI plus BamHI removes neither.
Recent studies with SV40 deletion mutants (55,56) and recombinant
viruses (57) suggest that a functional splice junction is required
for the formation of stable mRNA.

## Polyadenylation Site

The polyadenylated 3' termini of the SV40 late mRNAs lie at 0.17
map units (58).  This site is retained in vectors produced by cleavage with BamHI at 0.14 map units and is utilized in recombinant
viruses (30,41).

## Initiation Codons

The late region initiation codons lie at 0.94 map units for VP1,
0.76 map units for VP2 and 0.83 map units for VP3 (3,4).  These sites
can be removed or retained by using appropriate restriction endonucleases as described above.  There is evidence for preferential or
exclusive use of the first AUG in eukaryotic mRNAs containing more
than one potential initiation codon (1,59).  Accordingly, it may be
important to eliminate the viral initiation codons, while retaining
the transcription and initiation and processing sites, to obtain
expression of cDNA sequences.

## EXPRESSION OF FOREIGN GENES CLONED IN SV40

### Prokaryotic Genes

SV40 reiteration mutant vehicles have been used to clone two
different bacteriophage λ DNA segments, one of which contained a λ

promoter (27,28). No stable λ-specific RNA could be detected in monkey cells infected with these hybrids (D. Ganem, G. Fareed and G. Khoury, personal communication). This could be due to lack of a promoter, intrinsic instability of λ RNA in monkey cells or absence of necessary processing signals.

A bacteriophage λ DNA segment has also been cloned in an SV40 late region deletion vehicle (37). In this study, the poly(dA)-poly(dT) joining method was used to insert the λ fragment in both possible orientations into the SV40/HpaII-BamHI vehicle. In the initial report, no λ transcripts could be detected in monkey cells infected with either type of hybrid (37). However, subsequent experiments have demonstrated the presence of low levels of unstable, nuclear, λ-specific RNA (60).

Substantial quantities of bacterial RNA were found in monkey cells infected with the SV40-E. coli tRNA$^{Tyr}$su$^{+}$III hybrid described earlier (38,39). Analysis of the RNA synthesized during a one hour pulse with [$^{3}$H]uridine showed that about half of the bacterial RNA consisted of 19S transcripts containing covalently linked SV40 and bacterial sequences. However, no mature, functional tRNA$^{Tyr}$ was produced, perhaps because monkey cells lack processing enzymes required for the formation of bacterial tRNA.

## Eukaryotic cDNA Sequences

Two groups have constructed SV40 recombinants carrying rabbit β-globin coding sequences derived from a cloned cDNA. In the first study (30) (Figure 3), the SV40 vector extended clockwise from the BamHI site at 0.14 map units to the HindIII site at 0.945 map units. This vector retains the late region promoter, polyadenylation site and both the 16S and 19S mRNA splice junctions, but lacks all of the late region initiation codons. The rabbit β-globin cDNA fragment contained the complete globin coding sequence and termination codon together with most of the 5' untranslated region. It was inserted into the vector in the sense orientation relative to the SV40 promoter to allow transcriptional read-through of the globin coding strand. Monkey cells infected with this hybrid produced SV40-globin hybrid mRNAs analogous to both the 16S and 19S late mRNAs. Further, they synthesized rabbit β-globin in quantities comparable to that of the viral late proteins. Tryptic fingerprint analysis of the protein synthesized in monkey cells confirmed its identity as authentic rabbit β-globin.

Rabbit β-globin cDNA sequences have also been inserted in the SV40 late region deletion vehicles produced by cleavage with BamHI plus HaeII and with BamHI plus HpaII (40). Both of these vectors retain the late region promoter and polyadenylation site. They differ in that the first also retains the 19S mRNA splice junction whereas the second contains no known splice junctions. The first type of recombinant encoded a stable SV40-globin hybrid RNA analogous to the viral 19S mRNA. Monkey cells infected with this recombinant

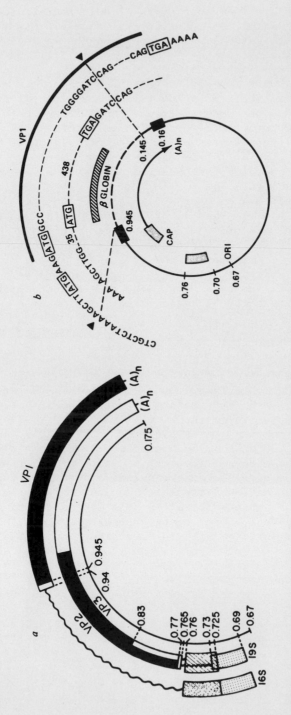

Figure 3. SV40 carrying rabbit β-globin coding sequences. Panel a shows the structures of the SV40 late gene region and of the late 16S and 19S mRNAs. Panel b shows the structure of the recombinant molecule and the nucleotide sequences at the SV40-globin gene junctions. This recombinant retains the promoter, splice junctions and polyadenylation site for the late mRNAs, but lacks all of the late protein initiation codons (30).

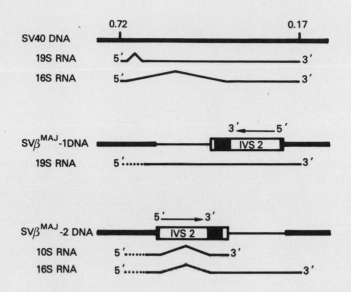

Figure 4. Transcription maps of SV40 recombinants carrying a segment
of the mouse genomic βMAJ-globin gene. These recombinants carry a
functional splice junction and polyadenylation site derived from the
3' portion of the mouse gene. When the nonsense strand of the mouse
DNA is transcribed (SVβMAJ-1), neither signal is utilized. When the
sense strand is transcribed (SVβMAJ-2), both signals are utilized (41)

synthesized rabbit β-globin as determined with a sensitive radio-
immunoassay. However, the amount of globin produced was very low,
perhaps because of the presence of a viral initiation codon upstream
of the globin start point. The globin sequences were also efficiently transcribed in the second virus, but the resulting RNA was
unstable, perhaps because it was not spliced.

## Eukaryotic Chromosomal Sequences

The first studies with eukaryotic chromosomal sequences have focused on the mouse β-globin and α-globin genes. Both of these genes are discontinuous, being divided into three coding segments separated by two intervening sequences (61-63). Hence, these experiments have provided an opportunity to test for utilization of the genomic signals for RNA splicing as well as for transcription initiation, polyadenylation and translation.

The initial experiments utilized a fragment derived from the 3' portion of the chromosomal $\beta^{MAJ}$-globin gene (41) (Figure 4). This fragment contained 18 base pairs from the 3' end of the second coding segment, the second intervening sequence, the third coding segment and transcribed but untranslated sequence, and 3' flanking sequences. Thus, it had potential signals for both RNA splicing and polyadenylation. It was cloned, in both possible orientations, in the SV40/BamHI-HaeII vector, which retains both the late region promoter and 19S mRNA splice junction. Both orientations of virus produced stable SV40-globin mosaic mRNAs. However, the mouse gene signals for splicing and polyadenylation were utilized only when the fragment was transcribed in the sense orientation (Figure 4). These results show that 18 base pairs is the maximum sequence length required to specify the 5' donor side of an RNA splice junction. Further, they show that symmetric secondary structures formed by perfect base pairing are not the sole basis for RNA splicing; clearly such structures, if they existed, would be present in the anti-sense as well as the sense RNA.

Recently, this fragment has also been cloned in the SV40/BamHI-HpaII vector, which retains the late promoter but does not carry any known splice junctions (57). When the fragment was transcribed in the sense orientation, so that the globin splice junction was functional, a stable SV40-globin hybrid RNA was produced. In contrast, when the fragment was transcribed in the anti-sense orientation, so that the foreign splice junction was not utilized, no stable globin transcripts could be detected. These data support the idea that splicing is a prerequisite for stable RNA formation (55,56).

Further information as to the expression of chromosomal sequences cloned in SV40 has been obtained by constructing a recombinant carrying the complete mouse $\beta^{MAJ}$-globin gene (42). The globin fragment, cloned in the sense orientation relative to the viral late promoter, included 280 base pairs of 5' flanking sequence and 600 base pairs of 3' flanking sequence as well as the complete coding and intervening sequences. Monkey cells infected with this hybrid produced globin mRNA that was spliced and polyadenylated at the appropriate globin gene sites. However, most (possibly all) of these transcripts were initiated at the upstream viral promoter rather than the globin promoter. The ability of this virus to direct the synthesis of mouse $\beta^{MAJ}$-globin was demonstrated by immunoprecipitation, two-dimensional gel electrophoresis and tryptic

fingerprint analysis. This shows that the mouse RNA splicing signals
are accurately recognized in monkey kidney cells despite the species
and cell-type differences. Further, the levels of globin mRNA and
protein produced in the monkey cells compared favorably with those of
the viral late gene products, suggesting that the mouse gene signals
function with high efficiency in this system.

To test more directly for the utilization of a chromosomal
promoter, the mouse α-globin gene has been inserted into SV40 in both
possible orientations relative to the viral promoter (M. Kaehler, D.
Hamer and P. Leder, unpublished data). Presumably, the anti-sense
hybrid would produce substantial quantities of globin only if the
globin promoter were active. In fact, both types of virus do produce
approximately 9S globin RNA that is translated into mouse α-globin.
The rate of globin synthesis is about 5-fold lower in the anti-sense
than the sense virus. However, a definite conclusion regarding the
activity of the mouse gene promoter will require precise mapping of
the 5' end of the mRNA.


ADVANTAGES OF SV40 RECOMBINANTS AS GENETIC AND BIOCHEMICAL TOOLS

Several methods have been developed for the introduction of
foreign genes into animal cells. These include microinjection of
toad germinal vesicles (65), microinjection of animal cells (66) and
direct DNA transformation (67-69) as well as the use of SV40
recombinants. Several unique features of the SV40 recombinant system
for such experiments should be noted.

First, the SV40 system requires neither tedious micro-
manipulation nor extensive tissue culture work. This is especially
true when the initial recombinants are constructed in E. coli rather
than monkey cells. Further, all the necessary reagents (except, of
course, the foreign DNA) are available from commercial sources.

Second, SV40 reaches levels of about 100,000 copies per cell
late in the lytic cycle. This tremendous gene amplification greatly
simplifies the detection and analysis of gene products encoded by the
foreign DNA.

Third, by varying the structure of the recombinant molecules,
it is possible to determine whether or not the foreign DNA will be
expressed under control of the viral signals for transcription
initiation, splicing, polyadenylation and translation initiation.
This may be especially useful if the insert lacks some or all of
these signals (e.g., chromosomal sequences altered by in vitro
mutagenesis).

Fourth, by using SV40 recombinants for transformation rather
than lytic infection, it should be possible to construct cell lines
with stably altered genotypes and phenotypes.

There is one difficulty with the direct DNA transformation
method: foreign DNA can integrate at a variety of sites on the host
chromosome. Therefore, if one prepares cell lines carrying a set of

mutated genes, it may be difficult to determine whether variations in
the expression of the gene reflect chromosomal position effects or
the induced mutations.  In contrast, a set of isogenic SV40 recombinants differing only at the desired site can readily be prepared.
By comparing the expression of the foreign gene in these hybrids, it
should be possible to map the fine structure of the gene.

Acknowledgments:  I thank P. Leder, M. Kaehler, M. Botchan, P.
Berg and G. Fareed for communicating their unpublished observations
and G. Khoury and M. Martin for their helpful comments on the
manuscript.

## REFERENCES

1    Kelly, T.J. and Nathans, D. (1977) in Advances in Virus Research
     (Lauffer, M.A., Bang, F.B., Maramorosch, K. and Smith, K.M.,
     eds.), Vol. 21, pp. 86-173, Academic Press, New York, NY.
2    Fareed, G.C. and Davoli, D. (1977) in Ann. Rev. Biochem. (Snell,
     E.S., Boyer, P.D., Meister, A. and Richardson, C.C., eds.) Vol.
     46, pp. 471-522, Palo Alto, CA.
3    Fiers, W., Contreras, R., Haegeman, G., Rogiers, R., Van de
     Voorde, A., Van Heuverswyn, H., Van Herreweghe, J., Volckaert,
     G. and Ysebaert, M. (1978) Nature 273, 113-120.
4    Reddy, V.B., Thimmappaya, B., Dhar, R., Subramanian, K.N., Zain,
     B.S., Pan, J., Ghosh, P.K., Celma, M.L. and Weissman, S.M.
     (1978) Science 200, 494-502.
5    Aloni, Y., Dhar, R., Laub, O., Horowitz, M. and Khoury, G.
     (1977) Proc. Nat. Acad. Sci. U.S.A. 74, 3686-3690.
6    Celma, M.L., Dhar, R., Pan, J. and Weissman, S.M. (1977) Nucl.
     Acids Res. 4, 2549-2560.
7    Hsu, M.-T. and Ford, J. (1977) Proc. Nat. Acad. Sci. U.S.A. 74,
     4982-4985.
8    Lavi, S. and Groner, Y. (1977) Proc. Nat. Acad. Sci. U.S.A. 74,
     5323-5327.
9    Lai, C.-J., Dhar, R. and Khoury, G. (1978) Cell 14, 971-982.
10   Ghosh, P.K., Reddy, V.B., Swinscoe, J., Choudary, P., Lebowitz,
     P. and Weissman, S.M. (1977) J. Biol. Chem. 253, 3643-3647.
11   Reddy, V.B., Ghosh, P.K., Lebowitz, P. and Weissman, S. (1978)
     Nucl. Acids Res. 5, 4195-4213.
12   Uchida, S., Yoshiike, K., Watanabe, S. and Furuno, A. (1968)
     Virology 34, 1-13.
13   Yoshiike, K. (1968) Virology 34, 391-402.
14   Uchida, S. and Watanabe, S. (1969) Virology 39, 721-728.
15   Tai, H.T., Smith, C.A., Sharp, P.A. and Vinograd, J. (1972)
     J. Virol. 9, 318-322.
16   Lavi, S. and Winocour, E. (1972) J. Virol. 9, 309-317.
17   Martin, M.A., Gelb, L.D., Fareed, G.C. and Milstein, J.B. (1973)
     J. Virol. 12, 748-759.

18    Brockman, W.W., Lee, T.N.H. and Nathans, D. (1973) Virology 54,
      384-395.

19    Fareed, G.C., Byrne, J.C. and Martin, M.A. (1974) J. Mol. Biol.
      87, 275-291.

20    Khoury, G., Fareed, G.C., Berry, K., Martin, M.A., Lee, T.N.H.
      and Nathans, D. (1974) J. Mol. Biol. 87, 289-301.

21    Lee, T.N.H., Brockman, W.W. and Nathans, D. (1975) Virology 66,
      53-65.

22    Ganem, D., Nussbaum, A., Davoli, D. and Fareed, G.C. (1976) J.
      Mol. Biol. 101, 57-74.

23    Davoli, D., Ganem, D., Nussbaum, A., Fareed, G.C., Howley, P.M.,
      Khoury, G. and Martin, M.A. (1977) Virology 77, 836-844.

24    Brockman, W.W., Gutai, M.W. and Nathans, D. (1975) Virology 66,
      36-44.

25    Brockman, W.W. and Nathans, D. (1974) Proc. Nat. Acad. Sci.
      U.S.A. 71, 942-947.

26    Mertz, J.E. and Berg, P. (1974) Proc. Nat. Acad. Sci. U.S.A. 71,
      4879-4885.

27    Ganem, D., Nussbaum, A., Davoli, D. and Fareed, G.C. (1976) Cell
      7, 349-358.

28    Nussbaum, A., Davoli, D., Ganem, D. and Fareed, G.C. (1976)
      Proc. Nat. Acad. Sci. U.S.A. 73, 1068-1074.

29    Davoli, D., Nussbaum, A.L. and Fareed, G.C. (1976) J. Virol. 19,
      1100-1107.

30    Mulligan, R.C., Howard, B.H. and Berg, P. (1979) Nature 277,
      108-114.

31    Mertz, J.E. and Davis, R.W. (1972) Proc. Nat. Acad. Sci. U.S.A.
      69, 3370-3374.

32    Hedgpeth, J., Goodman, H.M. and Boyer, H.W. (1972) Proc. Nat.
      Acad. Sci. U.S.A. 69, 3448-3452.

33    Heyneker, H.L., Shine, J., Goodman, H.M., Boyer, H.W. Rosenberg,
      J., Dickerson, R.E., Narang, S.A., Itakura, K., Lin, S. and
      Riggs, A.D. (1976) Nature 263, 748-752.

34    Scheller, R.H., Dickerson, R.E., Boyer, H., Riggs, A.D. and
      Itakura, K., (1977) Science 196, 177-180.

35    Jackson, D.A., Symons, R.H. and Berg, P. (1972) Proc. Nat. Acad.
      Sci. U.S.A. 69, 2904-2909.

36    Labban, P. and Kaiser, D. (1973) J. Mol. Biol. 78, 458-471.

37    Goff, S.P. and Berg, P. (1976) Cell 9, 695-705.

38    Hamer, D.H., Davoli, D., Thomas, C.A. Jr. and Fareed, G.C.
      (1977) J. Mol. Biol. 112, 155-182.

39    Hamer, D.H. (1977) in Recombinant Molecules: Impact on Science
      and Society (Beers, R.F. and Bassett, E.G., eds.), pp. 317-336,
      Raven Press, New York, NY.

40    Hamer, D.H., Smith, K.D., Boyer, S.H. and Leder, P. (1979) Cell
      17, 725-735.

41    Hamer, D.H. and Leder, P. (1979) Cell 17, 737-747.

42    Hamer, D.H. and Leder, P. (1979) Nature 281, 35-40.

43    Hamer, D.H. (1977) in Genetic Manipulation as it Affects the
      Cancer Problem (Schultz, J. and Brada, Z., eds.), pp. 37-47,
      Academic Press, New York, NY.
44    Upcroft, P. Skolnik, H. Upcroft, J.A., Solomon, D., Khoury, G.,
      Hamer, D.H. and Fareed, G.C. (1978) Proc. Nat. Acad. Sci. U.S.A.
      75, 2117-2121.
45    Muzyczka, N. (1979) Gene 6, 107-122.
46    Villarreal, L.P. and Berg, P. (1977) Science 196, 183-185.
47    Haegeman, G. and Fiers, W. (1978) Nucl. Acids Res. 5, 2359-2371.
48    Ghosh, P.K., Reddy, V.B., Swinscoe, J., Lebowitz, P. and
      Weissman, S.M. (1978) J. Mol. Biol. 126, 813-846.
49    Villarreal, L., White, R. and Berg, P. (1979) J. Virol. 29,
      209-219.
50    Bratosin, S., Horowitz, M. and Aloni, Y. (1979) Virology 92,
      310-323.
51    Canaani, D., Kahana, C., Mukamel, A. and Groner, Y. (1979) Proc.
      Nat. Acad. Sci. U.S.A. 76, 3078-3082.
52    Bina-Stein, M., Thoren, M., Salzman, N. and Thompson, J.A.
      (1979) Proc. Nat. Acad. Sci. U.S.A. 76, 731-735.
53    Mertz, J. and Berg, P. (1974) Proc. Nat. Acad. Sci. U.S.A. 71,
      4879-4883.
54    Shenk, T., Carbon, J. and Berg, P. (1976) J. Virol. 18, 664-671.
55    Lai, C-J. and Khoury, G.C. (1979) Proc. Nat. Acad. U.S.A. 76,
      71-75.
56    Gruss, P., Lai, C-J. and Khoury, G.C. (1979) Proc. Nat. Acad.
      Sci. U.S.A. (in press).
57    Hamer, D.H. and Leder, P. (submitted for publication).
58    Dhar, R., Zain, B.S., Weissman, S.M., Pan, J. and Subramanian,
      K. (1974) Proc. Nat. Acad. Sci. U.S.A. 71, 371-376.
59    Cancedda, R., Villa-Komaroff, L., Lodish, H.F. and Schlesinger,
      M. (1975) Cell 6, 215-222.
60    Goff, S. (1976) Ph.D. Thesis, Stanford Medical School.
61    Tilghman, S.M., Tiemeier, D.C., Seidman, J.G., Peferlin, B.M.,
      Sullivan, M., Maizel, J.V. and Leder, P. (1978) Proc. Nat. Acad.
      Sci. U.S.A. 75, 725-729.
62    Konkel, D.A., Tilghman, S.M. and Leder, P. (1978) Cell 15,
      1125-1132.
63    Leder, A., Miller, H.I., Hamer, D.H., Seidman, J.G., Norman, B.,
      Sullivan, M. and Leder, P. (1978) Proc. Nat. Acad. Sci. U.S.A.
      75, 6187-6191.
64    Schaffner, W., Topp, W. and Botchan, M. (1979) Alfred Benzend
      Symp. (in press).
65    Mertz, J.E. and Gurdon, J.B. (1977) Proc. Nat. Acad. Sci. U.S.A.
      74, 1502-1506.
66    Graessmann, M. and Graessmann, A. (1975) Virology 65, 591-594.
67    Graham, F.C., Abrahams, P.J., Mulder, C., Heijneker, H.C.,
      Warnaar, S.O., deVries, F.A.J., Fiers, W. and Van der Eb, A.J.
      (1974) Cold Spring Harbor Symp. Quant. Biol. 39, 637-650.
68    Wigler, M., Pellicer, A., Silverstein, S. and Axel, R. (1978)
      Cell 14, 725-731.

69   Mantei, N., Ooyen, A., Berg, J., Beggs, J., Boll, W., Weaver, R. and Weissmann, C. (1979) JCN–UCLA Symposium on Eucaryotic Gene Regulation (in press).

ADENOVIRUS-SV40 HYBRIDS: A MODEL SYSTEM FOR EXPRESSION OF FOREIGN

SEQUENCES IN AN ANIMAL VIRUS VECTOR

Joseph Sambrook and Terri Grodzicker

Cold Spring Harbor Laboratory

Cold Spring Harbor, New York   11724

While adequate methods have been developed to introduce segments of foreign DNA into cultured mammalian cells, no satisfactory means have been found to assure their expression. From this perspective, DNA animal viruses become attractive eukaryotic vectors. The genomes of several of these viruses have been completely sequenced; others nearly so. And during the last five years the arrangement and location have been determined not only of the principal viral genes, but also of the elements that control their expression. It has therefore become possible to insert pieces of foreign DNA into viral genomes at positions where they can be expressed. This approach is most successful when the coding sequences of a foreign gene are used to replace the sequences coding for a viral protein, leaving the flanking control sequences in place. Thus, cloned cDNA copies of mouse globin mRNA have been inserted into SV40 DNA at the position normally occupied by a late capsid gene (1,2). Transcription of the inserted sequences begins from the upstream viral late promoter and the resulting hybrid RNA is spliced, polyadenylated and capped at sites within the noncoding viral sequences that enclose the mRNA sequences coding for globin. Predictably, the mRNA is translated into globin. The substitution of one coding sequence for another is an extremely useful way to obtain expression of small genes for which DNA copies are available. Ultimately, however, one would not want to be restricted to cloning only coding regions nor to sequences that can be accomodated in relatively small vectors such as SV40.

To our knowledge, adenoviruses have not been used deliberately as vectors for foreign DNA. Ironically however, and as a result of a series of unlikely recombination events occurring several years ago, we know a good deal about the ways that foreign sequences inserted into adenovirus DNA are expressed. During the 1960s, human adenoviruses were grown in cultures of simian cells now known to

103

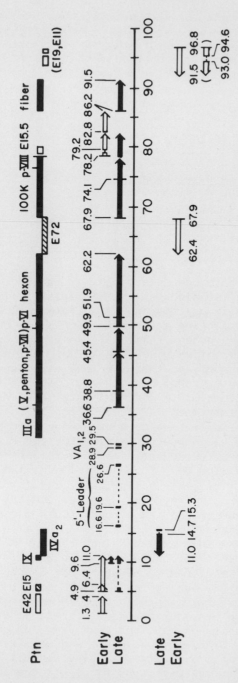

Figure 1. A physical map of the adenovirus genome. Vertical bars on the arrows indicate the map positions of 5' termini of RNA transcripts; arrowheads represent 3' termini. Dashed lines designate sequences absent from the mature mRNAs. The tripartite leader sequence is coupled to the 5' end of all late transcripts from 36 to 91.5. Viral associated (VA) RNAs extend to 0.45 map units from the 5' ends indicated. The map positions of the leader sequences and the VA RNAs have been determined by assuming coordinates of 0 to 29.1 for the BamHI B restriction fragment. RNA in the region 31 to 36 has not been mapped accurately. Proteins identified by cell-free translation of purified mRNAs are indicated. Hatched blocks represent proteins made both early and late. Taken from L.T. Chow, J.B. Lewis, J.M. Roberts, J.F. Atkins, C.W. Anderson, M.B. Mathews, U. Pettersson, R.E. Gelinas, R.J. Roberts, and T.R. Broker (see ref. 32).

have been contaminated with SV40. Human adenoviruses normally will
not replicate in monkey cells; however, for reasons that are not
entirely understood, SV40 codes for a helper function that renders
simian cells fully permissive to adenoviruses (3). Although the
genomes of the two viruses are different in topology and share no
detectable homology, their long-term cocultivation allowed recombinants to form whose genomes consist of SV40 sequences covalently
attached to adenovirus DNA (4,5). It thus became clear that SV40
can provide the helper function for adenovirus growth in monkey cells
when the relevant portion of its genome is inserted into adenoviral
DNA. The hybrid viruses are stable, replicate efficiently and are
packed into adenovirus coats; each contains a different quantity of
SV40 DNA inserted at one of a number of sites in the adenovirus
genome. In one sense, the several members of this group of adeno-
virus-SV40 hybrid viruses can be regarded as recombinants in which
the adenovirus moiety acts as a vector for the enclosed SV40 sequences. Studies of the expression of these SV40 sequences have
progressed rapidly and now are at a stage where they can be concisely
summarized. They hold lessons as well as promise for those who wish
to use adenovirus as vectors.

The genome of adenovirus 2 is divided into two functional domains
(Figure 1; for a review, see ref. 6). One comprises the five blocks
of early genes each associated with its separate promoter(s). The
immediate products of early transcription are believed to be largely
colinear with and only slightly longer than the gene blocks themselves. The expression of the early genes during the early phase of
infection causes the activation of the late genes. Most of the latter
are located on the "r" strand of the viral DNA in three noncontiguous
blocks separated from each other by intervening sets of early genes.
Despite this separation and the great distances involved, the late
genes are nevertheless transcribed from a single initiation point
located at about map position 16.4. The immediate product is believed to be a large nuclear RNA whose 5' end maps at position 16.4
and whose 3' end is copied, at least in some cases, from sequences at
or near the right-hand terminus of the viral DNA, some 28,000 nucleotides away. Exactly how cytoplasmic mRNAs are fashioned from such
precursors is unknown but the process involves internal cleavage,
addition of poly(A) to one newly generated 3' end and of capping
structures to the 5' end as well as the removal of internal sequences.
This splicing procedure causes the coding sequence of each of the major
late mRNAs to be coupled to a common tripartite leader derived from
coordinates 16.6, 19.6 and 26.6.

The best studied adenovirus-SV40 hybrids, those involving adenovirus 2, fall into three classes, in which the SV40 sequences:  a) are
placed under the control of an early adenovirus promoter; b) are
placed under the control of the major late adenovirus promoter; c)
retain autonomous control.

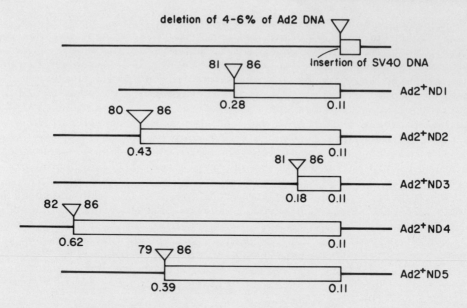

Figure 2.   Genome structures of nondefective adenovirus 2-SV40 hybrid
viruses.   The general structure is given at the top of the figure.
All hybrids contain an insertion of SV40 DNA (boxes) at the site from
which adenovirus 2 sequences (triangles) have been deleted.   All deletions of adenovirus 2 DNA have a common end point (position 86 on
the adenovirus genome) and all SV40 insertions share a common end
at position 0.11 on the SV40 physical map.   The orientation of SV40
sequences in the hybrids is counterclockwise on the SV40 physical map.

SV40 SEQUENCES UNDER CONTROL OF
AN EARLY ADENOVIRUS PROMOTER

A family of related nondefective adenovirus 2-SV40 hybrid
viruses contain varying amounts of SV40 DNA inserted into an early
region (E3, Figure 2) on the adenovirus genome.   Ad2[+]ND4, isolated
from monkey cells, expresses both the SV40 helper function and SV40
T-antigen in infected cells.   It is the parent of other nondefective
hybrid viruses, Ad2[+]ND1, Ad2[+]ND2, Ad2[+]ND3 and Ad2[+]ND5 that were
derived from it after growth and recombination in human cells (7,8).

All of the hybrid viruses have SV40 DNA inserted at a site
(86 map units) from which adenovirus DNA has been lost. The deleted
adenoviral DNA is clearly not required for virus growth in any of
the cell culture systems so far tested. The SV40 segments present
in the hybrids begin at the same position (0.11) on the SV40 genome
and comprise an overlapping series that extends into the SV40 early
region (5,9-11; Kelly and Lewis, unpublished data; Zain et al.,
unpublished data). Ad2[+]ND4 contains SV40 sequences that map from
position 0.62 to 0.11; Ad2[+]ND2 has a segment that maps from 0.43 to
0.11; Ad2[+]ND5 has sequences derived from 0.39 to 0.11; Ad[+]ND1 contains
SV40 DNA mapping from 0.28 to 0.11; and Ad2[+]ND3 has an insertion that
maps from position 0.18 to 0.11. Thus, the hybrid viruses contain
varying amounts of early SV40 sequences. The deletions of adenovirus 2 DNA that the hybrids have suffered also form an overlapping
set extending left from position 86 on the adenovirus genome.

The hybrid viruses synthesize early SV40 RNA under the control
of the early adenovirus E3 promoter. Transcription starting at
this promoter continues through the SV40 early region and terminates
at the normal early/late boundary (SV40 approximate map position 0.15)
that is present in all viruses. Thus, the SV40 early region is transcribed as part of a hybrid adenovirus 2-SV40 molecule (11-16,19;
Chow et al., unpublished data).

The SV40-specific transcript induced by each hybrid contains
varying lengths of the 5' end of the adenovirus E3 mRNA and of the
3' end of SV40 early mRNA. While the 3' end of the mRNA is determined by signals for transcription termination and poly(A) addition
coded for by the SV40 sequences in the hybrid, the leader sequences
present at the 5' end are determined by adenovirus controlling signals. It has been shown that Ad2[+]ND1 hybrid mRNA has the leader
sequences of the E3 early mRNAs at early times after infection while
mRNAs isolated late after infection contain the late adenovirus 2
tripartite leader (14,18; Chow et al., unpublished data). Thus,
Ad2[+]ND1 hybrid transcripts switch from an early to a late pattern of
splicing and processing as do authentic E3 transcripts themselves.

The hybrid transcripts are translated into a variety of proteins that share peptides with SV40 large T-antigen and with each
other (17,19-24). The size of these proteins is related to the size
of the insertion carried by the hybrid virus (Table 1); in fact, the
hybrids Ad2[+]ND4 and Ad2[+]ND2, which carry the largest insertions, form
several T-antigen related proteins. Only Ad2[+]ND3, which carries the
smallest insertion, fails to produce an SV40-related protein. SV40-
specific proteins made by the hybrids share common sequences from
the C' terminal end of SV40 T-antigen and the differences between
them map at the N' terminus or middle of the molecule. It has not
been shown directly that any of the hybrid proteins carry adenovirus
peptides, although we know that other hybrid viruses code for fusion
proteins that are biologically active (see below).

The multiple hybrid proteins produced by any one virus (Ad2[+]ND4
and Ad2[+]ND2) are probably produced from unique mRNA (Westphal et al.,
unpublished data, cited in ref. 25). However, it is not yet clear

Table 1

SV40-Specific Antigens and Proteins Produced by
Nondefective Adenovirus 2-SV40 Hybrid Viruses

| Hybrid viruses | Properties of hybrid viruses | |
| | Size of SV40-specific proteins | SV40 antigens induced |
| --- | --- | --- |
| Ad2$^+$ND1 | 30K | U,TSTA* |
| Ad2$^+$ND2 | 42K,56K | U,TSTA |
| Ad2$^+$ND3 | none | none |
| Ad2$^+$ND4 | 96K,74K,72K,64K, 60K,56K,42K | U,TSTA,T |
| Ad2$^+$ND5 | 42K | none |

*TSTA induced in Balb/c mice, but not in hamsters

whether these transcripts differ in leader and control sequences and/
or in the coding regions they contain nor is it known whether any
differences there may be in splicing patterns are determined by the
adenovirus and/or SV40 sequences they carry.

What is striking is that the hybrid virus proteins are biologically active for a variety of the functions associated with SV40
large T-antigen although most of them carry only truncated portions
of that molecule.  In fact, one can produce a map relating the SV40
T-antigen sequences in each hybrid, and the proteins they code for,
to the biological functions they induce:  1) Ad2$^+$ND4 is the only
hybrid virus that produces SV40 T-antigen as determined by complement fixation and immunofluorescence (8,26); 2) only Ad2$^+$ND4 and
Ad2$^+$ND2 induce the SV40 transplantation specific tumor antigen (TSTA)
in hamsters while Ad2$^+$ND1, in addition, is able to induce SV40 TSTA
in Balb/c mice (27); 3) Ad2$^+$ND1, Ad2$^+$ND2 and Ad2$^+$ND4 can induce SV40
U-antigen in infected cells.  U-antigen, a perinuclear antigen, is
thought to be a part of the SV40 large T-antigen molecule; 4) Ad2$^+$ND1,
Ad2$^+$ND2 and Ad2$^+$ND4 all express the SV40 helper function and are able
to grow efficiently on monkey cells as well as human cells; 5) Ad2$^+$ND3
and Ad2$^+$ND5 induce no SV40 antigens and fail to grow in monkey cells.

The only virus that does not fit into the general scheme is
Ad2$^+$ND5, which codes for a 42K related T-antigen protein, although
it fails to grow in monkey cells and induces no SV40 antigens. However, the Ad2$^+$ND5 42K protein is unique in that it is extremely
unstable and cannot be immunoprecipitated with SV40 anti-T serum.

These results taken together suggest that the functional domains
of SV40 large T-antigen are relatively autonomous and can function
even if the normal configuration of the wild-type protein is not
present.  If this turns out to be a common feature of viral and
other eukaryotic proteins, the fusion of a variety of genes to foreign

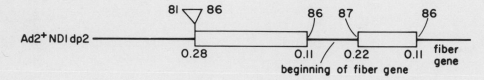

Figure 3.   The genome structure of Ad2+ND1 dp2, a revertant of a
host range mutant of Ad2+ND1 that has regained the ability to grow
on monkey cells.   Ad2+ND1 dp2 contains the original SV40 insertion
present in Ad2+ND1 (SV40 map positions 0.28 to 0.11).   It also contains a new insertion adjacent to the original SV40 segment.   The
new insertion consists of 1% of adenovirus 2 DNA (map positions 86-
87) and SV40 DNA that maps between coordinates 0.22 and 0.11 on the
SV40 map.

promoters and controlling elements will result in the translation
of many active proteins.

SV40 SEQUENCES UNDER CONTROL OF
A LATE ADENOVIRUS PROMOTER

Two examples are known in which the SV40 sequences integrated
into an adenovirus–SV40 hybrid are placed under the control of the
major late adenovirus promoter.

Ad2+ND1 dp2

This virus is a derivative of Ad2+ND1 whose genome, as we have
seen, contains a simple insertion of SV40 DNA approximately 900
nucleotides in length, in partial replacement of the segment of
adenovirus DNA that maps between positions 80.6 and 86.   Crodzicker
et al. (28) isolated a number of mutants of Ad2+ND1 that have lost
the ability to grow efficiently in simian cells because they contain
chain terminator mutations in the gene coding for helper protein.
From one such mutant, H71, containing an ochre codon, Eugene
Lukanidin selected a revertant that was again able to replicate in
simian cells (unpublished data).   The genome of Ad2+ND1 dp2 (Figure
3) retains the sequences of SV40 DNA that are present in H71 and
expressed under the control of the promoter associated with early
region 3 of adenovirus DNA.   However, an additional insertion of
SV40 sequences is present, separated from the first by approximately
350 nucleotide pairs of adenovirus DNA.   This short tract spans the
beginning of the late adenovirus gene, which codes for the structural
protein fiber.

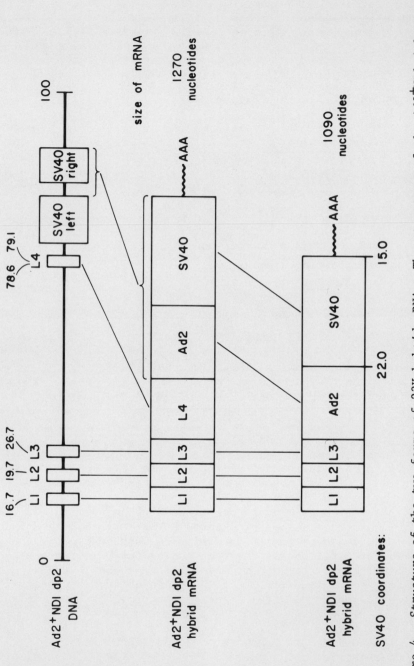

Figure 4. Structure of the two forms of 23K hybrid mRNA. The structure of the Ad2$^+$ND1 dp2 genome is shown at the top of the figure. A schematic representation of the two forms of the 23K hybrid mRNA containing and lacking the 4th leader is also included. The figure is based on work publish–ed by Dunn et al., 1978 (14).

At late times after infection (and only at late times), the cytoplasm of infected cells contains large quantities of two distinct and separable populations of mRNAs that are specific for Ad2+ND1 dp2 (i.e., they are not found in cells infected with Ad2+ND1) (14). The two mRNAs are hybrid molecules and contain SV40 as well as adenovirus sequences. Their structures are shown in Figure 4. At their 5' ends are the sequences of the traditional tripartite leaders. Then, in order, follow an optional fourth leader, a block of coding sequences and, finally, a tract of polyriboadenylic acid. The two mRNAs differ from each other only by the presence or absence of the optional fourth leader. Both late hybrid mRNAs direct the synthesis of a polypeptide of approximately 24K consisting of fiber sequences at its N-terminal end and SV40 sequences at its C-terminal end. This protein carries the helper function, which allows Ad2+ND1 dp2 to grow efficiently on monkey cells (29).

## Ad2+D2

AD2+D2 is a defective adenovirus 2-SV40 hybrid obtained by plaque isolation from a heterogeneous population of hybrids called Ad2++HEY (30). The structure of its genome is shown in Figure 5. It contains a set of SV40 sequences that are colinear with those mapping counter-

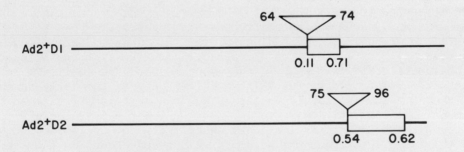

Figure 5. Genome structures of defective adenovirus 2-SV40 viruses. The single lines show adenovirus 2 sequences present in the hybrid viruses while the triangles extending above the maps show which adenovirus 2 sequences are missing from the hybrid virus. Coordinates on the adenovirus 2 physical maps are given as whole numbers. SV40 segments present in the hybrid viruses are shown as boxes on the map. SV40 coordinates are given as fractional numbers representing positions on the SV40 circular physical map. The orientation of SV40 sequences in Ad2+D1 is clockwise with respect to the conventional SV40 map and counterclockwise in the Ad2+D2 hybrids.

clockwise on the SV40 genome from position 54 to position 63. The
SV40 sequences are inserted into the adenovirus genome in partial
replacement of the DNA sequences mapping between position 76 and 96.
The left end of the SV40 DNA sequences of Ad2$^+$D2 are attached to the
segment of the adenovirus 2 genome coding for the late 100K protein.
At early times after infection with Ad2$^+$D2, no expression of SV40
DNA sequences is detectable; at late times, however, considerable
quantities are present of RNA that hybridizes to the early region
of SV40 DNA and of protein that cross reacts immunologically with
SV40 T-antigen. Therefore, in Ad2$^+$D2 as in Ad2$^+$ND1 dp2, expression
of SV40 early functions is placed quite firmly under adenovirus late
control. The SV40 protein coded by Ad2$^+$D2 is considerably larger
(115K) in size than SV40 T-antigen (96K) and consists of the N-
terminal portion of the 100K protein and the C-terminal sequences of
SV40 T-antigen. This protein, synthesized in very large amounts
late during Ad2$^+$D2 infection, has been purified virtually to homogeneity and has been shown to display many if not all of the properties both in vivo and in vitro of SV40 T-antigen (29,31).

The entire late region of SV40 DNA is present in the Ad2$^+$D2
genome but is not expressed to a significant degree at any time
during lytic infection (30).


## SV40 SEQUENCES THAT RETAIN AUTONOMY

The best example of this class of hybrid is Ad2$^+$D1, the structure of whose genome is shown in Figure 5 (30). The SV40 sequences
of Ad2$^+$D1 are colinear with those that map between positions 71 and
10 on the wild-type SV40 genome. The left end of the inserted SV40
DNA is joined to the adenovirus genome within the DNA sequences that
code for the 72,000 dalton early, DNA-binding protein; the right end
abuts the adenovirus gene coding for the late nonstructural 100K
protein. The SV40 sequences are transcribed from right to left and
SV40 specific RNA is formed in the cytoplasm at both early and late
times after infection. The RNA hybridizes only to the early strand
of SV40 DNA and is translated into a protein that in its size and
immunological reactivity is indistinguishable from SV40 large T-
antigen. The SV40 insertion contains not only the coding sequences
of the large T-antigen but also intervening sequences and the elements believed to control the expression of the early SV40 genes.
Because at least one of these gene products (large T-antigen) is
expressed normally, it would seem likely that the inserted sequence
is transcribed, processed and translated by pathways established
for SV40.


## CONCLUSIONS

Consideration of the above facts leads to the following
general conclusions: 1) insertion of a coding sequence in phase

into either a late or an early adenovirus gene leads to the production of a hybrid protein which consists of adenovirus sequences at
its N' terminal end and SV40 sequences at its C' terminal end; 2)
the hybrid proteins are translated from hybrid RNAs which are synthesized at times in infection and at rates that are appropriate
for the viral gene to which the insert is coupled.  Thus, the insertion of a coding sequence in phase into a late adenovirus gene
results in the production of very large quantities of a hybrid protein that, in some cases at least, retains the biological functions
coded by the inserted sequences; 3) the arrangements of the leader
sequences in the hybrid mRNAs are identical to those of authentic
adenoviral mRNAs; 4) adenovirus transcription at both early and
late times after infection is sensitive to signals for termination
(or poly(A) addition) that map within the inserted sequences; 5)
certain types of promoters carried by the inserted sequences function autonomously; others either remain inactive or give rise to
unstable transcription products.

The reason adenoviruses carry and express SV40 sequences so
efficiently is a consequence of the selective advantage donated
to the recombinant by the inserted sequences themselves.  However,
there would seem to be no theoretical or practical reasons why nonselectable genes should not function with equal efficiency.  Perhaps
the only difficulty is the possibility that a defective adenovirus
vector carrying a nonselectable gene would recombine with its obliga-
tory helper to form a wild-type adenovirus genome that might come to
dominate the population.  However, there are several ways to avoid
this problem.  For example, distantly related adenovirus serotypes
could be used as helpers; alternatively, the transacting functions
necessary for replication could be provided by a viral sequence integrated into the cellular genome.  Given these possibilities, it does
not seem too optimistic to believe that within the next few years
adenoviruses might come to occupy for eukaryotic systems the position
now held by bacteriophage lambda for prokaryotes.

REFERENCES

1   Hamer, D.H., Smith, K.D., Boyer, S.H. and Leder, P. (1979)
    Cell 17, 725-735.
2   Mulligan, R., Howard, B. and Berg, D. (1979) Nature 277,
    108-114.
3   Rabson, A.S., O'Conor, G.T., Berezesky, I.K. and Paul, F.J.
    (1964) Proc. Soc. Exp. Biol. Med. 116, 187-190.
4   Crumpacker, C.S., Levin, M.J., Wiese, W.H., Lewis, A.M. Jr.
    and Rowe, W.P. (1970) J. Virol. 6, 788-794.
5   Kelly, T.J. Jr. and Lewis, A.M. (1973) J. Virol. 12, 643-652.
6   The Molecular Biology of Tumor Viruses (1979) (Tooze, J., ed.),
    Second Edition, Cold Spring Harbor Laboratory, Cold Spring
    Harbor, NY.

7   Lewis, A.M. Jr., Levin, M.J., Wiese, W.H., Crumpacker, C.S. and
    Henry, P.H. (1969) Proc. Nat. Acad. Sci. U.S.A. 63, 1128-1135.
8   Lewis, A.M. Jr., Levine, A.S., Crumpacker, C.S., Levin, M.J.,
    Samaha, R.J. and Henry, P.H. (1973) J. Virol. 11, 655-664.
9   Morrow, J.F., Kelly, T.J. Jr. and Lewis, A.M. Jr. (1973) J.
    Virol. 12, 653-658.
10  Lebowitz, P., Kelly, T.J. Jr., Nathans, D., Lee, T.H.N. and
    Lewis, A.M. Jr. (1974) Proc. Nat. Acad. Sci. U.S.A. 71, 441-445.
11  Westphal, N., Lai, S.P., Lawrence, C., Hunter, T. and Walter, G.
    (1979) J. Mol. Biol. 130, 337-351.
12  Flint, S.J., Wewerka-Lutz, Y., Levine, A.S., Sambrook, J. and
    Sharp, P.A. (1975) J. Virol. 16, 662-673.
13  Dunn, A.R. and Hassell, J.A. (1977) Cell 12, 23-26.
14  Dunn, A.R., Mathews, M.B., Chow, L.T., Sambrook, J. and Keller,
    W. (1978) Cell 15, 511-526.
15  Khoury, G., Lewis, A.M. Jr., Oxman, M.N. and Levine, A.S. (1973)
    Nature New Biol. 246, 202-205.
16  Levine, A.S., Levin, M.J., Oxman, M.N. and Lewis, A.M. Jr.
    (1973) J. Virol. 11, 672-681.
17  Anderson, C.W., Lewis, J.B., Baum, P.R. and Gesteland, R.F.
    (1976) J. Virol. 18, 685-692.
18  Chow, L.T., Broker, T.R. and Lewis, J.L. (1979) J. Mol. Biol.
    (in press).
19  Lopez-Revilla, R. and Walter, G. (1973) Nature New Biol. 244,
    165-167.
20  Grodzicker, T., Anderson, C., Sharp, P.A. and Sambrook, J.
    (1974) J. Virol. 13, 1237-1244.
21  Walter, G. and Martin, H. (1975) J. Virol. 16, 1236-1247.
22  Deppert, W. and Walter, G. (1977) Proc. Nat. Acad. Sci. U.S.A.
    73, 2505-2509.
23  Deppert, W., Walter, G. and Linke, H. (1977) J. Virol. 21,
    1170-1186.
24  Mann, K., Hunter, T., Walter, G. and Linke, H. (1977) J. Virol.
    24, 151-169.
25  Linke, H., Hunter, T. and Walter, G. (1979) J. Virol. 29,
    390-394.
26  Lewis, A.M. Jr. and Rowe, W.P. (1970) J. Virol. 5, 413-420.
27  Jay, G., Jay, F.T., Chang, C., Friedman, R.M. and Levine, A.S.
    (1978) Proc. Nat. Acad. Sci. U.S.A. 75, 3055-3059.
28  Grodzicker, T., Lewis, J.B. and Anderson, C.W. (1976) J. Virol.
    19, 559-571.
29  Tjian, R., Fey, G. and Graessmann, A. (1978) Proc. Nat. Acad.
    Sci. U.S.A. 75, 1279-1283.
30  Hassell, J.A., Lukanidin, E., Fey, G. and Sambrook, J. (1978)
    J. Mol. Biol. 120, 209-247.
31  Tjian, R. (1978) Cell 13, 165-179.
32  Chow, L.T., Lewis, J.L., Broker, T.R. (1980) Cold Spring Harbor
    Symp. Quant. Biol. (in press).

# MOLECULAR CLONING IN BACILLUS SUBTILIS

D. Dubnau, T. Gryczan, S. Contente and A.G. Shivakumar

Department of Microbiology
The Public Health Research Institute of the City of
  New York
New York, New York  10016

## INTRODUCTION

Bacillus subtilis is the most studied prokaryote other than
Escherichia coli.  It is desirable that molecular cloning be available to the many laboratories presently engaged in research on the
physiology, biochemistry and genetics of this organism.  This technology will be a powerful aid in the study of such phenomena as sporulation and competence, which cannot be investigated in E. coli.
Other problems, such as the heterospecific barrier to gene expression,
manifested by B. subtilis (1-4), and the regulation of transcription
and translation in B. subtilis can be studied more easily with the
aid of a molecular cloning technology.  Finally, the fact that much
of the fermentation industry is based on the use of the Bacilli makes
the development of this technology desirable from the viewpoint of
applied science.
    In this review, we will summarize recent progress in this field.
Although several laboratories are attempting to develop bacteriophage
vectors in B. subtilis (see, for instance, ref. 5) and although
several plasmids indigenous to the Bacilli have been described (4,
6-14), we will restrict our discussion to the use of the plasmids
of Staphylococcus aureus, which have been introduced in B. subtilis
by transformation (1, 15-17).  Nevertheless, many of our comments
will have general applicability.

## PLASMID TRANSFORMATION IN B. SUBTILIS

Ehrlich (15) first reported that competent cultures of B.
subtilis can take up plasmid (CCC) DNA derived from S. aureus and

that this DNA is often capable of autonomous replication and expression.  Further investigation in several laboratories has yielded
the following information.  The development of competence for transformation by plasmid and chromosomal DNA follows a similar time
course (18).  Transformation of competent cultures with plasmid DNA
is first order with respect to DNA concentration, unlike transfection
using bacteriophage DNA.  This suggests that a single plasmid molecule is sufficient for a successful transformation event, although
$10^3$ to $10^4$ molecules are taken up per event (18).  Thus, the process
is inefficient, unlike chromosomal transformation.  When linearized
by cleavage at unique restriction endonuclease sites, plasmid DNA
loses its ability to transform (15,18).  Similarly, when nicked by
the random action of pancreatic DNase, CCC plasmid DNA can no longer
transform.  It appears that although covalent continuity is required
for plasmid transformation, superhelicity is not, since treatment
with nicking-closing enzyme does not affect transforming efficiency
(18).  Transformation of competent cultures by CCC DNA occurs readily
in recE strains (16).  The recE product is required for both chromosomal transformation and transduction and seems to be required for
the integration of bacterial DNA (19,20).

     Canosi et al. (21) reported that plasmid transformation is due
largely (if not entirely) to the presence of oligomeric forms in
native plasmid DNA preparations.  CCC monomer isolated following
agarose gel electrophoresis has little or no transforming activity.
The requirement for oligomeric, CCC DNA can be reasonably explained
in terms of our present understanding of DNA uptake.  During uptake
of duplex DNA, one strand is degraded and the other enters the cell
(22-25).  Some material is lost from the end of this strand (or is
otherwise made unavailable for subsequent recombination events) (26,
27).  It is possible that the uptake of oligomeric strands by this
process permits the occurrence of recE-independent intracellular
annealing and repair that can precisely regenerate a unit length
CCC plasmid molecule.  This process is analogous to that proposed
to explain transfection (28).  However, since plasmid transformation
exhibits a first order dependence on DNA concentration, the opposite
strands with overlapping regions of homology must derive from the
same molecule.  (This requirement is due to the reasonable assumption
that the monomer units in the active oligomers are joined head-to-
tail.  If not, then snap-back of a single strand might generate a
repairable duplex molecule.)

     In contrast, when a homologous resident plasmid is present,
markers carried on either linear, nicked or monomer plasmid DNA can
yield transformants (27).  This rescue process is recE-dependent and
requires homology between donor and resident plasmids.  When the
resident plasmid is homologous to a discrete portion of the donor
molecule, certain rules govern the transforming activity of a marker
carried on the nonhomologous donor moiety.  A linearizing cleavage
cannot occur within the nonhomologous sequence, nor can it occur at
the junction between the homologous and nonhomologous portions.  The
transforming activity of such a marker is directly dependent on the

distance between this junction and a unique cleavage site within the homologous moiety.  These properties of plasmid transformation are directly useful in designing certain cloning experiments in B. subtilis (see below).

Chang and Cohen (29) have adapted the polyethylene glycol cell fusion technique (30,31) to obtain efficient plasmid transformation in protoplasts prepared from noncompetent cultures of B. subtilis. These can be readily regenerated to the bacillary form.  Protoplasts can be transformed by the CCC monomer at about the same efficiency as by dimers and higher oligomers.  Linear plasmid DNA transforms protoplasts about 1% as well as do circular molecules, while nicked monomers transform as well as covalently continuous forms (32).

Based on this discussion, it is instructive to compare transformation of plasmid-free competent cultures and protoplasts by plasmid DNA that has been cleaved by a restriction endonuclease at a unique site and then religated.  In the case of competent cultures, ligation at high DNA concentration will increase the recovery of transforming activity since transformation is due to the formation of oligomers (21,33).  As a result, the recovery of transforming activity is often well over 100%.  Transformation of protoplasts by ligated DNA is not dependent on oligomer formation.  In fact, poor recovery of transforming activity has been reported using ligated DNA (29,32).  This may indicate a preference for lower molecular weight DNA in protoplast transformation.  Oligomerization of plasmid DNA by T4 ligase has similarly been reported to decrease transformation in Escherichia coli (33).  Thus, ligation of DNA should perhaps be performed under conditions that do not favor oligomerization when protoplasts are to be used and under conditions that do favor oligomerization when competent cultures are to be employed.


RESTRICTION AND MODIFICATION IN B. SUBTILIS

Although the competent strain of B. subtilis has been reported to restrict the growth of certain bacteriophage grown on heterologous hosts, the effect of this restriction activity is weak (34).  No more than about a tenfold effect was observed.  A restriction deficient mutant was isolated that did not display this effect (34).  It was reported that another wild-type isolate of B. subtilis specifies a restriction endonuclease (Bsu) that has the same specificity as HaeIII (GGCC) (35,36).  The genes that specify this restriction-modification system were transferred into the competent strain of B. subtilis and were reported to have little or no effect on transformation by bacterial DNA (36).  We have used this strain to test the effect of restriction on plasmid transformation (37).  Unlike transformation by bacterial DNA, transformation by unmodified pUB110, pC194 and pSA2100 DNA is reduced about tenfold compared to modified DNA or compared to transformation of the R⁻ (wild-type) culture.  Transformation using DNA from plasmids that do not contain HaeIII sites was not affected even when the DNA was isolated from nonmodifying

Table 1

Partial List of Plasmids Transferred to
Bacillus subtilis from Staphylococcus aureus

| Plasmid | Antibiotic resistance | Mechanism of resistance | Molecular weight | References |
|---------|----------------------|------------------------|-----------------|------------|
| pC194 | Cm$^r$ | chloramphenicol acetyl transferase | $2.0 \times 10^6$ | 15,40,45,69 |
| pE194 | Em$^r$ | methylation of 23S rRNA (inducible) | $2.3 \times 10^6$ | 17 |
| pSA0501 | Sm$^r$ | | $2.7 \times 10^6$ | 16,40 |
| pSA2100 | Cm$^r$Sm$^r$ | | $4.7 \times 10^6$ | 16,40 |
| pUB110 | Km$^r$ | kanamycin nucleo-tidyl transferase | $3.0 \times 10^6$ | 16,40 |
| pT127 | Tc$^r$ | | $2.9 \times 10^6$ | 15 |

strains. It appears, then, that restriction can affect transformation
by plasmid but not by bacterial DNA, raising the possibility that
cloning of foreign DNA in B. subtilis may be limited by restriction
as it is in E. coli. However, the weakness of the restriction effect
exerted by the wild-type 168 strain on bacteriophage infection suggests that this may not be an important obstacle. In fact, no
difference was detected in transformation of the 168 R⁻ mutant mentioned above compared to transformation of the R⁺ (wild-type) 168
strain, with plasmid DNA isolated from modifying (B. subtilis) and
presumably nonmodifying (S. aureus) organisms (37).


VECTORS

Following Ehrlich's (15) initial report, several additional
S. aureus plasmids have been transferred to B. ꞌbtilis (16,17).
Table 1 summarizes some properties of several of these plasmids.
Our list is not intended to be exhaustive since progress is rapid
and much of the available information is unpublished. In general,
these plasmids are stable in B. subtilis, except that pE194, pSA0501
and pSA2100 are temperature-sensitive and segregate at temperatures
above about 35⁰C. The plasmid pC194 and its derivatives (including
pSA2100) suffer frequent deletions (38,39). It is likely that pC194
is a transposon-like element. Indeed, pSA2100 was originally isolated by Iordanescu (40) as a spontaneous recombinant of pSA0501
and pC194. It was subsequently shown to contain all the sequences
of the parental plasmids (16) and to exhibit a stem-loop structure
at the junction of the double- and single-stranded moieties in
heteroduplexes prepared from pSA0501 and pSA2100 with the stem-loop
contained within the single-stranded (pC194) portion (41). We have
observed that chimeric plasmids constructed from pC194 are unstable

and generate frequent deletions by a recE-independent process (39).
Stable derivatives can, however, be obtained (see below).  Physical
maps (16,39) of several of the plasmids listed in Table 1 are presented in Figure 1.

Chimeras that confer multiple antibiotic resistance have been
constructed using these plasmids (38).  Restriction endonuclease
cleavage sites have been identified in which insertion of foreign
DNA inactivates one or another of these antibiotic resistance markers
(42,43).  Table 2 presents a partial list of chimeras that are potentially useful for cloning, describes some of their properties and
lists the available unique cleavage sites and the resistance characters that are inactivated by insertion.  In each case the unique
sites that do not exhibit insertional inactivation of antibiotic
resistance have been shown not to occur in essential plasmid genes,
since the sites can be removed by deletion or receive foreign DNA
inserts without interfering with replication of the chimera (39).
Some of these same sites, however, may be essential in the parental
plasmids listed in Table 1, since these are not multiple replicons.
The plasmid pBD64 is a spontaneous deletion derivative of pBD12 that
has lost 1.3 megadaltons of DNA entirely from within the pC194 moiety
of pBD12 (42).  It confers both Km$^r$ and Cm$^r$ and appears to be quite
stable, unlike its parent.  The temperature-sensitive (ts) character
of pSA0501 is also expressed in pBD6 and pBD8 (44).  These plasmids
thus presumably replicate using the pSA0501 system.  The other chimeras listed in Table 2 are not temperature-sensitive, although
several of them are derived from pE194, a ts-plasmid.  The ts character of pBD6 and pBD8 is potentially useful as a quick screen to demonstrate that a cloned fragment is indeed carried on the plasmid,
since any phenotype expressed by the fragment should be irreversibly
lost after growth at elevated temperature.  Growth of ts-plasmids at
subinhibitory temperatures allows the adjustment of plasmid copy
number to intermediate levels (45).  This may permit titration of
cloned gene products and a study of gene dosage effects.  Physical
maps of the chimeras listed in Table 2 are presented in Figure 2
(38,39).


## Shotgun Cloning in B. subtilis

Following the initial success of Keggins et al. (46), who used
pUB110 as a cloning vector for the isolation of EcoRI-generated
Bacillus trp fragments, most workers have had great difficulty in
shotgun cloning chromosomal fragments in B. subtilis competent
cultures, although R. Rudner (47) has successfully repeated the construction of a pUB110-B. pumilus trp clone as reported by Keggins
et al. (46).  In contrast, it has been simple to clone plasmid fragments as evidenced by the list of chimeric plasmids in Table 2.
Analysis of several of our experiments demonstrated that when foreign
DNA is ligated to a plasmid vector, the yield of transformants
selected for a plasmid resistance marker drops markedly compared to

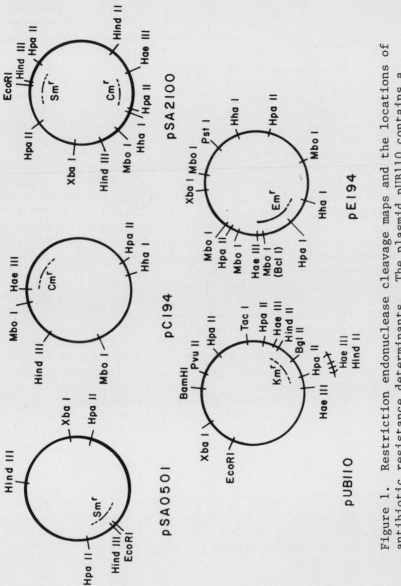

Figure 1. Restriction endonuclease cleavage maps and the locations of antibiotic resistance determinants. The plasmid pUB110 contains a fourth HaeIII site that has not been mapped. As indicated, the positions and orientation of one HaeIII and HindII on pUB110 are uncertain with respect to a HpaII site. A second MboI site exists on pSA2100, but its position is uncertain.

Table 2

Partial List of Plasmid Chimeras

| Plasmid | Parental plasmids | Ref. | Antibiotic resistance | Molecular weight | Single sites* | Inactivated resistance marker |
|---------|-------------------|------|----------------------|------------------|---------------|-------------------------------|
| pBD6 | pSA0501, pUB110 | 38 | $Km^r, Sm^r$ | $5.8 \times 10^6$ | BamHI | none |
|  |  |  |  |  | TacI | none |
|  |  |  |  |  | HindIII | $Sm^r$ |
|  |  |  |  |  | BglII | $Km^r$ |
| pBD8 | pUB110, pSA2100 | 38 | $Km^r, Cm^r, Sm^r$ | $6.0 \times 10^6$ | HindIII | $Sm^r$ |
|  |  |  |  |  | BamHI | none |
|  |  |  |  |  | XbaI | none |
|  |  |  |  |  | BglII | $Km^r$ |
|  |  |  |  |  | EcoRI | $Sm^r$ |
| pBD9 | pE194, pUB110 | 38 | $Em^r, Km^r$ | $5.4 \times 10^6$ | PstI | none |
|  |  |  |  |  | EcoRI | none |
|  |  |  |  |  | BamHI | none |
|  |  |  |  |  | TacI | none |
|  |  |  |  |  | BglII | $Km^r$ |
|  |  |  |  |  | HpaI | $Em^r$ |
|  |  |  |  |  | BclI | $Em^r$ |
| pBD10 | pBD8, pE194 | 38 | $Km^r, Cm^r, Em^r$ | $4.4 \times 10^6$ | BglII | $Km^r$ |
|  |  |  |  |  | BamHI | none |
|  |  |  |  |  | XbaI | none |
|  |  |  |  |  | BclI | $Em^r$ |
|  |  |  |  |  | HpaI | $Em^r$ |
| pBD11 | pBD8, pE194 | 38 | $Km^r, Em^r$ | $4.4 \times 10^6$ | BamHI | none |
|  |  |  |  |  | XbaI | none |
|  |  |  |  |  | BglII | $Km^r$ |
|  |  |  |  |  | HpaI | $Em^r$ |
|  |  |  |  |  | BclI | $Em^r$ |
| pBD12 | pUB110, pC194 | 38 | $Km^r, Cm^r$ | $4.5 \times 10^6$ | HindIII | none |
|  |  |  |  |  | EcoRI | none |
|  |  |  |  |  | XbaI | none |
|  |  |  |  |  | BamHI | none |
|  |  |  |  |  | TacI | none |
|  |  |  |  |  | BglII | $Km^r$ |
| pBD64 | pBD12 | 42 | $Km^r, Cm^r$ | $3.2 \times 10^6$ | EcoRI | none |
|  |  |  |  |  | XbaI | none |
|  |  |  |  |  | BamHI | none |
|  |  |  |  |  | TacI | none |
|  |  |  |  |  | BglII | $Km^r$ |

*These unique restriction sites are available for cloning since their removal by insertion of foreign DNA or by deletion does not interfere with replication of the chimeric plasmid (T. Gryczan and D. Dubnau, unpublished data).

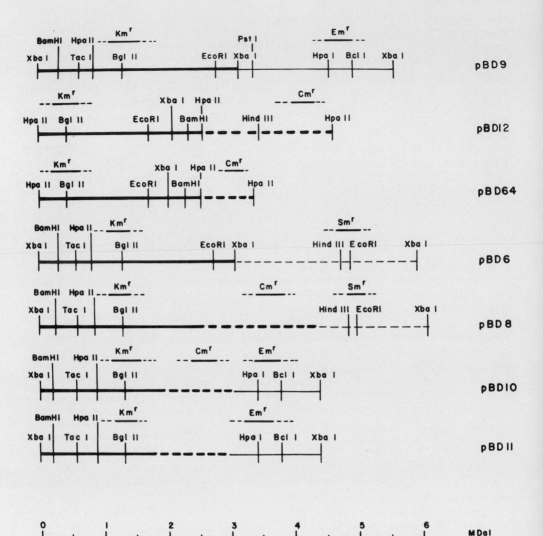

Figure 2.   Restriction endonuclease cleavage maps.
Maps of the chimeric plasmids used in this study
are shown in a linear form cleaved at a specific
restriction enzyme site.  Derivation of chimeric
plasmid DNA is as follows: pUB110 (——), pE194(——),
pC194(---), pSA0501(---).  In pBD8, pBD10 and pBD11,
boundaries between parental plasmid DNAs are uncertain and are given as approximate values.  Locations of specific antibiotic resistance markers are
listed above the linear restriction enzyme cleavage
maps.

the recovery of this marker when the plasmid is ligated to itself.
The extent of this loss is much greater than expected solely from
competition at the level of DNA uptake by competent cells (32).  A
reasonable explanation for this effect is provided by the observation of Canosi et al. (21) that only plasmid multimers transform
competent B. subtilis.  As the ratio of foreign to vector DNA is
elevated during ligation, the yield of vector oligomers must decrease.  Attempts to increase the frequency of hybrid ligates will
always decrease the yield.  The difficulty involved in shotgun
cloning is, therefore, due to an inherent property of plasmid transformation in competent B. subtilis--the requirement for oligomers.
      Several stratagems may be considered to overcome this problem.
Two have been tested successfully.


                        Cloning by Recombination

      As described above, plasmid transformation in a strain carrying
a homologous resident plasmid can be accomplished using linear plasmid
DNA.  This type of transformation therefore does not require oligomeric DNA but is recE-dependent (27).  Figure 3 shows how plasmid
transformation by recombination can be used in a shotgun cloning
experiment.  Ligated DNA damaged during or prior to uptake, even
if monomeric, can be rescued by recombination with the homologous
resident plasmid, if the damage occurs within the homologous sequence.  This procedure has been tested using B. licheniformis
chromosomal DNA cleaved with BglII or with BclI (32).  The vector
used was pBD64 (Table 2 and Figure 2).  Selection was simultaneously
for $Cm^r$ and for complementation of trpC2, hisH2 and aroB2.  The yield
of recombinant plasmids was 6 to 200 per µg of chromosomal DNA.  The
procedure has proven to be highly reproducible.  Agarose gel electrophoresis patterns of DNA samples from several representative recombinant plasmids are shown in Figure 4.  Given the variety of plasmids
and restriction sites now available in B. subtilis, this simple procedure holds considerable promise.  One drawback to the use of pBD64
and pUB110 as the donor (vector) and resident plasmids is the weak
incompatibility expressed by these plasmids toward one another.  As
indicated in Figure 3, the transformed cell contains both the hybrid
plasmid and several copies of the resident.  Thus, inactivation of
$Km^r$ cannot be used as a screen for insertion in the BglII site of
pBD64 unless segregation of the two plasmids occurs.  In our experiments, $Km^s$ clones were isolated only after several growth cycles or
after retransformation of the hybrid plasmids into plasmid-free hosts.
Many other combinations of donor (vector) and resident plasmids can
be utilized depending on the needs of the investigator.  Several of
them express strong incompatibility, thus permitting the use of insertional inactivation as an immediate screen for hybrid plasmids.
For instance, pBD9 and pE194 are strongly incompatible (37) and could
be used as donor and resident plasmid with selection for $Km^r$ and
screening for $Em^s$.  The BclI or HpaI sites of pBD9 would be available

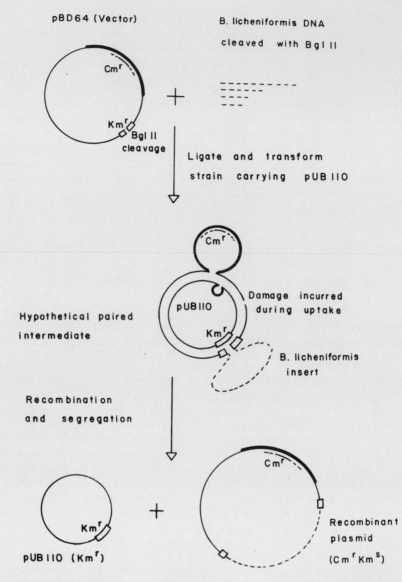

Figure 3.   Cloning by recombination.  The plasmid pBD64 is cleaved
with BglII in the Km$^r$ gene (open bar) and ligated to BglII-cleaved B.
licheniformis DNA.  A competent recipient that carries pUB110 is
used as a recipient for transformation.  The heavy lines indicate
the positions and extents of two regions of inhomology between the
vector and the resident plasmid.  B. licheniformis DNA is indicated
by dashed lines.  If a recombinant molecule produced by ligation is
damaged during uptake, recombination on either side of this damage
will repair this damage.  Damage within a nonhomologous segment will
not be repaired.  Since pUB110 is a multicopy plasmid, the cell
bearing a recombinant plasmid will also contain several pUB110
molecules.

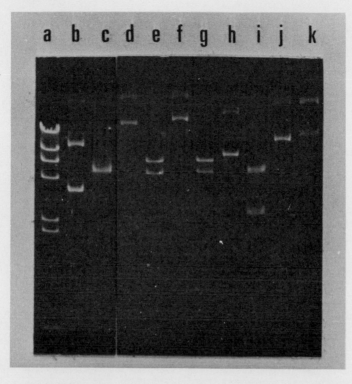

Figure 4. BglII cleavage patterns of recombinant plasmids on a 0.8% agarose gel. (a) λ DNA cleaved with HindIII, (b) pBD64, uncut, (c) pBD64, BglII, (d) pBD81, uncut (pBD81 complements trpD, trpC, trpF, trpB, trpA and hisH and contains a 4.1 megadalton BglII insert), (e) pBD81, BglII, (f) pBD83, uncut (pBD83 is identical to pBD81 but contains an additional 0.6 megadalton BglII fragment), (g) pBD83, BglII, (h) pBD84, uncut (pBD84 complements aroF and aroB and contains a 1.8 megadalton BglII insert), (i) pBD84, BglII, (j) pBD85, uncut (pBD85 complements trpC, trpF, trpB and trpA and contains a 2.8 megadalton insert), (k) pBD85, BglII.

for such an experiment. A more general drawback of cloning by recombination is its recE-dependence, since an attempt to clone B. subtilis DNA would result in transformation by chromosomal integration (48,49). Recently, we have succeeded in introducing pUB110 into B. licheniformis by transformation in the hope that a reciprocal system can be established in that organism to permit cloning of B. subtilis fragments. These experiments are in progress.

## Escherichia coli as an Intermediate Host

Many investigators have isolated B. subtilis chromosomal, plasmid or bacteriophage DNA fragments by first cloning in E. coli, identifying clones carrying the desired fragment and then introducing the hybrid plasmid into B. subtilis (2,3,12,49-59). In this type of experiment, double-replicon vectors that can replicate in both B. subtilis and E. coli are obviously of great utility. Several such vectors have already been described (2,3,29,57,60). One has been used to construct a B. subtilis gene bank in E. coli that is reported to represent 80% of the B. subtilis chromosome (57). B. subtilis chromosomal gene banks have also been constructed in E. coli by other investigators (61,62). Although several B. subtilis DNA fragments have been shown to complement E. coli auxotrophic mutations, some B. subtilis genes will most likely not be readily detected in E. coli because they will not express. For instance, many sporulation genes belong to this category. In these cases, specific RNA probes (58) or brute force screening procedures involving retransformation of B. subtilis by pooled plasmid samples (57) can be employed to detect desired clones.

## Other Cloning Stratagems

Several other approaches to cloning in B. subtilis hold promise. An obvious one is to use protoplast transformation since this occurs with high efficiency and oligomers are not required. However, two obstacles need to be overcome. It has been observed that ligated DNA transforms protoplasts poorly (29,32). This may be due, as suggested above, to a preference for lower molecular weight DNA. If this does not place too great a limitation on the size of the cloned fragment, then adjustment of DNA concentration during ligation may increase the efficiency of transformation. We have noted that recE protoplasts do not transform with high efficiency (39). Thus, it may be difficult to clone homologous fragments, although Chang and Cohen (29) have reported that chromosomal transformation (by integration) does not occur in protoplasts. A third approach is to use plasmid-free competent cultures but to adjust ligation conditions to optimize the yield of recombinant plasmid oligomers. This approach is currently under investigation in our laboratory.

## Gene Expression from Cloned Fragments

Two methods have been developed that are useful for the study of the expression of plasmid genes including those carried on cloned fragments: the minicell system and plasmid amplification.
B. subtilis minicell strains were isolated by Reeve and coworkers (63) and have been used to study bacteriophage and more recently plasmid-specific protein synthesis (2,64-67). Minicells, which can be readily isolated by differential centrifugation, contain no detectable chromosomal DNA. All plasmids tested to date segregate into minicells, thus permitting the study of plasmid gene expression in the absence of host macromolecular synthesis (44,67).

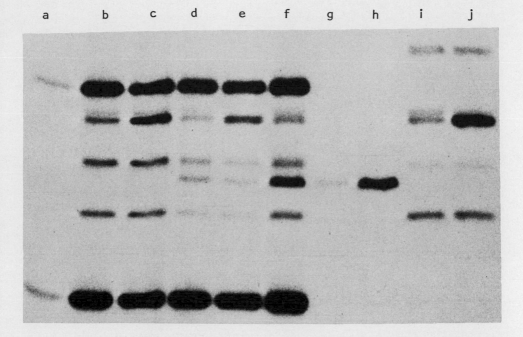

Figure 5.  SDS-polyacrylamide (15%) gel electrophoresis of [$^3$H]
leucine-labeled minicell extracts, (a) pUB110, (b) pBD9, uninduced,
(c) pBD9, Em-induced, (d) pBD10, uninduced, (e) pBD10, Em-induced,
(f) pBD10, Cm-induced, (g) pC194, uninduced, (h) pC194, Cm-induced,
(i) pE194 (copy control mutant), uninduced, (j) pE194 (copy control
mutant), Em-induced.


We have found that washed preparations of <u>B. subtilis</u> minicells are
stable for well over a year at -70°C, retaining full protein synthetic ability.  It is important to isolate the minicells from the late
exponential stage of growth.  Older cultures produce much greater
yields but the minicells have low activity (44).  Upon incubation,
the plasmid-specific, protein-synthetic capacity of minicells de-
creases rapidly, with a half-life of about 30 minutes at 37°C (68).
The reason for this is unclear.  Figure 5 shows a typical SDS-
acrylamide gel obtained using [$^3$H]leucine-labeled extracts of minicells carrying plasmids pC194, pE194 (copy control mutant), pUB110
and pBD10.  The chimera pBD10 produces several of the polypeptides
specified by its three parental plasmids.  Also illustrated is
the inducibility of the 22K chloramphenicol acetyl transferase
(45,67,69) and the 29K Em$^r$ protein of pE194.  The slower moving of
the two pUB110 polypeptides is the 38K kanamycin nucleotidyl transferase, responsible for the drug resistance conferred by this plasmid

(42,70).  The minicell system has recently been used to study the
barriers to heterospecific gene expression in B. subtilis (2).

Another approach to the study of proteins specified by plasmid
genes is to amplify the plasmid and to analyze proteins produced
under conditions of elevated gene dosage.  Temperature-sensitive DNA
replication mutant strains carrying plasmid are incubated for several
hours at a nonpermissive temperature.  Chromosome replication is
retarded and in many cases the plasmid continues to replicate.  This
has been accomplished using pUB110, with dnaA, dnaB, dnaC, dnaD, dnaE
and dnaI mutants of B. subtilis (71).  Recently, we have extended
these observations to the plasmids pC194 and pBD9 and have shown,
using SDS-polyacrylamide gels and enzymatic assay of the chloramphenicol acetyl transferase, that several of the plasmid-specific polypeptides continue to be synthesized at the elevated temperature for
several hours, proportionally to gene dosage (44).  Thus, on SDS-
polyacrylamide gels, plasmid-specific proteins can be detected
against a relatively diminished background of host protein synthesis.
The polypeptides detected in this way are identical in mobility to
those seen in minicell extracts, thus tending to allay concerns that
the minicell environment may generate aberrant polypeptides.

Acknowledgments:  We wish to acknowledge valuable discussions
with Y. Kozlov, I. Smith, E. Dubnau, L. Mindich and R.P. Novick, as
well as the expert secretarial assistance of A. Howard.  This work
was supported by National Institutes of Health grant AI-10311 and
American Cancer Society grant VC300, awarded to D. Dubnau.

REFERENCES

1    Ehrlich, S.D. (1978) Proc. Nat. Acad. Sci. U.S.A. 75, 1433-1436.
2    Goebel, W., Kreft, J. and Burger, K.J. (personal communication).
3    Kreft, J., Bernhard, K. and Goebel, W. (1978) Mol. Gen. Genet.
     162, 59-67.
4    Tanaka, T. and Sakaguchi, K. (1978) Mol. Gen. Genet. 165,
     269-276.
5    Dean, D.H., Perkins, J.B. and Zarley, C.D. (1978) in Spores VII
     (Chambliss, G. and Vary, J.C., eds.), pp. 144-149, American
     Society for Microbiology, Washington, D.C.
6    Bernhard, K., Schrempf, H. and Goebel, W. (1978) J. Bacteriol.
     133, 897-903.
7    Le Hégarat, J.-C. and Anagnostopoulos, C. (1977) Mol. Gen.
     Genet. 157, 167-174.
8    Lovett, P.S. (1973) J. Bacteriol. 115, 291-298.
9    Lovett, P.S. and Bramucci, M.G. (1975) J. Bacteriol. 124,
     484-490.
10   Lovett, P.S. and Burdick, B.D. (1973) Biochem. Biophys. Res.
     Commun. 54, 365-370.
11   Lovett, P.S., Duvall, E.J. and Keggins, K.M. (1976) J.
     Bacteriol. 127, 817-828.

12    Niaudet, B. and Ehrlich, S.D. (1979) Plasmid 2, 48–58.
13    Tanaka, T., Kuroda, M. and Sakaguchi, K. (1977) J. Bacteriol.
      129, 1487–1494.
14    Tanaka, T. and Koshikawa, T. (1977) J. Bacteriol. 131, 699–701.
15    Ehrlich, S.D. (1977) Proc. Nat. Acad. Sci. U.S.A. 74, 1680–1682.
16    Gryczan, T.J., Contente, S. and Dubnau, D. (1978) J. Bacteriol.
      134, 318–329.
17    Weisblum, B., Graham, M.Y., Gryczan, T. and Dubnau, D. (1979) J.
      Bacteriol. 137, 635–643.
18    Contente, S. and Dubnau, D. (1979) Mol. Gen. Genet. 167, 251–
      258.
19    Hoch, J.A., Barat, M. and Anagnostopoulos, C. (1967) J.
      Bacteriol. 93, 1925–1937.
20    Dubnau, D., Davidoff-Abelson, R., Scher, B. and Cirigliano, C.
      (1973) J. Bacteriol. 114, 273–286.
21    Canosi, U., Morelli, G. and Trautner, T.A. (1978) Mol. Gen.
      Genet. 166, 259–267.
22    Davidoff-Abelson, R. and Dubnau, D. (1973) J. Bacteriol. 116,
      154–162.
23    Dubnau, D. and Cirigliano, C. (1972) J. Mol. Biol. 64, 9–29.
24    Lacks, S. (1962) J. Mol. Biol. 5, 119–131.
25    Piechowska, M. and Fox, M.S. (1971) J. Bacteriol. 108, 680–689.
26    Guild, W.R., Cato, A. Jr. and Lacks, S. (1968) Cold Spring
      Harbor Symp. Quant. Biol. 33, 643–645.
27    Contente, S. and Dubnau, D. (1979) Plasmid 2, 555–571.
28    Porter, R.D. and Guild, W.R. (1978) J. Virol. 25, 60–72.
29    Chang, S. and Cohen, S.N. (1979) Mol. Gen. Genet. 168, 111–115.
30    Fodor, K. and Alföldi, L. (1976) Proc. Nat. Acad. Sci. U.S.A.
      73, 2147–2150.
31    Schaeffer, P., Cami, B. and Hotchkiss, R.D. (1976) Proc. Nat.
      Acad. Sci. U.S.A. 73, 2151–2155.
32    Gryczan, T., Contente, S. and Dubnau, D. (1979) Mol. Gen. Genet.
      (in press).
33    Mottes, M., Grandi, G., Sgaramella, V., Canosi, U., Morelli, G.
      and Trautner, T.A. (1979) Mol. Gen. Genet. 174, 281–286.
34    Uozumi, T., Hoshino, T., Miwa, K., Horinouchi, S., Beppu, T. and
      Arima, K. (1977) Mol. Gen. Genet. 152, 65–69.
35    Bron, S. and Murray, K. (1975) Mol. Gen. Genet. 143, 25–33.
36    Trautner, T.A., Pawlek, B., Bron, S. and Anagnostopoulos, C.
      (1974) Mol. Cen. Genet. 131, 181–191.
37    Contente, S. and Dubnau, D. (unpublished data).
38    Gryczan, T.J. and Dubnau, (1978) Proc. Nat. Acad. Sci. U.S.A.
      75, 1428–1432.
39    Gryczan, T.J. and Dubnau, D. (unpublished data).
40    Iordanescu, S. (1975) J. Bacteriol. 124, 597–601.
41    Novick, R.P., Edelman, I., Della Latta, P., Swanson, E.C. and
      Pattee, P.A. (1978) in Contributions to Microbiology and
      Immunology (Hertman, I., ed.), Vol. 6, Karger, Basel,
      Switzerland (in press).

42    Gryczan, T.J., Shivakumar, A.G. and Dubnau, D. (1979) J.
      Bacteriol. (in press).
43    Löfdahl, S., Sjöström, J.E. and Philipson, L. (1978) Gene 3,
      161-172.
44    Hahn, J. and Dubnau, D. (unpublished data).
45    Shivakumar, A.G. and Dubnau, D. (unpublished data).
46    Keggins, K.M., Lovett, P.S. and Duvall, E.J. (1978) Proc. Nat.
      Acad. Sci. U.S.A. 75, 1423-1427.
47    Rudner, R. (personal communication).
48    Duncan, C.H., Wilson, G.A. and Young, F.E. (1978) Proc. Nat.
      Acad. Sci. U.S.A. 75, 3664-3668.
49    Duncan, C.H., Wilson, G.A. and Young, F.E. (1977) Gene 1,
      153-167.
50    Bonamy, C., Keryer, E. and Szulmajster, J. (1978) in Spores VII
      (Chambliss, G. and Vary, J.C., eds.), pp. 139-143, American
      Society for Microbiology, Washington, D.C.
51    Chi, N.-Y.W., Ehrlich, S.D. and Lederberg, J. (1978) J.
      Bacteriol. 133, 816-821.
52    Ehrlich, S.D., Bursztyn-Pettegrew, H., Stroynowski, I. and
      Lederberg, J. (1976) Proc. Nat. Acad. Sci. U.S.A. 73, 4145-4149.
53    Ehrlich, S.D., Bursztyn-Pettegrew, H., Stroynowski, I. and
      Lederberg, J. (1977) in Recombinant Molecules: Impact on Science
      and Society (Beers, R.F. Jr. and Bassett, E.G., eds.), pp.69-80,
      Raven Press, New York, NY.
54    Horinouchi, S., Uozumi, T., Hoshino, T., Ozaki, A., Nakajima,
      S., Beppu, T. and Arima, K. (1977) Mol. Gen. Genet. 157,
      175-182.
55    Mahler, I. and Halvorson, H.O. (1977) J. Bacteriol. 131, 374-
      377.
56    Nagahari, K. and Sakaguchi, K. (1978) Mol. Gen. Genet. 158,
      263-270.
57    Rapoport, G., Klier, A., Billault, A., Fargette, F. and
      Dedonder, R. (1979) Mol. Gen. Genet. (in press).
58    Segall, J. and Losick, R. (1977) Cell 11, 751-761.
59    Young, F.E., Duncan, C. and Wilson, G.A. (1977) in Recombinant
      Molecules: Impact on Science and Society (Beers, R.F. Jr. and
      Bassett, E.G., eds.), pp. 33-43, Raven Press, New York, NY.
60    Stepanov, A.I. (personal communication).
61    Hutchinson, K. (personal communication).
62    Steinberg, W. (personal communication).
63    Reeve, J.N., Mendelson, N.H., Coyne, S.I., Hallock, L.L. and
      Cole, R.M. (1973) J. Bacteriol. 114, 860-873.
64    Reeve, J.N. (1977) Mol. Gen. Genet. 158, 73-79.
65    Reeve, J.N. and Cornett, J.B. (1975) J. Virol. 15, 1308-1316.
66    Reeve, J.N., Mertens, G. and Amann, E. (1978) J. Mol. Biol. 120,
      183-207.
67    Shivakumar, A.G., Hahn, J. and Dubnau, D. (1979) Plasmid 2,
      279-289.
68    Shivakumar, A.G., Kozlov, Y. and Dubnau, D. (unpublished data).

69   Fitton, J.E., Packman, L.C., Harford, S., Zaidenzaig, Y. and
     Shaw, W.V. (1978) in Microbiology-1978 (Schlessinger, D., ed.),
     pp. 249-252, American Society for Microbiology, Washington, D.C.
70   Sadaie, Y. (personal communication).
71   Shivakumar, A.G. and Dubnau, D. (1978) Plasmid 1, 405-416.

# BACTERIAL PLASMID CLONING VEHICLES

H.U. Bernard and D.R. Helinski

Department of Biology
University of California at San Diego
La Jolla, California  92093

Plasmids are defined as extrachromosomal, self-replicating and stably-inherited nucleic acid molecules. Without exception, plasmids have been isolated from bacteria as double-stranded circular DNA molecules. These genetic elements are maintained in bacterial cells in low copy number (stringent control of replication) or high copy number (relaxed control of replication). A variety of host properties are specified by plasmid elements including antibiotic and heavy metal resistance, bacteriocin production and toxin production. In addition many but not all plasmids of Gram-negative bacteria and several Gram-positive species carry a functional set of genes that specifies bacterial sexuality. These conjugative plasmids promote the conjugal transfer of themselves, nonconjugative plasmids present in the same cell and regions of the host chromosome. Not all plasmid elements can coexist in a bacterial cell. Two different plasmids that cannot be stably maintained together in a bacterium are considered to be members of the same incompatibility group. A large number of incompatibility groups have been identified for naturally occurring plasmids within the Gram-negative group of bacteria (1).

Naturally occurring plasmids can be modified by in vitro recombinant DNA techniques to facilitate their use as cloning vehicles in bacteria. A bacterial plasmid is considered a suitable cloning vehicle if it has the following properties: (a) it is readily isolated from cells; (b) it possesses a single restriction site for one or more restriction endonucleases; (c) the insertion of a linear DNA molecule at one of these sites does not alter its replication properties; and (d) it can be reintroduced into a bacterial cell and cells carrying the plasmid with or without the insert can be selected or identified.

The characterization of plasmids and the bacteriophage lambda of the bacterium Escherichia coli, the development of methods to

manipulate biochemically the structure of plasmid and phage DNA and
the derivation of procedures for the reintroduction of these elements
into bacteria have led to the revolutionary new technology termed
recombinant DNA, gene cloning and genetic engineering.  These fundamental developments along with the societal impact resulting from
the applications of this technology have been treated in detail in
a number of review articles (2-7).  In this article, we will attempt
to describe recent progress in the construction of plasmids as
vehicles for gene cloning in bacteria.  The majority of the available plasmid cloning vehicles will be listed in tables to facilitate
their comparison.  Certain plasmid elements will be described in
detail when they exhibit a particularly unusual property.  Emphasis
will be on the description of plasmid cloning vehicles that have the
following properties:  (a) contain single restriction enzyme sites
of a type that permit the direct cloning of DNA fragments that are
produced by cleavage with the same restriction enzyme; (b) contain
single restriction enzyme sites of a nature that insertion of a DNA
fragment at that site results in the inactivation of an identifiable
gene and, thus, allows selection or screening for a DNA insertion;
(c) allow testing of the properties of cis-acting genetic elements
or DNA segments after insertion of these elements into particular
restriction sites on the vehicle; (d) allow the establishment of
several plasmid vehicles in a cell and, thus, facilitate the study
of the interaction in vivo of specific DNA segments cloned on different plasmid vehicles; (e) extend the host-range for the cloning
of DNA fragments, either by their ability to be maintained in hosts
different from E. coli or by their ability to replicate in E. coli
as well as in other bacterial hosts; (f) contain regulatory genes
or nucleotide sequences that promote the expression of genes on a
DNA insert; and (g) promote high frequency transformation or selectively enrich for large DNA inserts.


PLASMID CLONING VEHICLES OF E. coli DERIVED FROM
HIGH COPY NUMBER AND AMPLIFIABLE REPLICONS

     E. coli plasmid cloning vehicles have been constructed from a
variety of naturally occurring plasmids.  Depending on their physical
structure, their incompatibility group and their copy number, different vehicles may vary in their applicability to a specific DNA
cloning problem.  In many cases, the principle aim of the cloning
experiment is the insertion of a particular restriction fragment
into the vehicle and its amplification in E. coli.  For this type
of experiment, a particularly useful vehicle is one that contains
a single restriction site in the plasmid molecule that is cleaved
by the restriction enzyme used in generating the DNA fragment to be
cloned.  An additional advantage is realized if the vehicle possesses
this restriction site near or within the sequence of a plasmid-coded
gene that is inactivated by the insertion event, thus permitting
screening or enrichment for transformants carrying the hybrid plasmid.

This type of vehicle should carry at least one additional selective marker for the selection of transformants.  In general, for purposes of biological containment, the vehicle should not carry a functional set of genes for conjugal transfer (EK1 vector in E. coli (8)) or be readily mobilizable (EK2 vector in E. coli (8)) by a conjugative plasmid.

Since it is often desirable to obtain the cloned restriction fragment in large quantities, efforts have been concentrated especially on the construction of cloning vehicles that are derived from the naturally occurring plasmids ColE1 (9), pMB1 (10) and P15A (11).  These E. coli plasmid replicons offer several advantages over other plasmid types.

(a)    The plasmid with or without a DNA insert is maintained in logarithmically growing cells at a copy number of greater than 10 per chromosome equivalent for the naturally occurring plasmid and considerably higher levels for low molecular weight derivatives of these plasmids.

(b)    This copy number can be amplified many fold by growth of the cells in the presence of the protein synthesis inhibitor chloramphenicol, thus facilitating the isolation of large amounts of the vehicle and its DNA insert.

(c)    Only a small segment of the naturally occurring plasmid is required for its replication in vivo.  This facilitates reconstruction of the plasmid in order to develop even more effective vehicles.

Table 1 lists properties of the majority of the presently available E. coli cloning vehicles of this group as well as including plasmid vehicles that will be treated in the following section.

When it is possible to carry out positive selection of a gene on the inserted fragment, all vehicles of the high copy number replicons having a single restriction site in a nonessential region should be similarly useful.  If no phenotypic trait is expressed by the inserted fragment, vehicles may be preferred that permit the screening of a gene inactivated by the insertion event.  Several plasmids that provide this property for different restriction sites are pACYC184 (11), pBR322 (14), pMK16 (21)(Figure 1), pMK20 (21) and pMK2005 (21).  While normally it will be advantageous to use any one of the prototype vehicles for a specific approach, an alternative vehicle may be chosen depending on the number or nature of its selective markers or to the particular arrangement of restriction sites. For example, if it is necessary to clone a restriction fragment with one PstI and one SalI end, both selective markers of pBR322 would be lost upon insertion, while the use of pMK2004 (21) (Figure 2) retains the kanamycin resistance.  In other cases, the nature of the genetic information encoded by a plasmid may make a vehicle particularly suitable as in the case of the plasmid pHA105, which codes for polypeptides no larger than 16,000 daltons.  The use of pHA105 facilitates the identification of larger products of cloned genes in E. coli minicells (12).

Table 1

E. coli Plasmid Cloning Vehicles

Table 1 lists plasmid cloning vehicles that are either at present widely used or which potentially are very useful, e.g. due to the occurrence of single restriction sites not known to be present in other vehicles. Vehicles derived from the natural plasmids ColE1, pMB1, P15A and R6K are considered as high copy number plasmids and are normally maintained at greater than 10 copies per chromosome. ColE1 (34,35) and pMB1 (10) derivatives can be greatly amplified in the presence of chloramphenicol while P15A derivatives have moderate amplification properties (11). R6K derivatives cannot be amplified. RK6 derivatives are present at approximately 5 to 8 copies and F-derivatives at 1 to 2 copies per chromosome (21). No attempt has been made to list isoschizomers of restriction endonucleases recognizing a particular restriction site. In particular, note that the sequence cleaved by SmaI, which produces blunt ends, is also cut by XmaI, which generates cohesive sticky ends. Cloning vehicles contain single sites mostly for those restriction endonucleases that are specific for hexanucleotide sequences. While most cloning vehicles are cleaved by restriction endonucleases recognizing tetranucleotide sequences into many fragments, it should be noted that a number of fragments produced by 4-cutters can be cloned into 6-cutter sites and recovered with 4-cutters, e.g. MboI fragments can be cloned into BamHI and BglII sites. An additional approach extensively used is the cloning of oligo(dC)-tailed DNA fragments into an oligo(dG)-tailed PstI site (36,37) or the cloning of a poly(dA)-tailed fragment into a HindIII site that has been filled in with DNA-polymerase I and tailed with poly(dT). In both cases, the original restriction site is restored.

| Plasmid | Replicon | Size[a] (kb) | Selective markers | Single restriction sites | Remarks | Refs. |
|---|---|---|---|---|---|---|
| pAC105 (pHA105) | ColE1 | 2.4 | ColE1 imm. | EcoRI | Useful for characterizing cloned gene products in minicells. pAC105 does not express polypeptides larger than 16K | 12 |
| pACYC177 | P15A | 3.7 | Amp[r] Kan[r] | BamHI, SmaI, HincII,XhoI, HindIII,PstI | PstI insertions inactivate Amp[r]. HindIII, SmaI and XhoI insertions inactivate Kan[r]. | 11 |

| Plasmid | Replicon | Markers | Restriction sites | Comments | Ref. |
|---|---|---|---|---|---|
| pACYC184 | P15A | Cam$^r$ Tet$^r$ | BamHI,EcoRI, HindIII,SalI | EcoRI insertions inactivate Cam$^r$. BamHI, HindIII$^c$ and SalI insertions inactivate Tet$^r$ [b]. | 11 |
| pBR313 | pMB1 | Amp$^r$ Tet$^r$ | BamHI,SalI EcoRI,SmaI, HindIII,HpaI | BamHI, HindIII and SalI insertions inactivate Tet$^r$ [b]. | 13 |
| pBR322 | pMB1 | Amp$^r$ Tet$^r$ | AvaI,PstI, BamHI,PvuII, ClaI,SalI, EcoRI,HindIII | Insertions into the PstI site inactivate Amp$^r$. BamHI, HindIII and SalI insertions inactivate Tet$^r$. Complete nucleotide sequence is known. | 14–16 |
| pBR324 | pMB1 | Amp$^r$ Tet$^r$ Col E1 imm | BamHI,EcoRI, HindIII,SalI | EcoRI insertions inactivate the colicin E1 gene and BamHI, HindIII and SalI insertions inactivate Tet$^r$. | 17 |
| pBR325 | pMB1 | Amp$^r$ Cam$^r$,Tet$^r$ | BamHI,EcoRI, HindIII,PstI | EcoRI insertions inactivate Cam$^r$. | 17 |
| pCRI | ColE1 | Col E1 imm Kan$^r$ | EcoRI, (BstEII)$^d$, HindIII, SalI,XhoI | HindIII and XhoI insertions inactivate Kan$^r$. | 18,19 |
| pCT7 | ColE1 | Col E1 imm Kan$^r$ | HincIII,SstI, (HindIII), (SmaI)(XhoI) | | 20 |
| pDF41 | F | trpE | BamHI,EcoRI, HindIII,SalI | | 21 |
| pKB158 | pMB1 | Tet$^r$ Phage λ imm | BamHI,PstI, BglII,SalI, EcoRI,HpaI | | 22 |

Table 1 (contd.)

| Plasmid | Replicon | Size[a] (kb) | Selective markers | Single restriction sites | Remarks | Refs. |
|---|---|---|---|---|---|---|
| pKN80 | ColEl | 15.9 | Amp$^r$ Col El imm | HincII,HpaI | Insertions into HincII and HpaI sites can be selected positively. | 23 |
| pKC7 | pMB1 | 5.8 | Amp$^r$ Tet$^r$ | BamHI,HindIII, BclI,PvuI,BglII, (SalI),BstEII, SmaI,EcoRI,XhoI | | 24 |
| pKN410 | RI | 15.0 | Amp$^r$ | BamHI,EcoRI, HindIII | Cloning vehicle that, due to temperature-dependent copy number, allows controlled amplification of genes and gene products. | 25 |
| pMB9 | pMB1 | 5.8 | Tet$^r$ | BamHI,EcoRI, HindIII,SalI | Mainly useful for EcoRI cloning since BamHI,HindIII and SalI insertions inactivate Tet$^r$. | 13 |
| pMK16 | ColEl | 4.6 | Col El imm Kan$^r$ Tet$^r$ | BamHI,SmaI, EcoRI,XhoI, HincII,d (BstEII),SalI | BamHI, HincII and SalI insertions inactivate Tet$^r$. SmaI and XhoI insertions inactivate Kan$^r$. | 21 |
| pMK20 | ColEl | 4.1 | Col El imm Kan$^r$ | EcoRI,XhoI, HindIII,d (BstEII), PstI,SmaI | HindIII, SmaI and XhoI insertions inactivate Kan$^r$. | 21 |
| pMK2004 | pMB1 | 5.2 | Amp$^r$ Kan$^r$ Tet$^r$ | BamHI,SmaI, EcoRI,XhoI, PstI,SalI | PstI insertions inactivate Amp$^r$. BamHI and SalI insertions inactivate Tet$^r$ and SmaI and XhoI insertions inactivate Kan$^r$. | 21 |

| Plasmid | Replicon | Mol. wt. | Markers | Restriction sites | Comments | Ref. |
|---|---|---|---|---|---|---|
| pMK2005 | ColE1 | 6.9 | Col E1 imm, trpE | BglII, SmaI, EcoRI, (BstEII)[d], HindIII, HpaI | | 21 |
| pRK248 | RK2 | 9.6 | Tet[r] | BglII, EcoRI, SalI. | SalI insertions inactivate Tet[r]. | 21 |
| pRK2501 | RK2 | 11.1 | Kan[r] Tet[r] | BglII, XhoI, EcoRI, HindIII, SalI | SalI insertions inactivate Tet[r]. SmaI, HindIII and XhoI insertions inactivate Kan[r]. | 21 |
| pRK353 | R6K | 11.0 | trpE | BamHI, EcoRI | | 21,26 |
| pRK646 | R6K | 3.4 | Amp[r] | BamHI, BglII, PstI | High containment of cloned DNA since pRK646 can only be maintained in E. coli strains carrying the R6K pir gene. | 21 |
| pRK702 | R6K | 2.2 | lacO, Amp[r] | HaeII, TaqI, | Same containment as for pRK646 above. | 27 |
| pSC101 | pSC101 | 8.8 | Tet[r] | BamHI, EcoRI, HindIII, SalI | Mainly useful for EcoRI cloning since BamHI, HindIII and SalI inactivate Tet[r]; 1st plasmid described as cloning vehicle. | 28,29 |
| pRSF2124 | ColE1 | 11.0 | Amp[r] Col E1 imm | BamHI, EcoRI | | 30 |
| pVH51 | ColE1 | 3.5 | Col E1 imm | EcoRI, HincIII | | 31 |
| pVH153 | ColE1 | 10.3 | trpED | KpnI | | 32,33 |

[a]Molecular weights published in Megadalton have been transferred to kilobases using the equation 1 kb = 0.66 Megadalton.

[b]Tetracycline-sensitive clones can be enriched in a medium containing cycloserine plus tetracycline (2,38).

Table 1 (contd.)

[c]HindIII fragments inactivate tetracycline resistance only if they do not contain a functional E. coli promoter (39,40).

[d]Cloning into the BstEII site may interfere with plasmid replication (21).

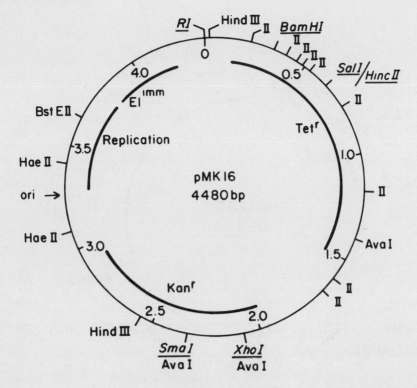

Figure 1.   Plasmid pMK16, a vehicle useful for cloning <u>Bam</u>HI, <u>Eco</u>RI, <u>Hinc</u>II, <u>Sal</u>I, <u>Sma</u>I and <u>Xho</u>I fragments (21).

PLASMID CLONING VEHICLES FROM PLASMIDS F, RK2, R1 AND R6-5

In some cases, it may be advantageous to use cloning vehicles derived from  plasmids other than the multi-copy ColEl type.  For instance, it is at times desirable to study the trans-acting effect of the product of a gene cloned into one plasmid on a DNA segment inserted into another plasmid.  To obtain stable coexistence of the two cloning vehicles in E. coli cells, it is necessary that these plasmids be members of different incompatibility groups.  Of the plasmid cloning vehicles described in the previous section that are multicopy and amplifiable, derivatives of the extensively used ColEl and pMBl replicons are incompatible.  The P15A replicon, however, is

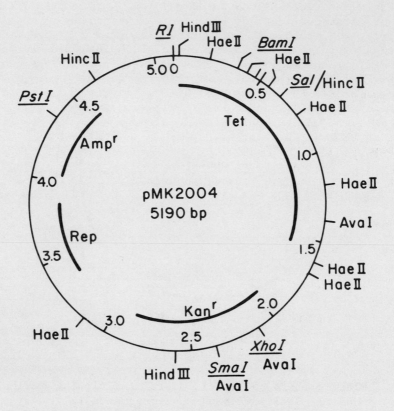

Figure 2.  Plasmid pMK2004, a vehicle useful for cloning BamHI, EcoRI, PstI, SalI, XmaI and XhoI fragments (21).

compatible with ColE1 (11).  Plasmid pSC101 and cloning vehicles derived from F, RK2 and R6-5 are compatible with the ColE1 and pMB1 replicons.

In some instances, it also may be desirable to have the DNA insert present at a low gene dosage level.  For this purpose, a low copy number cloning vehicle derived from the F replicon is preferable (21).  Vehicles derived from the broad host range plasmid RK2, which is stably maintained in virtually all Gram-negative bacteria including species distantly related to E. coli (41,42), are potentially useful for cloning of DNA in organisms other than E. coli and will be

discussed in more detail in a later section.  Other plasmids may
provide an increased level of biological containment of cloned DNA.
For example, the plasmid vehicle pRK646, derived from the naturally
occurring antibiotic resistance plasmid R6K, contains only the
replication origin of R6K (21)  Plasmid vehicle pRK646 is completely
dependent for its replication on a trans-acting function (the π protein) specified by the pir gene of R6K (43).  The pir gene of R6K has
been inserted into a lambda bacteriophage and integrated into the E.
coli chromosome.  The E. coli strain carrying the pir gene inserted
into its chromosome will maintain plasmid pRK646 at a copy number of
10 to 15 copies per chromosome.  This dependency of pRK656, and,
therefore, any DNA insert in this cloning vehicle, on a host strain
carrying the R6K pir gene provides a high level of biological containment for recombinant DNA studies.


                              COSMIDS


     Based on the properties of specialized transducing phages carry-
ing part of the phage lambda genome and ColEl plasmid DNA, Shimada,
Takagi and collaborators (44,45) could show that the presence of a
small segment of the phage lambda DNA containing the cohesive end
site cosλ on the plasmid molecule is a sufficient prerequisite for in
vivo packaging of this DNA into infectious particles.  The in vivo
packaging mechanism seemed to select DNA molecules of the full size
of the lambda genome since small hybrid plasmids with cosλ were
packaged as dimers or trimers.
     This system has been developed into a particularly useful series
of plasmid cloning vehicles designated cosmids by Collins, Bruning
and Hohn (46,47).  These cloning vehicles, consisting of a ColEl type
replicon joined to the cosλ site and several selectable genes, combine the uniquely advantageous properties of the plasmid and phage
systems for recombinant DNA work.  For example, although cosmids are
stably inherited at a copy number of approximately five copies per
cell, they can be amplified by growth of the cells in the presence of
chloramphenicol.  In addition, since cosmids carry the cosλ site
which is necessary for the assembly of phage lambda proteins to form
infectious viral packages in vivo or in vitro (48), in vitro ligated
and packaged cosmid hybrid DNA can be established by transfection
with a much higher frequency in E. coli cells than is possible using
hybrid DNA and a standard bacterial transformation procedure.  Cosmids also can be employed to greatly enrich for large DNA inserts due
to the selection of the packaging system for DNA molecules of a
length equivalent to the phage lambda genome (47 kb).  The DNA insert
must be of sufficient size to compensate for the difference in lambda
DNA length and the size of the cosmid DNA (10.9 to 24.3 kb).  The
cosmid vehicles described to date contain single restriction sites
for several different restriction enzymes (BamHI, BglII, EcoRI,
HindIII, KpnI, PstI, SalI, XmaI) (46,47; this volume, pp.169-183).

## PLASMID CLONING VEHICLES USEFUL FOR DETECTION OF cis-ACTING GENETIC ELEMENTS

Most plasmid cloning vehicles have been constructed for the purpose of optimizing the properties of these vehicles for the insertion, amplification and expression of cloned genes. Plasmid cloning vehicles also have been constructed in order to test for the presence of promoter and transcription termination sequences and functional origins of replication on a cloned DNA segment.

A system of wide applicability for detecting the presence of promoter or transcription termination sequences on given DNA fragments has recently been introduced by Casadaban and Cohen (49). Plasmid pMC81 and several related vehicles contain a fusion of the ara promoter to the lacZ gene of E. coli. Restriction fragments can be inserted into single HindIII, KpnI, SmaI, HincII or HpaI sites between these two genetic regions in the vehicles. If an inserted fragment contains a DNA sequence that functions as a promoter in E. coli in vivo, lacZ expression occurs in the absence of the inducer arabinose. If the insert contains a DNA sequence that terminates transcription, transformants will exhibit the lac⁻ phenotype even in the presence of arabinose.

The inability of the ColE1 replicon to be stably maintained in a polA mutant of E. coli (50) can be used to identify and clone DNA segments containing a polA-independent functional origin of replication in E. coli (51). Restriction fragments containing such an origin of replication can be cloned into ColE1-derived cloning vehicles in an E. coli polAts mutant at a temperature permitting the replication of the hybrid plasmid from the ColE1 origin. In a nonconditional polA mutant strain or at a nonpermissive temperature in an E. coli polAts mutant only plasmids carrying on the DNA insert a polA independent functional origin of replication are maintained. Using this approach, fragments carrying the replication origins of the E. coli (52 and Salmonella typhimurium (53) chromosomes, the plasmids R6K (43,54), RK2 (55,56) and phage lambda (57) have been isolated and characterized.

## PLASMID CLONING VEHICLES THAT PROMOTE EXPRESSION IN E. coli OF GENES ON A DNA INSERT

It has been generally observed that genes of E. coli cloned into high copy number cloning vehicles are readily expressed in E. coli cells. Normally, a gene dosage effect is observed, i.e. the high copy number of a given gene is reflected in a substantial increase of the gene product over those concentrations that are found in a cell carrying only one copy of the gene on the chromosome (9,58-71). In contrast to these self-cloning experiments and the finding that many genes from related or unrelated bacteria are expressed when present in plasmid cloning vehicles in E. coli (72-74), genes from lower eukaryotes are inactive or only weakly expressed in E. coli (75,76).

In addition, the expression of genes from higher eukaryotes under the control of natural eukaryotic sequences that determine correct transcription and translation has not been observed as yet in E. coli. Since the genetic code and in particular the AUG translation start and the three stop codons are similar or identical in bacteria and eukaryotes, it is likely that one or more of the following factors are responsible for the failure of gene expression in heterologous systems in vivo.

1.  The E. coli RNA polymerase may not recognize DNA sequences that function as promoter regions in eukaryotic cells.

2.  The transcript of a eukaryotic gene containing intervening sequences cannot be properly processed in E. coli.

3.  E. coli transcripts may contain structural features that assure an adequate half-life for translation and these structural properties may be absent in transcripts of cloned eukaryotic genes.

4.  A nucleotide sequence in most prokaryotic mRNAs (Shine-Dalgarno sequence (SD-sequence)) appears to be essential for the E. coli ribosome to function in the initiation of translation (77) and this sequence normally is not observed in eukaryotic mRNAs (78).

5.  Certain eukaryotic proteins require a posttranslational modification for biological activity and it is unlikely that this activation process can be carried out by E. coli cells.

6.  A eukaryotic protein may be particularly sensitive to E. coli proteases.

Relatively few observations have been made on the stability of different heterologous transcripts and translation products in E. coli. It has been observed recently that the level of transcript corresponding to the gene for the enzyme dihydroquinase of Neurospora crassa, which has been cloned in the plasmid pBR322, is elevated several fold in an E. coli pnp⁻ mutant (polynucleotide phosphorylase) when compared with a pnp⁺ strain (79). The elevated concentration of the transcript results in a considerable increase in the level of the dihydroquinase enzyme in the pnp⁻ mutant.

The possibility that small eukaryotic proteins are unstable in E. coli was raised from studies on the product of a fused gene consisting of the N-terminal codons on the E. coli β-galactosidase and a nucleotide sequence coding for the 14 amino acids of the hormone somatostatin. When the fused gene contained only 30 N-terminal codons of β-galactosidase, the hybrid somatostatin protein could not be detected in vivo (80). However, detectable levels of the hybrid protein were found when the somatostatin gene replaced only a few C-terminal codons of the β-galactosidase genes.

Considerable effort has been extended in the construction of plasmid cloning vehicles where the transcription of inserted genes is under the control of defined promoters on the plasmid. Four well-defined prokaryotic promoters have been used for this purpose. These include the promoters of the lac operon of E. coli with its lacZ (β-galactosidase) gene (81), the N-gene of phage lambda (82), the tryptophan operon of E. coli (83 and the Amp$^r$ gene coding for the enzyme β-lactamase of the plasmid pBR322 (37,84).

In addition to those constructions directed at overcoming transcriptional barriers, attempts have been made to bring the translation of transcripts of gene inserts under the control of translation initiation signals on the plasmid. Various approaches have been employed including fusing the gene insert to the N-terminal portion of a plasmid gene (37,80,85-90), fusing the gene insert to an E. coli ribosome-binding site (91-93) or fusing the gene insert to a nucleotide sequence that can function as an artificial ribosome-binding site (84).

## lac Promoter ($p_{lac}$)

Plasmids useful for the controlled transcription of inserted prokaryotic or eukaryotic genes initiating at the promoter ($p_{lac}$) of the lac operon of E. coli have been constructed by several inde-pendent approaches. In initial constructions using $p_{lac}$, an EcoRI fragment encompassing 6.8 kb of the phage λplac5 carrying the lac promoter, operator and most of the lacZ gene coding for the enzyme β-galactosidase was inserted into the plasmid pRSF2124 (81). From this vehicle the plasmid pBGP120 could be derived by deletion of the EcoRI site upstream of the transcription direction of the lac pro-moter. Plasmid pBGP120 can be selected by the ampicillin resistance conferred by the pRSF2124 component and screened for the expression of the β-galactosidase enzyme. The enzyme is functional despite the loss of several C-terminal amino acids due to the removal of the C-terminal codons by the EcoRI cleavage carried out to generate this fragment from λplac5. Since the mobility of the protein corres-ponding to the β-galactosidase enzyme showed no difference in mobil-ity to the wild-type enzyme by SDS-gel electrophoresis, it was con-cluded that translation of this gene was terminated several codons downstream from the EcoRI site (94).

To show transcriptional read-through of RNA polymerase beyond the EcoRI site in the lacZ gene, a fragment containing the gene for the 28s r-RNA of Xenopus was inserted into the EcoRI site of pBGP120 (81,94). RNA hybridizing to Xenopus r-DNA was specified by this plasmid and the production of this RNA was stimulated in cells by induction of the lac operon with cyclic AMP plus isopropylthio-galactoside. Hybrid proteins corresponding to β-galactosidase showed increases in their molecular weights by 2000 and 6000 daltons, res-pectively, depending on the orientation of the insert and presumably the availability and position of a terminator codon in the inserted fragment.

A cloning vehicle as pBGP120 may be particularly useful for the expression of fused genes consisting of the β-galactosidase gene and sequences coding for polypeptides with biological function, as in the the case of the chemically synthesized DNA segment coding for the hormone somatostatin (80). For this purpose, a derivative of pBR322, pSomII-3, carrying the same 6.8 kb $p_{lac}$ promoter fragment from λplac5, was used to promote transcription and translation of the

fused β-galactosidase and somatostatin genes (80). The somatostatin
gene was joined to the β-galactosidase gene in such a way that its
codons were arranged in the same frame as the translational read-
through from the latter gene. The chemically synthesized somatostatin gene contained two stop codons at the C-terminus for translational termination. The β-galactosidase-somatostatin fusion protein
produced in E. coli could be cleaved with cyanogen-bromide at the
N-terminal methionine residue of the somatostatin polypeptide in
order to generate a full-length somatostatin polypeptide chain. The
applicability of this method is restricted to genes coding for polypeptides not containing methionine residues.

A similar construction was used to obtain fusion proteins
between β-galactosidase and the A or B chain of human insulin (88).
In these experiments, up to 20% of total cellular protein was produced as the fusion protein and the yield of the cyanogen-bromide
cleaved insulin polypeptides was 10-fold higher than obtained with
somatostatin.

In an independent series of constructions, small fragments
containing the 5'-terminal portion of the lac operon have been
inserted into different ColEl-type cloning vehicles and the expression of prokaryotic and eukaryotic genes examined. The gene inserts
were under the transcriptional control of the lac-promoter with
translation controlled by their own ribosome binding site, by the
transcriptional and translational controlling regions of the lac
promoter and the β-galactosidase ribosome-binding site, or fused to
the N-terminal codons of the β-galactosidase gene. A 203 base-pair
HaeIII fragment of lac DNA was isolated encompassing part of the lac
i gene, the lac UV5 promoter and operator and 7 N-terminal codons of
the β-galactosidase gene (95,96). It was possible to create EcoRI
sites at both ends of this fragment by cleaving plasmid pMB9 with
EcoRI, filling in the sticky ends with DNA polymerase I and blunt-end
ligation of the HaeIII fragment into the flush-ended plasmid molecule. With this procedure, the full EcoRI recognition sequences were
restored (96,97). LacZ bacteria, containing a plasmid carrying the
lac operator, can be easily identified by their characteristic color
on XG-plates since the high dosage of the lac operator present in the
plasmids competes with the lac operator on the bacterial chromosome
for binding of repressor proteins and this results in constitutive
expression of the lac operon (96-100). In the case of one of these
constructions, plasmid pOP203-3, the EcoRI site upstream of the lac
promoter was fortuitously lost (97). Into the remaining EcoRI site,
a TaqI fragment of chicken ovalbumin cDNA was inserted with the use
of EcoRI linker molecules (87). It was assumed that in this hybrid
plasmid, designated pUC1001, a new cistron was generated encoding for
an ovalbumin-like protein that was extended by 18 amino acids at its
N-terminus, consisting of the 7 N-terminal amino acids of the
β-galactosidase and a stretch of naturally untranslated DNA 5'
terminal to the ovalbumin translation start. An E. coli strain
harboring pUC1001 produced 1.5% of its total cell protein as a

protein which was ovalbumin-like in its molecular weight and antigenic properties (87).

In a similar construction, a HindIII-EcoRI fragment containing
Plac and originally derived from pOP203, was inserted into pBR322
to obtain the plasmid vehicle pOMPO. This vehicle was cleaved at its
single EcoRI site and after S1 treatment, blunt-end ligated to an
S1-treated HhaI fragment encoding for most of the ovalbumin protein
(85). It was found that 30,000 to 90,000 molecules per cell were
expressed from pOMP2 in E. coli as a β-galactosidase-ovalbumin fusion
protein--approximately the same amount as the corresponding protein
produced by cells carrying pUC1001.

The concept of creating a fusion product between the lac-
promoter, the β-galactosidase-ribosome-binding site, several
N-terminal codons of the β-galactosidase gene and a DNA segment
containing the eukaryotic gene has led to the construction of plasmid
cloning vehicles of potentially general applicability. Plasmid
cloning vehicles pPCφ1, pPCφ2 and pPCφ3 carry on a pBR322 cloned
replicon a single EcoRI site allowing the fusion of DNA sequences on
an EcoRI fragment with N-terminal sequences of the β-galactosidase in
all three translational reading frames (86).

With a similar vehicle, pKB252, the expression of the phage λcI
gene was brought under the transcriptional control of the lac promoter with translation being facilitated by the ribosome-binding site
of cI (96). E. coli strains harboring pKB252 produced 10,000 cI
protein molecules per cell (corresponding to 0.8% of the total cell
protein). Extending these studies, a 95 base-pair EcoRI-Alu sub-
fragment of the 203 base-pair HaeIII (EcoRI) fragment lacking any of
the β-galactosidase codons, was fused at an Hph site with the cI
cistron (91). This construction (pKB280) deleted the ribosome-
binding site of cI and brought the expression of cI under transcriptional and translational control of the lac promoter and the
β-galactosidase SD-sequence and resulted in a 3-fold higher production of the cI protein in vivo in comparison to pKB252. Considering that a single copy of the β-galactosidase gene on the bacterial
chromosome is capable of specifying the production of 5000 β-galactosidase molecules then only about 10% of the theoretical expression
of β-galactosidase (assuming 50 copies of the ColEl vehicle per
chromsome) was observed with the pKB280 vehicle. The efficiency of
expression was higher when derivatives of plasmid pBR322 were used
that contained the phage lambda cro gene and the same 95 base-pair
lac fragments used to construct pKB280 (92). In the case of two of
these plasmid constructs, pTR213 and pTR214, 190,000 and 120,000
molecules of cro protein (corresponding to 1.6% and 1.0% of the total
cell protein, respectively) were produced in vivo when cro expression
occurred from the lac promoter and the cro SD-sequence, respectively. Unexpectedly, similar derivatives that, according to their
nucleotide sequence, should express cro from the lac promoter and
the β-galactosidase SD-sequence gave considerably less expression.

Recently, the expression of a complete eukaryotic gene was
brought under control of an E. coli promoter and ribosome-binding

site (93).  An HaeIII fragment coding for amino acids 24-191 of the
human growth hormone, including its C-terminus, was fused to a chemically synthesized DNA fragment encoding for the N-terminal amino
acids of this protein.  This DNA segment was then inserted into a
plasmid carrying the lac promoter and the β-galactosidase SD-
sequence.  The hybrid plasmid contained the proper spatial arrangement of the regulatory elements and the AUG starting codon since
186,000 protein monomers of the growth hormone were produced per E.
coli cell carrying the recombinant plasmid (95).

It is clear that continued studies on plasmid cloning vehicles
containing small restriction fragments carrying $p_{lac}$ and the
β-galactosidase-SD-sequence potentially can greatly increase our
understanding of the effect of spacing and framing of these regulatory sequences with respect to the AUG start codon on the expression
of a gene.

## Lambda $p_L$ Promoter

The usefulness of the promoter for early leftward transcription
in phage lambda, $p_L$, for the construction of cloning vehicles that
promote a high level of expression of a gene insert, was suggested by
the observation that both in vivo and in vitro constructed phage
lambda derivatives carrying genes of the trp operon of E. coli
efficiently expressed in vivo genes under the control $p_L$ (101,102).
It was also observed that $p_L$ present on fragments obtained from different lambda phages was functional when inserted into a plasmid
element (103-105).  Two derivatives of plasmid pMK2004, pHUB2 and
pHUB4 (Figure 3), have been constructed carrying a PstI-EcoRI and a
PstI-SalI fragment, respectively, from λtrp43 that contained the $p_L$
promoter (82,106).  These showed the following properties in vivo.
    1.   With a cI mutant gene that encodes for a temperature-
sensitive repressor protein, $p_L$ is shut off at 30°.  In these
constructions, the temperature-sensitive cI gene is either present in
one copy on the bacterial chromosome, or inserted into the low copy
number plasmid pRK248 which is compatible with pHUB2 and pHUB4, or
inserted into the high copy number vehicle that carries $p_L$ itself.
    2.   Promoter $p_L$ can be activated by heat shock with the
result that transcription from $p_L$ proceeds continuously for several
hours at 42°C.
    3.   When plasmid pRK248cIts is used to provide the repressor
protein, the cloning vehicles pHUB2 and pHUB4 can be used in a
variety of E. coli hosts.
    4.   Vehicles pHUB2 and pHUB4 carry single HpaI, EcoRI, BamHI
and SalI sites downstream from $p_L$ that are available for the insertion of genes on different restriction fragments.
    5.   The expression of the trpA genes of Salmonella typhimurium
and Shigella dysenteria promoted by $p_L$ after insertion into the
EcoRI, BamHI and SalI sites of these plasmids reached 2%, 6.6% and
1.2% of total cell protein, respectively, as estimated by determining

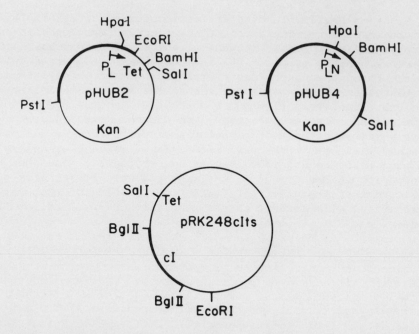

Figure 3. Plasmids pHUB2 and pHUB4, vehicles that promote the transcription of genes on fragments inserted into HpaI, EcoRI, BamHI and SalI sites. The cIts gene on the compatible plasmid pRK248cIts allows control of transcription from p<sub>L</sub> (82).

trpA protein enzymatic activity after heat induction. With pHUB4, the expression of the gene for the trp repressor protein was stimulated 100-fold (107). Since both pHUB2 and pHUB4 contain the 5' terminal part of the N-gene, they can conceivably be further developed for the expression of eukaryotic genes, either by constructing a fused gene between N-gene sequences and the insert, or by making use of the SD-sequence of N.

## trp Promoter ($P_{trp}$)

In an earlier study (9), it was shown that the expression of the five genes of the tryptophan operon shows a gene dosage effect when this operon is present on a high copy number (ColE1) plasmid. With

these hybrid plasmids, the trp enzymes accounted for at least 20 to
25% of the cellular proteins after stimulation of transcription from
the trp promoter by the tryptophan analogue 3-β-indolylacrylic acid.
The results of this study also indicated that in the absence of inducer the concentration of trp repressor in trpR$^+$ E. coli is sufficient to repress totally the trp operon despite the high copy number
of the trp promoters.

These observations have led to the construction of a plasmid
vehicle where expression of inserted genes is under the control of
regulatory elements of the trp operon (83). A 5.4 kb fragment of the
trp operon carrying the trp promoter and operator region, the trpE
gene and part of the trpD gene was inserted into the HindIII site of
pBR322. The resulting hybrid plasmid, designated ptrpED3, was then
cleaved at its single EcoRI site, incubated with exonuclease III,
digested with S1 nuclease to form blunt ends and recircularized with
T4 DNA ligase. Transformation of E. coli was carried out with the
treated ptrpED3 DNA with selection of transformants for trpE and
Amp$^r$. One transformant carried a 6.7 kb plasmid that had a single
HindIII site at the end of the trpD segment. Densitometer tracings
of protein gels obtained from extracts of cells that expressed the
trpE protein and a trpD protein fragment from ptrpED5-1 for 3 hours
after induction with 3-β-indolylacrylic acid suggested that 30% of
the total cell protein consisted of these two proteins (83)

This trp plasmid vehicle can be used to obtain fusion proteins
consisting of the N-terminal part of the trpD protein and polypeptides and proteins encoded by DNA sequences inserted at the single
HindIII site of ptrpED5-1. In the case of a plasmid construct containing sequences of the human growth hormone gene inserted into the
HindIII site of ptrpED5-1, 3% of total cell protein could be obtained
in vivo as a trpD-human growth hormone fusion protein (89).


## β-Lactamase Promoter ($p_{β-lact}$)

A fourth defined system used for promoting the expression of
genes on fragments inserted into a cloning vehicle is the lactamase
gene which is expressed constitutively. The nucleotide sequence of
this gene is known (108) and its product, β-lactamase, is a periplasmic protein whose transport properties are determined by its N-terminal sequence. A single PstI site is located at the position of 181
to 182 codons from the N-terminal end of the gene. A sequence encoding for 214 amino acids of the C-terminal part of the rat growth hormone gene have been directly fused in frame with the β-lactamase gene
at its single PstI site (90). In addition, cDNA sequences encoding
for rat preproinsulin (37) and mouse dihydrofolate reductase (84) have
been inserted into the β-lactamase gene by the oligo(dG)-oligo(dC)
joining procedure. In all three experiments, antigenically active
fusion proteins were expressed from the hybrid plasmids in E. coli.
The presence of the fusion protein in the periplasmic space was

Table 2

E. coli Plasmid Cloning Vehicles Promoting the Expression of Inserted Genes

This table is an attempt to list the information presented in the text in such a way that a comparison can be made of the effectiveness of the different vehicles.

| Promoter | Replicon | Cloning vehicle | Properties of cloning vehicle | Inserted gene | Plasmid carrying insert | Expression of inserted gene | Refs |
|----------|----------|-----------------|-------------------------------|---------------|-------------------------|----------------------------|------|
| P̲l̲a̲c̲ | pRSF2124 | pBGP120 | DNA can be inserted into the single EcoRI site downstream from P̲l̲a̲c̲ in C-terminal part of the β-galactosidase gene. β-galactosidase is expressed with a modified C-terminus. | Xenopus 28srDNA | pBGP123 | Qualitative proof of transcription from P̲l̲a̲c̲ through the E̲c̲o̲RI site into 28srDNA. | 81,94 |
| P̲l̲a̲c̲ | pBR322 | — | Fragment with P̲l̲a̲c̲ was inserted into pSOMII carrying the somatostatin gene. Relative arrangement of β-galactosidase and somatostatin gene in pSOMII-3 is similar to insertion of 28srDNA in pBGP123. | Somatostatin | pSOMII-3 | Expression of β-galactosidase-somatostatin fusion protein, from which somatostatin can be recovered by cyanogen-bromide treatment and can be detected radio-immunologically. Somatostatin constituted 0.001-0.03% of total cell protein. A similar construction was used to obtain β-galactosidase-insulin A and B-chain fusion proteins constituting ∿20% of total cellular protein. | 80,88 |

| Promoter | Vector | Plasmid | Construction | Gene / Plasmid | Description | Ref. |
|---|---|---|---|---|---|---|
| $P_{lac}$ | pMB9 | pOP203-3 | 203 base-pair HaeIII fragment carrying $P_{lac}$ was inserted into EcoRI site of pMB9, filled in with DNA polymerase I, generating 2 EcoRI sites. EcoRI site upstream of $P_{lac}$ was then deleted. | Chicken ovalbumin  pUC1001 | Chicken ovalbumin gene on TaqI fragment was cloned with EcoRI linkers into pOP203-3. β-galactosidase-ovalbumin fusion protein = 1.5% of total cell protein. | 87,96 97 |
| $P_{lac}$ | pBR322 | pOMP0 | Derivative of pOP203 obtained by recloning a HindIII-EcoRI fragment carrying $P_{lac}$ into pBR322. | Chicken ovalbumin  pOMP2 | Chicken ovalbumin gene on HhaI fragment cloned by blunt-end ligation into S1-treated EcoRI site of pOMP0. 30,000-90,000 copies of the fusion proteins were expressed/cell. | 85 |
| $P_{lac}$ | pMB9 | — | 203 bp HaeIII fragment carrying $P_{lac}$ was cloned into the EcoRI site of pKB158, a pMB9 derivative carrying phage λ cI gene. | cI  pKB252 | 2 lac promoters cloned in tandem in pKB252 stimulated the production of 0.8% of total cell protein as cI-protein; a 36-fold increase over the cI-protein concentration in a lysogen. | 96 |
| $P_{lac}$ | pMB9 | — | Fusion of a 95 bp EcoRI-Alu fragment with the cI gene bringing transcription and translation under $P_{lac}$ and β-galactosidase SD control. | cI  pKB280 | The product of the cI gene constituted 2.5% of total cell protein. | 91 |

Table 2 (contd.)

| Promoter | Replicon | Cloning vehicle | Properties of cloning vehicle | Inserted gene | Plasmid carrying insert | Expression of inserted gene | Refs |
|----------|----------|-----------------|-------------------------------|---------------|-------------------------|------------------------------|------|
| $P_{lac}$ | pBR322 | — | 95 bp EcoRI-Alu fragment was inserted into pTR151, a pBR322 derivative carrying the cro gene of phage lambda. | cro | pTR313 pTR314 | cro is expressed under control of $P_{lac}$ and the cro SD-sequence. 1.6 and 1% of total cell protein was the product of the cro gene (as determined by a cro protein radioimmune assay). | 92 |
| $P_L$ | pMK2004 | pMK2004 λtrp43 | Used only to test the expression of the trpB gene fused by in vivo recombination to $P_L$. 5% of total cell protein was expressed as the product of the trpB gene. | — | — | — | 82, 106 |
| $P_L$ | pMK2004 | pHUB2 | pHUB4 has single HpaI, EcoRI, BamHI and SalI sites downstream from $P_L$ available for cloning. | trpA (Salmonella) | pHUB2 trpA (EcoRI) | trpA expression is fully repressed by cIts at 30°C and constituted 2% of total cell protein after heat inactivation of the cIts-protein (determined on basis of enzyme activity). | 82, 106 |

| $P_L$ | pMK2004 | pHUB4 | pHUB4 has single HpaI, BamHI and SalI sites downstream from $P_L$, within a continuous phage λ DNA segment available for cloning. | trpA (Shigella) | pHUB4 trpA (BamHI) pHUB4 trpA (SalI) | trpA protein is 6.6% and 1.2% of total cell protein after heat inactivation of cIts protein (determined on the basis of enzyme activity), respectively. | 82, 106 |
|---|---|---|---|---|---|---|---|
| Ptrp | pBR322 | ptrpED5 -1 | Single HindIII, BamHI and SalI sites are available for cloning downstream from $P_{trp}$. On activation of $P_{trp}$ the product of the trpE gene is 30% of total cell protein (via autoradiography). | human pre- growth hormone | ptrpED50 -chGH800 | Expression of a fusion protein between trpD protein and human growth hormone was equivalent to ∿3% of total cell protein. | 83,89 |
| $P_{\beta-lact}$ | pBR322 | pBR322 | pBR322 contains a single PstI site within the sequence coding for β-lactamase (Amp^r). Insertion of a gene into this site may lead to constitutive production of fusion protein carrying 182 N-terminal amino acids of β-lactamase. This fusion protein may be processed and transported into the periplasmic space like β-lactamase. | mouse dihydro- folic acid reductase | pDHFR7 (and others) | Cloning of DHFR-cDNA with oligo(dG)-oligo(dC)-tailing into the PstI site of pBR322. Oligo(dG)-oligo(dC)-tails may have provided an E. coli ribosome binding site leading to production of DHFR enzyme of natural size rather than a fusion protein in the case of certain hybrid plasmids. | 84 |

Table 2 (contd.)

| Promoter | Replicon | Cloning vehicle | Properties of cloning vehicle | Inserted gene | Plasmid carrying insert | Expression of inserted gene | Refs |
|---|---|---|---|---|---|---|---|
| P$_{\beta lact}$ | pBR322 | pBR322 | Same as for pBR322 above. | rat prepro- insulin | pI47 | Cloning of rat proinsulin cDNA with oligo(dG)-oligo(dC)- tailing into PstI site of pBR322. One clone expressed a β-lactamase-proinsulin fusion protein, which is also trans- ported into the periplasmic space (radioimmunoassay). It was estimated that 100 molec- ules were produced/cell. | 37 |
| P$_{\beta lact}$ | pMB9 | — | Same as for pBR322 above. | rat pre- growth hormone | pEX-RGH | Insertion of a BamHI-PstI fragment from pBR322 encom- passing the N-terminal part of β-lactamase into a pMB9 derivative carrying RGH-cDNA bearing a PstI site in the RGH coding region. pEX-RGH, a fusion protein between the 181 N-terminal β-lactamase and 214 residues of rat pre- growth hormone, was expressed in E. coli minicells. | 90 |

observed in the case of the proinsulin hybrid protein. The nucleotide sequence of a region of one of the dihydrofolate reductase
hybrid plasmids that expressed detectable levels of the enzyme in E.
coli cells suggested that the GC-tails used for constructing the
hybrid plasmid may function as a ribosome-binding site in vivo,
possibly as a result of their similarity to the four G-residues in
SD-sequences (84). Most of the observations presented in this section are summarized in Table 2.


PLASMID CLONING VEHICLES FOR BACTERIAL HOSTS OTHER THAN E. coli

On the basis of observations with the ColEl plasmid, it is
likely that cloning vehicle derivatives of ColEl, pMBl and P15A will
be restricted by their host range replication properties to E. coli
and related bacteria. Since in vitro recombinant DNA techniques provide unique opportunities to extend our understanding of the molecular genetics of all organisms, it is important to have available
plasmid cloning vehicles for bacterial hosts other than E. coli. The
availability of such cloning vehicles should facilitate the use of
recombinant DNA techniques in the manipulation of a variety of bacteria of medical and agricultural importance. In an effort to obtain
cloning vehicles for Gram-negative bacteria other than E. coli, extensive work has been carried out on the P-1 incompatibility group of
plasmids. P-1 group plasmids are unusual in their ability to replicate and be maintained stably in virtually all Gram-negative bacteria. Similar efforts have been directed to developing plasmid cloning
vehicles for Gram-positive bacteria. For example, naturally occurring plasmids in Staphylococcus aureus have been modified by recombinant DNA techniques for use as cloning vehicles capable of replication in both Staphylococcus aureus and several Bacillus species (this
volume, pp. 115-131.)


Broad Host Range Cloning Vehicles

To construct plasmid cloning vehicles for Gram-negative bacteria
distantly related to E. coli, particular advantage was taken of the
Inc P-1 group plasmid RK2 which is similar if not identical to plasmids RP1 and RP4 of this group. These plasmids can be transferred
and stably maintained in a variety of Gram-negative bacteria, including the agriculturally important genera Rhizobium, Azotobacter and
Agrobacterium. It has been shown that the RK2 plasmid contains
single restriction sites for the endonucleases EcoRI, HindIII, BamHI
and BglII (109). A broad host range cloning system has recently been
developed from RK2 by incorporating separately the transfer and
replication regions of this plasmid into two different plasmids. The
three regions of the naturally occurring RK2 plasmid that are essential for replication, designated oriV, trfA and trfB, have been
incorporated into the plasmid cloning vehicle designated pRK290
(109a). This plasmid is a 20-kb tetracycline-resistant deletion

derivative of RK2. It is mobilizable but nonconjugative and has two
unique restriction enzyme sites, BglII and EcoRI. Both of these
sites have been used successfully for cloning large DNA inserts.
Vehicle pRK290 can be complemented in trans for conjugal transfer by
the helper plasmid pRK2013 (55). The latter plasmid is a kanamycin-
resistant hybrid containing ColE1 linked to a segment of RK2 DNA that
specifies the transmissibility properties of this plasmid. Helper
plasmid pRK2013 is self-transmissible in E. coli, but cannot be
stably maintained in other Gram-negative hosts due to the narrow host
range of the ColE1 replicon. High frequency transfer (1 to 10% per
recipient) of foreign DNA inserted into pRK290 can be achieved from
E. coli to other distantly-related Gram-negative bacteria with this
binary vehicle system. Such a high rate of transfer is particularly
valuable where it is necessary to operate against an active restriction system in the intended recipient. The nonself-transmissible
nature of the vector itself provides a high level of biological
containment.


### Plasmid Cloning Vehicles for Gram-Positive Bacteria

The classical distinction between Gram-negative and Gram-
positive bacteria as two well-separated classes of prokaryotes perhaps draws further support from the observation that to date no
plasmid replicon has been found to be stably maintained in both a
Gram-positive and Gram-negative bacterium (72-74). For this reason,
it has been necessary to construct vehicles for gene cloning in Gram-
positive bacteria that are distinct from the vehicles effective in E.
coli. Naturally occurring plasmids of Gram-positive bacteria have
been modified for this purpose. After it was observed that certain
S. aureus antibiotic resistance plasmids could be transformed and
stably maintained in B. subtilis (110), a variety of S. aureus plasmids with this property were identified and different species of
Bacillus were shown to be transformable with these plasmids (see this
volume, pp. 115-131). Several of these naturally occurring plasmids
possessed the desirable cloning vehicle properties of relatively
small size, the presence of antibiotic resistance genes and single
restriction sites. By manipulations in vitro and in vivo, cloning
vehicles carrying several selectable markers and single restriction
sites for several restriction enzymes were derived. In several
cases, it was shown that the insertion in vitro of DNA fragments at
these sites did not affect their replication properties. Table 3
summarizes the properties of some of these plasmid constructs.
By fusion of each of the two of these vehicles, pC194 and
pBS161-1, with the E. coli vehicles pBR322, pACYC184 and pWL7, hybrid
plasmids could be obtained that replicate and are maintained in both
Gram-positive and Gram-negative bacteria, presumably by virtue of
replication initiating at either of the two origins of these hybrid
replicons (73,74). Some of these constructions generated hybrid
plasmids (pVH13 and pVH14) that contained single restriction sites

Table 3

S. aureus and Bacillus Cloning Vehicles

Naturally Occurring Plasmids Useful as Cloning Vehicles

| Plasmid | Natural occurrence | Used in | Size (kb) | Selective markers | Single restriction sites | Restriction sites available for cloning | Refs. |
|---|---|---|---|---|---|---|---|
| pE194 | S. aureus | B. subtilis | 3.6 | Ery$^r$ | PstI, XbaI | XbaI | 111,112 |
| pC194 (pCM194) | S. aureus | B. subtilis | 3.0 | Cam$^r$ | HindIII, HpaII | HindIII, HpaII | 73,112, 113,114 |
| pUB110 | S. aureus | B. subtilis, B. pumilus, B. licheniformis | 4.2 | Kan$^r$ (Nm$^r$) | BamHI, BglII, EcoRI, XbaI | EcoRI, XbaI | 112,115, 116 |
| pSA0501 | S. aureus | B. subtilis | 4.2 | Str$^r$ | EcoRI, HindIII, XbaI | XbaI | 112,113 |
| pSA2100 | S. aureus | B. subtilis | 7.0 | Cam$^r$ | EcoRI, XbaI | XbaI | 112,113 |
| pCW6 | S. aureus | S. aureus | 3.9 | Cam$^r$ | BstEII, HindIII, HpaII, XbaI | | 116 |
| pCW8 | S. aureus | S. aureus | 2.9 | Cam$^r$ | HindIII, HpaII | HindIII | 116 |
| pC221 | S. aureus | S. aureus | 4.4 | Cam$^r$ | BstEII, EcoRI, HindIII | EcoRI, HindIII | 116 |

Table 3 (contd.)

In vitro Constructed Cloning Vehicles

| Plasmid | Parent plasmids | Used in | Size (kb) | Selective markers | Single restriction sites | Restriction sites available for cloning | Refs. |
|---|---|---|---|---|---|---|---|
| pSC194 | pS194 pC194(in vivo recombination) | S. aureus | 7.4 | Cam$^r$ Str$^r$ | EcoRI | EcoRI(insertion inactivates Str$^r$) | 113,114 |
| pBS161-1 | pBC16,pBS1 | B. subtilis | 3.65 | Tet$^r$ | EcoRI,HindIII | EcoRI,HindIII | 74 |
| pCW13 | pC221,pI228 | S. aureus | 8.6 | Cam$^r$ Ery$^r$ | BglII,BstEII EcoRI,HaeIII PstI | BglII,EcoRI, PstI | 116 |
| pCW14 | pCW8 | S. aureus | 7.1 | Cam$^r$ Ery$^r$ | BglII,PstI | BglII | 116 |

and, therefore, were suitable for examining the expression of DNA
inserts in both a Gram-positive and Gram-negative bacterial species.


## Streptomyces Plasmids

Among the Gram-positive bacteria, the genus Streptomyces has
commanded much attention with regard to possibilities of gene cloning
because of the considerable importance of species within this genus
as producers of a variety of valuable antibiotics.

Early genetic studies provided evidence for two sex plasmids in
S. coelicolor (117). One of these plasmids (SCP2) has been isolated
in the form of a covalently-closed circular DNA molecule of a molecular weight of 18 to 20 x $10^6$ (118-120). A high fertility variant of
SCP2, designated SCP2*, also has been obtained and shown to be of an
identical or similar size (121). Both SPC2 and SCP2* are maintained
at 1 to 4 copies per chromosome of S. coelicolor (119,120) and contain single restriction sites for the cohesive-end generating enzymes
EcoRI and HindIII. It is clear that SCP2 and SCP2* possess certain
of the essential properties of size, recoverability and single restriction sites for use as a plasmid cloning vehicle. Manipulation of
these plasmids in vitro should result in the construction of derivatives that are improved in their cloning vehicle properties. This is
particularly likely in view of the development of a highly efficient
transformation procedure for S. coelicolor, S. parvulus and S.
lividans with SCP2* DNA (122). The procedure employed utilizes
polyethylene glycol to facilitate uptake of the covalently-closed circular DNA form of this plasmid by protoplasts of these organisms and
the screening for transformants after regenerating the protoplasts by
visual detection of plasmid-containing spores since SCP2* determines
a "lethal zygosis" phenotype in a lawn of plasmid-minus cells.

Similarly high transformation frequencies with S. lividans have
been observed for plasmids SLP.1 and SLP.2 which are two of a series
of six plasmids (SLP1.1 to SLP1.6) isolated from spontaneous variants
to S. lividans strain JI1326 (121,123). Each of these six plasmids
differ only in a single region of their genomes. SLP1.2, the largest
of the plasmids (molecular weight of 8.23 x $10^6$) contains an additional 2 x $10^6$ dalton segment in this region when compared with
SLP1.6, the smallest plasmid of this series. The other plasmid
members contain smaller segments of DNA in this region. These
plasmids are maintained in S. lividans at a copy number of 3 to 4
copies per genome (120). A single BamHI site in a nonessential
region of SLP1.2 and a single PstI site in the nonessential regions
of SLP1.1 and SLP1.5 make these plasmids particularly attractive as
potential cloning vehicles. In vitro manipulation of the SLP1 family
of plasmids should improve upon their cloning vehicle properties at
least for S. lividans.

## CONCLUDING REMARKS

After the initial construction of plasmid cloning vehicles that would stably maintain DNA inserts in E. coli cells and the finding that genes of eukaryotic origin were expressed poorly or not at all in E. coli, considerable effort has been extended to develop plasmid cloning vehicles that facilitated the expression of a gene insert. While all of the problems of expression of "foreign" DNA in E. coli cells are far from resolved, much progress has been made in attempting to realize the promise of using bacterial cells and recombinant DNA techniques for the production of high levels of products specified by cloned prokaryotic and eukaryotic genes. Similarly, considerable progress has been made in extending the use of recombinant DNA techniques to bacteria distantly related to E. coli including Gram-positive species. In view of their great medical, industrial and agricultural importance, there is little question as to the potential benefits to be derived from recombinant DNA work with these bacteria and, therefore, the need for further development of plasmid vehicles in bacteria other than E. coli.

## REFERENCES

1    DNA Insertion Elements, Plasmids and Episomes (1972) (Bukhari, A.I., Shapiro, J.A. and Adhya S.L., eds.), Appendix B, Cold Spring Harbor Laboratory, Cold Spring Harbor, NY.
2    Collins, J. (1977) Curr. Topics Microbiol. Immunol. 78, 121-170.
3    Sinsheimer, R.L. (1977) Ann. Rev. Biochem. 46, 415-438.
4    Timmis, K., Cohen, S.N. and Cabello, F. (1978) Progr. Molec. Subcell. Biol. 6, 1-58.
5    Recombinant Molecules: Impact on Science and Society (1977) (Beers, R.F. and Bassett, E.G., eds.), Raven Press, New York, NY.
6    Molecular Cloning of Recombinant DNA (1977), 9th Miami Winter Symp., Academic Press, New York, NY.
7    Research with Recombinant DNA (1977) National Academy of Sciences, Washington, DC.
8    Federal Register, U.S.A., Vol. 43, No. 247.
9    Hershfield, V., Boyer, H.W., Yanofsky, C., Lovett, M.A. and Helinski, D.R. (1974) Proc. Nat. Acad. Sci. U.S.A. 71, 3455-3459.
10   Betlach, M., Hershfield, V., Chow, L., Brown, W., Goodman, H.M. and Boyer, H.W. (1976) Fed. Proc. 35, 2037-2043.
11   Chang, A.C.Y. and Cohen, S.N. (1978) J. Bacteriol. 134, 1141-1156.
12   Avni, H. and Markovitz, A. (1979) Plasmid 2, 225-236.
13   Bolivar, F., Rodriguez, R.L., Betlach, M.C. and Boyer, H.W. (1977) Gene 2, 75-93.

14    Bolivar, F., Rodriguez, R.L., Greene, P.J., Betlach, M.C.,
      Heyneker, H.L., Boyer, H.W., Crosa, J.H. and Falkow, S. (1977)
      Gene 2, 95-113.
15    Sutcliffe, J.G. (1979) Cold Spring Harbor Symp. Quant. Biol. 43,
      77-90.
16    Mayer, H., Schwartz, E., Melzer, M., Grosschedl, R., Schutte, H.
      and Hobom, G. (1978) Abstracts XII Internat. Congr. Microbiol.,
      p. 106, Munich.
17    Bolivar, F. (1978) Gene 4, 121-136.
18    Covey, C., Richardson, D. and Carbon, J. (1976) Mol. Gen. Genet.
      145, 155-158.
19    Armstrong, K.A., Hershfield, V. and Helinski, D.R. (1972)
      Science 196, 172-176.
20    Thomas, C. (personal communication).
21    Kahn, M., Kolter, R., Thomas, C., Figurski, D., Meyer, R.,
      Remaut, D. and Helinski, D.R. in Methods in Enzymology (Colowick
      S.P. and Kaplan, N.O., eds.) (in press).
22    Backmann, K., Hawley, D. and Ross, M.J. (1977) Science 196,
      182-183.
23    Schumann, W. (1979) Mol. Gen. Genet. 174, 221-224.
24    Rao, R.N. and Rogers, S.G. (1979) Gene 7, 79-82.
25    Uhlin, B.E., Molin, S., Gustavson, P. and Nordstrom, K. (1979)
      Gene 6, 91-106.
26    Kolter, R. and Helinski, D.R. (1978) Plasmid 1, 571-580.
27    Kolter, R. (personal communication).
28    Cohen, S.N. and Chang, A.C.Y. (1973) Proc. Nat. Acad. Sci.
      U.S.A. 70,1293-1297.
29    Cohen, S.N. and Chang, A.C.Y. (1977) J. Bacteriol. 132, 734-737.
30    So, M., Gill, R. and Falkow, S. (1976) Mol. Gen. Genet. 142,
      239-249.
31    Hershfield, V., Boyer, H.W., Chow, L. and Helinski, D.R. (1976)
      J. Bacteriol. 126,447-453.
32    Helinski, D.R., Hershfield, V., Figurski, D. and Meyer, R.M.
      (1977) in Recombinant Molecules: Impact on Science and Society
      (Beers, R.F. and Bassett, E.G., eds.), pp. 151-165, Raven Press,
      New York, NY.
33    Kahn, M. and Bernard, H.U. (unpublished data).
34    Bazaral, M. and Helinski D.R. (1968) Biochemistry 7, 3513-3517.
35    Clewell, D.B. and Helinski, D.R. (1972) J. Bacteriol. 110,
      1135-1146.
36    Ohtsuka, T. (personal communication).
37    Villa-Komaroff, L., Efstratiadis, A., Broome, S., Lomedico, P.,
      Tizard, R., Naber, S.P., Chick, W.L. and Gilbert, W. (1978)
      Proc. Nat. Acad. Sci. U.S.A. 75, 3727-3731.
38    Rodriguez, R.L., Bolivar, F., Goodman, H.M., Boyer, H.W. and
      Betlach, M. (1976) in Molecular Mechanisms in Control of Gene
      Expression, ICN-UCLA Symp. Mol. Cell Biol. (Nierlich, D.P.,
      Rutter, W.J. and Fox, C.F., eds.) Vol. 5, pp. 471-477, Academic
      Press, New York, NY.

39  Rodriguez, R.L., Tait, R., Shine, J., Bolivar, F., Heyneker, H.,
    Betlach, M. and Boyer, H.W. (1977) in Molecular Cloning of
    Recombinant DNA (Scott, W.A. and Werner, R., eds.) pp. 73-85,
    Academic Press, New York, NY.

40  Widera, G., Gautier, F., Lindenmayer, W. and Collins, J. (1978)
    Mol. Gen. Genet. 163, 301-305.

41  Datta, N. and Hedges, R.W. (1972) J. Gen. Microbiol. 70, 453-460.

42  Olsen, R. and Shipley, P. (1973) J. Bacteriol. 113, 772-780.

43  Kolter, R., Inuzuka, M. and Helinski, D.R. (1978) Cell 15, 1199-
    1208.

44  Fukumaki, Y., Shimada, K. and Takagi, Y. (1976) Proc. Nat. Acad.
    Sci. U.S.A. 73, 3238-3242.

45  Umene, K., Shimada, K. and Takagi, Y. (1978) Mol. Gen. Genet.
    159, 39-45.

46  Collins, J. and Hohn, B. (1978) Proc. Nat. Acad. Sci. U.S.A. 75,
    4242-4246.

47  Collins, J. and Bruning, H.J. (1978) Gene 4, 85-107.

48  Hohn, B. and Murray, K. (1977) Proc. Nat. Acad. Sci. U.S.A. 74,
    3259-3263.

49  Casadaban, M.J. and Cohen, S.N. (1980) J. Mol. Biol. (in press).

50  Kingsbury, D.T. and Helinski, D.R. (1970) Biochem. Biophys. Res.
    Commun. 41, 1538-1544.

51  Cabello, F., Timmis, K. and Cohen, S.N. (1976) Nature 259, 285-
    290.

52  Hirota, Y., Yasuda, S., Yamada, M., Nishimura, A., Sugimoto, K.,
    Sugisaki, H., Oka, A. and Takanami, M. (1979) Cold Spring Harbor
    Symp. Quant. Biol. 43, 129-138.

53  Zyskind, J. and Smith, D. (personal communication).

54  Kilter, R. and Helinski, D.R. (1978) J. Mol. Biol. 124, 425-441.

55  Figurski, D.H. and Helinski, D.R. (1979) Proc. Nat. Acad. Sci.
    U.S.A. 76, 1648-1652.

56  Thomas, C.M., Stalker, D., Guiney, D. and Helinski, D.R. (1979)
    in Symposium on Plasmids of Medical, Environmental and Commer-
    cial Importance (Puhler, A. and Timmis, K., eds.), pp. 375-385,
    Elsevier Press, Amsterdam.

57  Lusky, M. and Hobom, G. (1979) Gene 6, 137-172.

58  Nagahari, K., Tanaka, T., Hishinuma, F., Kuroda, M. and
    Sakaguchi, K. (1977) Gene 1, 141-152.

59  Wickner, S.H., Wickner, R.B. and Raetz, C.R.M. (1976) Biochem.
    Biophys. Res. Commun. 70, 389-396.

60  Uhlin, B.E. and Nordstrom, K. (1977) Plasmid 1, 1-2.

61  Gelfand, D.M., Shepard, H.M., O'Farrell, P.M. and Polisky, B.
    (1978) Proc. Nat. Acad. Sci. U.S.A. 75, 5869-5873.

62  Steffen, D. and Schleif, R. (1977) Mol. Gen. Genet. 137, 341-344.

63  Thomson, J.A. (1977) Gene 1, 347-356.

64  Thomson, J., Gerstenberger, P.D., Goldberg, D.E., Gociar, E.,
    Orozco de Silva, A. and Fraenkel, D.G. (1979) J. Bacteriol. 137,
    502-506.

65  Crabeel, M., Charlier, D., Cunin, R. and Glansdorff, N. (1979)
    Gene 5, 207-231.

66    Vapnek, D., Alton, N.K., Bassett, C.L. and Kushner, S.R. (1976)
      Proc. Nat. Acad. Sci. U.S.A. 73, 3492-3496.
67    Young, I.G., Jaworowski, A. and Poulis, M.I. (1978) Gene 4,
      25-36.
68    Fallow, A.M., Jinks, C.S., Yamamoto, M. and Nomura, M. (1979) J.
      Bacteriol. 138, 385-396.
69    Klinkert, J. and Klein, A. (1979) Mol. Gen. Genet. 171, 219-227.
70    Sancar, A. and Rupert, C.S. (1978) Gene 4, 295-308.
71    Hare, D.L. and Sadler, J.R. (1978) Gene 3, 269-278.
72    Ehrlich, S.D., Brusztyn-Pettegrew, U., Stroynowski, I. and
      Lederberg, J. (1976) Proc. Nat. Acad. Sci. U.S.A. 73, 4145-4149.
73    Ehrlich, S.D. (1978) Proc. Nat. Acad. Sci. U.S.A. 75, 1433-1436.
74    Kreft, J., Bernhard, K. and Goebel, W. (1978) Mol. Gen. Genet.
      162, 59-62.
75    Struhl, K., Cameron, J. and Davis, R. (1976) Proc. Nat. Acad.
      Sci. U.S.A. 73, 1471-1475.
76    Walz, A., Ratzkin, B. and Carbon, J. (1978) Proc. Nat. Acad. Sci.
      U.S.A. 75, 6172-6176.
77    Shine, J. and Dalgarno, L. (1974) Biochem. J. 141, 609-615.
78    Kozak, M. (1978) Cell 15, 1109-1123.
79    Kushner, S. (personal communication).
80    Itakura, K., Hirose, T., Crea, R., Riggs, A.D., Heyneker, H.L.,
      Bolivar, F. and Boyer, H.W. (1977) Science 198, 1056-1063.
81    Polisky, B., Bishop, R.J. and Gelfand, D.H. (1976) Proc. Nat.
      Acad. Sci. U.S.A. 73, 3900-3904.
82    Bernard, H.U., Remaut, E., Hershfield, M.V., Das, H.K., Helinski,
      D.R., Yanofsky, C. and Franklin, N. (1979) Gene 5, 59-76.
83    Hallewell, R.A. and Emtage, S. (1980) Gene (in press).
84    Chang, A.C.Y., Nunberg, J.H., Kaufman, R.J., Erlich, H.A.,
      Schimke, R.T. and Cohen, S.N. (1978) Nature 275, 617-624.
85    Mercereau-Puijalon, O., Royal, A., Cami, B., Garapin, A., Krust,
      A., Gannon, F. and Kourilsky, P. (1978) Nature 275, 505-510.
86    Charnay, P., Perricandet, M., Galibert, F. and Tiollais, P.
      (1978) Nucl. Acids Res. 5, 4479-4494.
87    Fraser, T.H. and Bruce, B.J. (1978) Proc. Nat. Acad. Sci. U.S.A.
      75, 5936-5940.
88    Goeddel, D.V., Kleid, D.G., Bolivar, F., Heyneker, H.L., Yansura,
      D.G., Crea, R., Hirose, T., Kraszewski, A., Itakura, K. and
      Riggs, A.D. (1979) Proc. Nat. Acad. Sci. U.S.A. 76, 106-110.
89    Martial, J.A., Hallewell, R.A., Baxter, J.D. and Goodman, H.M.
      (1979) Science 205, 602-607.
90    Seeburg, P.H., Shine, J., Martial, J.A., Ivarie, R.D., Morris,
      J.A., Ullrich, A., Baxter, J.D. and Goodman, H.M. (1978) Nature
      276, 795-798.
91    Backman, K. and Ptashne, M. (1978) Cell 13, 65-71.
92    Roberts, T.M., Kacich, R. and Ptashne, M. (1979) Proc. Nat. Acad.
      Sci. U.S.A. 76, 760-764.
93    Goeddel, D.V., Heyneker, H.L., Hozumi, T., Arentzen, R., Itakura,
      K., Yansura, D.G., Ross, M.J., Miozzari, G., Crea, R. and
      Seeburg, P.H. (1979) Nature 281, 544-548.

94   O'Farrell, P.H., Polisky, B. and Gelfand, D.H. (1978) J.
     Bacteriol. 134, 645-654.
95   Gilbert, W., Gralla, J., Majors, J. and Maxam, A. (1975) in
     Protein-Ligand Interactions (Suno, H.A. and Blauer, G., eds.),
     pp. 193-206, Walter de Gruyter, Berlin.
96   Backman, K., Ptashne, M. and Gilbert, W. (1976) Proc. Nat. Acad.
     Sci. U.S.A. 73, 4174-4178.
97   Fuller, F., Johnsrud, L. and Gilbert, W.(personal communication).
98   Sadler, J.R., Tecklenburg, M., Betz, J.L., Goeddel, D.V.,
     Yansura, D.G. and Caruthers, M.H. (1977) Gene 1, 305-321.
99   Marians, K.J., Wu, R., Stawinski, J., Hozumi, T. and Narang, S.A.
     (1976) Nature 263, 744-748.
100  Heyneker, H.L., Shine, J., Goodman, H.M., Boyer, H.W., Rosenberg,
     J., Dickerson, R.E., Narang, S.A., Itakura, K., Lin, S. and
     Riggs, A.D. (1976) Nature 263, 748-752.
101  Franklin, N.C. (1971) in Bacteriophage Lambda (Hershey, A.D.,
     ed.), pp. 621-638, Cold Spring Harbor Laboratory, NY.
102  Moir, A. and Brammer, W.J. (1976) Mol. Gen. Genet. 149, 87-99.
103  Helinski, D.R. (1977) in Genetic Engineering for Nitrogen Fixa-
     tion (Hollaender, A., ed.) pp. 19-46, Plenum Press, New York, NY.
104  Helinski, D.R., Hershfield, V., Figurski, D. and Meyer, R.L.
     (1977) in Recombinant Molecules: Impact on Science and Society
     (Beers, R.F. and Bassett, E.G., eds.), pp. 151-165, Raven Press,
     New York, NY.
105  Hedgpeth, J., Ballivet, M. and Eisen, H. (1978) Mol. Gen. Genet
     163, 197-203.
106  Bernard, H.U. and Helinski, D.R. (1980) in Methods in Enzymology
     (Colowick, S. and Kaplan, N., eds.) Academic Press, London and
     New York (in press).
107  Gunsalus, R.P., Zurawski, G. and Yanofsky, C. (1979) J.
     Bacteriol. 140, 106-113.
108  Sutcliffe, J.G. (1978) Proc. Nat. Acad. Sci. U.S.A. 75,
     3737-3741.
109  Meyer, R.J., Figursky, D. and Helinski, D.R. (1977) in DNA
     Insertion Elements, Plasmids and Episomes, pp. 559-566, Cold
     Spring Harbor Laboratory, NY.
109a Ditta, G. and Stansfield, S. (personal communication).
110  Ehrlich, S.D. (1977) Proc. Nat. Acad. Sci. U.S.A. 74, 1680-1682.
111  Iordanescu, S. (1977) J. Bacteriol. 129, 71-75.
112  Gryczan, T.J., and Dubnau, D. (1978) Proc. Nat. Acad. Sci. U.S.A.
     75, 1428-1432.
113  Iordanescu, S. (1975) J. Bacteriol. 124, 597-601.
114  Lofdahl, S., Sjostrom, J.E. and Philipson, L. (1978) Gene 3,
     161-172.
115  Keggins, K., Lovett, P.S. and Duvall, E.J. (1978) Proc. Nat.
     Acad. Sci. U.S.A. 75, 1423-1427.
116  Wilson, C.R. and Baldwin, J.N. (1978) J. Bacteriol. 136, 402-413.
117  Hopwood, D.A., Chater, K.F., Dowding, J.E. and Vivian, A. (1973)
     Bacteriol. Rev. 37, 371-405.

119   Schrempf, H. and Goebel, W. (1977) J. Bacteriol. 131, 251-258.
120   Bibb, M.J., Freeman, R.F. and Hopwood, D.A. (1977) Mol. Gen.
      Genet. 154, 155-166.
121   Bibb, M.J., Ward, J.M. and Hopwood, D.A. (1980) Developments in
      Industrial Microbiology 21 (in press).
122   Bibb, M.J., Ward, J.M. and Hopwood, D.A. (1978) Nature 274,
      398-400.
123   Hopwood, D.A., Bibb, M.J., Ward, J.M. and Westpheling, J. (1979)
      in Plasmids of Medical, Environmental and Commercial Importance
      (Timmis, K.N. and Puhler, A., eds.), pp. 245-257, Elsevier/North-
      Holland Biomedical Press, Amsterdam.

# CLONING WITH COSMIDS IN E. COLI AND YEAST

Barbara Hohn and A. Hinnen

Friedrich Miescher-Institut
P.O. Box 273
CH-4002 Basel, Switzerland

## INTRODUCTION

Several problems concerning the organization of genetic information can more easily be solved if larger pieces of DNA can be
cloned. First, large genes can be isolated and studied intact;
second, gene libraries consisting of a relatively small number of
clones can be constructed; third, genetic linkage studies can be made
at the molecular level.

To meet these demands a new generation of vectors has been
constructed. These vectors consist of plasmids containing the
cohesive end site (cos) from bacteriophage lambda in addition to
conventional markers required for cloning. This region of the
cohered cohesive ends of lambda, together with some sequences adjacent to it (1,2), is recognized by the lambda-specific packaging
proteins and lambda-head precursors (for a review of lambda morphogenesis, including DNA packaging, see ref. 3). These plasmids,
termed cosmids (4), can be packaged into a lambda-capsid, provided
the DNA meets the following requirement: cos sequences have to be
located on the DNA molecule, in parallel arrangement and 37 to 52
kilobases apart, a distance corresponding to 73 to 105% of the
50-kilobase lambda genome (5). This requirement can be used in
cloning (Figure 1).

A small cosmid is linearized by a suitable restriction enzyme
and ligated to large DNA fragments. Thus, concatenated molecules are
formed resembling the natural packaging substrate. In vitro packaging selects them and converts them into DNase-resistant phage-like
particles. The packaging endonuclease converts the cos sequences
into cohesive ends which, upon transduction into E. coli, become
covalently closed to form a circular molecule. Regulation of replication is effected by the plasmid genes again and transductants can
be selected using a resistance marker carried by the vector cosmid.

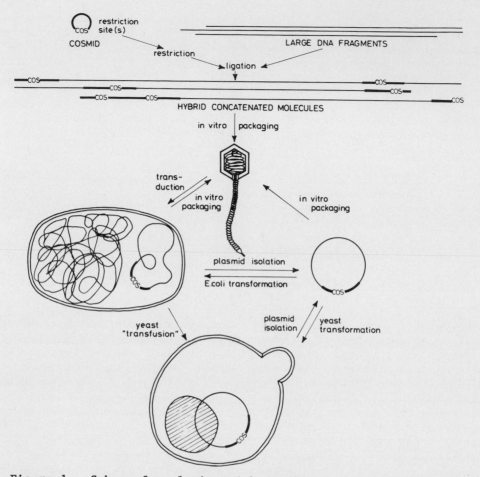

Figure 1.   Scheme for cloning with cosmids in E. coli and yeast.


     The use of these packageable plasmids not only allows effective
recovery of large hybrids--which was difficult before due to the
decrease in E. coli transformation efficiency with increasing plasmid
size (4,6)--but also enforces their selective recovery; vector
molecules and small hybrids are not packaged into transducing
particles.  This property, together with the advantages mentioned
above, allows the application of cloning with cosmids to other
organisms for which transformation systems are available.
     In this chapter, we describe this approach for cloning in yeast.
Saccharomyces cerevisiae, as a microbial eukaryote with a wealth of
genetically identified loci, has only recently emerged as an organism
amenable to modern molecular genetics.  This is due to the development of a yeast transformation system (7).  With stable auxotrophic
mutants, spheroplasts can be transformed with the corresponding
wild-type yeast genes by selection for prototrophy.  For technical
reasons, the transforming DNA molecules are hybrids containing

selectable genetic markers for both organisms. Due to the presence
of DNA sequences of the yeast 2μ plasmid or of sequences of chromosomal replication origins, these hybrids replicate autonomously in
yeast and can transform yeast at high frequency (7-10). The hybrid
nature of the yeast vehicles allows full utilization of the E. coli
cloning system.

The lower part of Figure 1 shows how the cosmid cloning system
can be applied in yeast. Since, for DNA amplification, recovery of
hybrid plasmids has to be preceded by a cloning step in E. coli,
cosmid cloning can effectively be used to select for large DNA
fragments.

In the following section, we discuss advantages and disadvantages of cosmids for particular cloning purposes, and which cosmids
offer the widest application for cloning in E. coli and yeast.
Finally, we include a practical methods section for in vitro
packaging, transduction and yeast transformation.


## WHEN TO USE A COSMID

Several aspects of the different vector systems--plasmids,
cosmids and bacteriophage lambda--will be compared. Table 1 summarizes the most important points and can serve as a guide. The
table is not intended to be self-explanatory but should facilitate
evaluation of the advantages and disadvantages of the three main
classes of vectors in several cloning operations. In the following
discussion of the table, as a matter of convenience, a noncosmid
plasmid will be termed plasmid to distinguish it from the package-
able cosmid.


### Enrichment of DNA Fragments to be Cloned

In preparation for cloning in a plasmid or in phage lambda, a
complete or partial digest of the DNA to be cloned can be enriched
for the size fraction hybridizing to a particular probe. If partially digested or sheared DNA, sized to 35-45 kilobases, is used to
clone in a cosmid (11), fractionation according to hybridization is
not possible. An exception would be a case in which a partial or
complete restriction digest yields a 35 to 45 kilobase fraction hybridizing to the probe. On the other hand, exploitation of the largest possible cloning capacity of phage lambda vectors in combination
with shearing of the DNA to be cloned excludes enrichment procedures
as well.


### Size of Clonable Fragments

In a plasmid, theoretically, any size fragment can be inserted.
Recovery of the large hybrid upon transformation, however, is low
(see cloning step 4, Table 1). The size range of DNA clonable in a
cosmid is given for pHC79 (11); larger cosmids can take only smaller

Table 1

Comparison of Cloning Operations with Plasmid,
Cosmid and Lambda Vectors

| Cloning step | Cloning vector | | |
|---|---|---|---|
| | Plasmid | Cosmid | Lambda |
| 1  Enrichment of DNA fragments to be cloned | + | (-) | + |
| 2  Size of clonable fragments | theoretically any | 30 to 45 kb | 0 to 20 kb |
| 3  Cloning enzymes | many | many | EcoRI,HindIII, BamHI,(SalI) |
| 4  Efficiency of transformation or in vitro packaging of hybrid | ++[a] <br> +[b] | ++ | ++ |
| 5  Screening by in situ hybridization | + | (+) | ++ |
| 6  Hybrid selection indirect | + | | + |
|     direct | (+) | ++ | ++ |
| 7  Hybrid stability | + | (+) | ++ |
| 8  Insert DNA recovery | ++ | ++ | + |

[a]small plasmids
[b]larger plasmids

loads.  The figures given for lambda summarize all possibilities:  if
isolated left and right arms of a lambda vector are used, there is a
lower and an upper limit for the size of the insert DNA required for
plaque formation (12).  If the production of a hybrid bacteriophage
is monitored by the inactivation of a particular, easily testable
gene (e.g., refs. 13,14), there is only an upper limit for the size
of the insert DNA.


Cloning Enzymes

     Since vector plasmids and cosmids are usually small molecules,
single sites for important restriction endonucleases are easily
introduced.  Bacteriophage lambda requires many genes for its lytic
cycle and thus it is much more difficult to manipulate this large DNA
molecule for cloning purposes.  Nevertheless, EcoRI and HindIII
vectors have been constructed (12-14) that may also be used as SalI,

XhoI or SstI vectors for special purposes. A newly constructed
lambda BamHI vector enriches the choice by an important member (B.
Klein and K. Murray, personal communication). The use of linkers or
adaptors extends the range of enzymes that one can use with a
particular vector.

The DNA to be cloned can be cleaved with an enzyme other than
the one used to restrict the vector DNA molecule. As will be discussed below, this is particularly useful for the production of a
random fragment population, which is best achieved by partial restriction with a four-base recognition enzyme cutting frequently and
by insertion into the vector molecule that has been restricted with
a six-base recognition enzyme generating the same ends.

## Efficiency of Transformation or of In Vitro Packaging of Hybrid DNA

The efficiency with which bacteria are transformed by plasmid
DNA is greatly dependent on the size of the DNA: larger (lambda size)
plasmids transform orders of magnitude less efficiently than small (5
to 10 kilobase size) plasmids (4,6,15). In vitro packaging yields
$10^7$ to $10^8$ plaque-forming units per µg nonrestricted lambda DNA and
in cloning experiments, 5 x $10^4$ to 5 x $10^5$ plaque-forming hybrid
phage particles per µg insert DNA and $10^4$ to $10^5$ hybrid cosmids per
µg insert DNA (16). This is two to three orders of magnitude above
transformation efficiencies of hybrid plasmids of similar size (4).

## Screening by In Situ Hybridization

Adaptation of the powerful colony hybridization procedure (17)
to screening of large numbers of colonies (18,19) almost brings it
into the efficiency range of screening of plaques (20). Since the
copy number of large hybrid cosmids is reduced compared to the one of
smaller plasmids, the signal to noise ratio of probes hybridizing to
recombinant cosmids is not optimal. This situation is greatly
improved by amplifying the hybrid cosmids in the colonies in situ
(18; J. Collins, personal communication).

The hybrid nature of the vector cosmid itself demands that the
hybridization probe contain neither sequences of the plasmid part of
the cosmid (pBR322 sequences in the case of pHC79) nor lambda sequences. Hybridization probes therefore should be separated from the
vector in which they were cloned or a plasmid should be chosen that
does not cross-hybridize with the cosmid.

## Hybrid Selection

In plasmids which carry two antibiotic resistance genes, the
inactivation due to DNA insertion into one of them can be tested.
Direct selection for hybrid plasmids is possible only after treatment of the restricted vector molecule with alkaline phosphatase, a

procedure which prevents ring closure of the vector molecules but
allows ligation to other DNA fragments. Also poly(dA)-poly(dT) or
poly(dG)-poly(dC) tailing of vector and insert DNA leads to direct
selection of hybrid molecules. Recombinant lambda bacteriophages can
be screened, as described in step 2 of Table 1, by the inactivation
of a gene carrying the cloning site or they can be directly selected
in cases where the immediate vector DNA is too small to yield a
plaque-forming phage particle. With the cosmid cloning system,
selection for hybrids is possible only by the direct method.

## Hybrid Stability

Two factors affecting stability of hybrid molecules are to be
considered: a) plasmids, cosmids and their derivatives depend on a
vector-host relationship for reproduction. Cloned genes, the expression of which in E. coli might have a deleterious effect on the
host, can be maintained more stably in a virulent phage cloning
vehicle because survival of the host cell is not required; b) as
noted previously (11), hybrid cosmids may be segregated from the
cells harboring them. This may in part be due to expressional instability, simply because in a larger hybrid the chance of carrying a
gene disadvantageous to its host is increased.

## Insert DNA Recovery

The production of large quantities of plasmid DNA, especially
after chloramphenicol amplification, is accomplished more easily than
production of bacteriophage lambda DNA. Moreover, the ratio of
insert to vector DNA, especially for large insert fragments, is much
greater in the case of plasmids or cosmids.
Another important point in this context is the possibility of
recovering vector-free insert DNA. Obviously this is a problem for
any large hybrid molecule and especially for hybrid cosmids.

In summary, it seems justified to conclude that cosmids should
be used for particular cloning purposes only: when large distances on
a genome need to be covered by large successive cloning steps, when,
in a shot-gun cloning experiment, the smallest possible number of
clones needs to be produced and screened (Table 2), when an
especially large restriction fragment must be cloned, when cosmid
cloning is considered to be the most convenient hybrid selection
device. All of these criteria also apply for cloning with yeast
cosmids.

## COSMIDS

Apart from the cosmids used to develop the new cloning procedure
(4), a series of cosmids has been developed that were still relatively limited in their general application (21,22). Recently the

Table 2

DNA Content of Various Organisms and Number of Cosmid Clones
Required for Gene Libraries

The number of cosmid clones required to represent a genome
are calculated on the basis of a published equation (21).
It is assumed that the average size of the inserted DNA is
between 35 and 40 kb.   Corrections are necessary if a
desired DNA sequence approaches the average insert size.

| Organism | DNA content per haploid genome (kb) | Ref. | No. of clones representing 95% of a genome |
|---|---|---|---|
| E. coli | $4.1 \times 10^3$ | 22 | $3.3 \times 10^2$ |
| S. cerevisiae | $1.5 \times 10^4$ | 23 | $1.2 \times 10^3$ |
| Drosophila | $1.7 \times 10^5{}_1$ | 24 | $1.4 \times 10^4$ |
| Sea urchin | $8.1 \times 10^5$ | 25 | $6.5 \times 10^4$ |
| Chicken | $1.2 \times 10^6$ | 26 | $9.6 \times 10^4$ |
| Mammals | $3.0 \times 10^6$ | 24 | $2.4 \times 10^5$ |
| Nicotiana tabacum | $3.9 \times 10^6$ | 27 | $3.1 \times 10^5$ |
| Xenopus laevis | $4.0 \times 10^6$ | 24 | $3.2 \times 10^5$ |
| (Lilium longiflorum)* | $(2.5 \times 10^8)$* | 27 | $(2.0 \times 10^7)$* |

*Ploidy level unknown.

production and use of the cosmid pHC79 has been described (11).  As
shown in the upper left part of Figure 2, it consists of the pBR322
moiety into which a BglII site has been introduced.  Into this site,
located in a nonessential part of the plasmid, a BglII fragment
carrying the cohered cohesive ends of the lambda vector charon 4A has
been inserted.

The possibility of moving this BglII fragment to other plasmids
has been exploited for the production of the yeast cosmid pYc1
(Figure 2).  The hybrid yeast-E. coli vector YEp6 (a gift from K.
Struhl; 10) was cut with restriction enzyme BamHI, which removes a
1.7 kilobase yeast DNA fragment carrying the his3 gene from yeast.
This digest was then ligated to BglII-cut cosmid pHC79.  By transforming E. coli strain B463 (hisB⁻) to His⁺, plasmids were recovered
containing the yeast his3 gene.  By checking the tet[R] marker, pHC79
derivatives could be distinguished from derivatives of YEp6.  Amp[R],
tet[S] colonies were screened by colony hybridization with a lambda
probe for the presence of the cos fragment.  Since BamHI and BglII
ends have the same sticky sequences, the BglII fragment from pHC79
could effectively be cloned into one of the BamHI sites of YEp6.
The yeast cosmid pYc1 constructed in this way retains only one

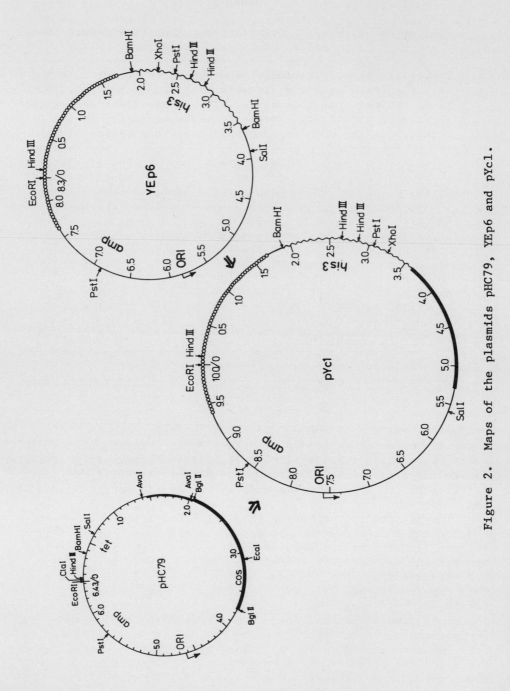

Figure 2. Maps of the plasmids pHC79, YEp6 and pYc1.

of the BamHI sites of YEp6 (hybrid BamHI/BglII sites are not cut by either of the two restriction enzymes). A vector was generated that has a unique BamHI site in addition to the cloning sites of YEp6 (EcoRI, XhoI and SalI).

The vector pYc1 is a true hybrid between a classical yeast vector and a cosmid. It can replicate autonomously in yeast and in E. coli since it has sequences of the yeast 2µ plasmid and contains the origin for replication of E. coli plasmid pBR322. In addition, selectable markers are present for both organisms (amp$^R$, his3 gene). The presence of the cos sequences of bacteriophage lambda allows this new plasmid to be packaged in vitro and thereby adds all the advantages of the cosmid cloning technique to the yeast cloning vector.

We have extended this new class of yeast vectors by making another derivative of pHC79. By introducing an EcoRI fragment of yeast DNA that codes for the yeast trp1 gene and carries a yeast chromosomal replication origin (10), we have constructed the yeast cosmid pYc2. Other yeast cosmids will be available soon (Hartmann and Collins, personal communication).

## HOW TO USE A COSMID

### Restriction, Sizing and Ligation of DNAs

Due to the requirement to fill a phage head with approximately 37 to 52 kilobase DNA (5), only those concatenated cosmid-insert-DNA molecules in which the cos sequences are 37 to 52 kilobases apart are packaged. Ligation of complete digests of cosmid DNA and DNA to be cloned yields a random assortment of cosmid molecules and insert sequences. If there is sufficient space in the phage-like particle for several ligated restriction fragments, these will be packaged and thereby cloned. The procedure, therefore, selects for a certain size but of course cannot distinguish a long inserted fragment from several smaller ones that are ligated together creating a nonauthentic gene arrangement. This should be avoided. If a random assortment of large fragments is collected, it is necessary to produce a narrow size range of clonable DNA fragments. Since the random distribution of break points in the DNA will generate a wide spectrum of randomly distributed size classes, fractionation of the molecules according to size is essential.

A gentle procedure for the isolation of genomic DNA has to be chosen to reduce the number of unclonable molecular ends created by shearing. For the production of DNA fragments of about 40 kb, the use of partial digests with enzymes cutting frequently, like Sau3A or MboI (for BamHI or BglII vectors), AvaI (for SalI vectors) and TaqI and HpaII (for ClaI vectors) is recommended (29). Mechanical shearing of the DNA in combination with linker addition or tailing also leads to random clonable fragments. The use of restriction enzymes

with six base-pair recognition sites might not allow the cloning of
exceptionally large fragments and fragments adjacent to them.
Fractionation of the DNA fragments into size classes can be accomplished using the differential electrophoretic mobilities or sedimentation differences.  For cloning in a 6 kilobase (kb) cosmid, the
insert DNA should be 30 to 45 kb long; for cloning in a 10 kb cosmid
the DNA fragments should be 25 to 40 kb in length.

The optimal concentrations for DNA ligation reflect the need for
the production of long concatenated molecules and the rejection of
the possibility for ring closure of the fragments.  The concentration
found most useful for cloning in pHC79 is 300 µg/ml each of restricted cosmid DNA and fragmented DNA to be cloned (11).

## In Vitro Packaging

The procedure for in vitro packaging was first developed as part
of a study on lambda morphogenesis (30).  It was later adapted to the
use of recombinant DNA (31) and used in the development of the cosmid
cloning system (4).  (For a detailed description see ref. 16.)  A
similar system involving a higher purification of the packaging
components has also been developed (32).

The packaging mixes are prepared by growing, inducing and
concentrating two E. coli strains whose prophages are defective in
different steps in the lambda morphogenetic pathway (3).  In the
presence of lambda or cosmid DNA, the phage head proteins, which have
been accumulating separately in the two strains upon induction of the
prophages, complement each other to produce infective lambda particles.  A combination of genetic and physical tricks prevents the
packaging of biologically active endogenous lambda DNA (16).  In
contrast to the packaging of lambda DNA, cosmid packaging does not
require UV inactivation of endogenous DNA molecules.

Cosmid ligation mixes can be added directly to the packaging
mixes, but care should be taken not to shear the highly viscous
ligated DNA preparation.  The resulting DNase-resistant particles,
freed from the cell debris by short centrifugation, can be stored
refrigerated like a phage lysate.  The ability to transduce recipient
bacteria drops down to 50%-70% in three months (11).

## Transduction

Bacteria to be transduced should be treated as any indicator for
bacteriophage lambda, i.e., the production of their phage receptors
should be induced by the addition of maltose to the medium.  The time
necessary for the expression of a resistance gene depends on whether
one uses resistance to ampicillin or tetracycline as the selective
marker.  As Figure 3 shows, the expression of the proteins required
for full resistance against 10 µg/ml tetracycline takes about 60 min
at 37°C.  Table 3 shows a comparison of frequencies of transduction

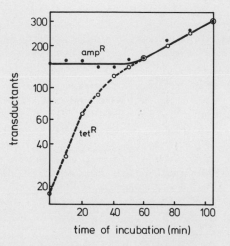

Figure 3. Kinetics of expression of the ampicillin and tetracycline resistance genes after transduction into SF8 of packaged pHC79– EcoRI/E. coli hybrids. SF8 competent cells were prepared as described in Table 2. Time 0 is the time of addition of L-broth to the bacteria-particle mixture. At given times after the incubation at 37°C samples were plated on agar plates containing 40 µg/ml ampicillin or 10 µg/ml tetracycline.

of one batch of hybrid packaged cosmids into E. coli strains with different recombination defects. (Strain 1400, provided by P. Kourilsky, can be used for a colony hybridization procedure in which partial lysis of the colonies is effected by inducing the prophage (20).)

Yeast Transformation

A yeast transformation protocol and some modifications have been described previously (7-10). The procedure is based on the coprecipitation of calcium-treated yeast spheroplasts and transforming DNA by polyethylene glycol 4000 (PEG 4000). If auxotrophic strains are transformed with the cloned wild-type genes, regeneration of the

Table 3

Efficiency of Transduction of Packaged Hybrid Cosmids into
Different E. coli Strains

The indicated E. coli strains were grown to late exponential phase in L-broth supplemented with 0.4% maltose, spun
down and resuspended in one-tenth the original volume in
10 mM $MgSO_4$. To 0.1 ml of these cells samples of the
packaged hybrid cosmids (E.coli EcoRI clones in pHC79)
were allowed to adsorb at room temperature for 15 min.
Expression of the ampicillin resistance gene was allowed
after addition of 1 ml L-broth and aeration at 37°C for 20
min for the first three strains and for 40 min at 32°C for
the fourth strain. Plating was done on plates containing
40 µg/ml ampicillin.

| E. coli strain | Normalized transduction frequency |
|---|---|
| SF8 lopl1 lig supE recB⁻C⁻ $r_K^-$ $m_K^-$ | 100 |
| HB101 recA⁻ $r_K^-$ $m_K^-$ | 80 |
| A0922 supE recA⁻ $r_K^-$ $m_K^+$ | |
| (A. Oppenheim) | 62 |
| 1400 = 803 $r_K^-$ $m_K^-$ recA⁻ supE⁺ supF⁺ | |
| (λIts857 ∇b2-cIII) (P. Kourilsky) | 66 |

spheroplast and selection of transformed colonies can take place
simultaneously on selective medium.

Yeast spheroplasts can be prepared by several cell wall de-
grading enzymes. Crude enzyme mixes (glusulase (7) Endo Laboratories, helicase (9) Industrie Biologique Francaise) and partially
purified enzymes (lyticase (10), zymolyase (60,000 units per g, Kirin
Brewery, Japan)) have been used with similar efficiencies.

In order to test yeast transformation with pYcl carrying large
inserts, yeast DNA was cut incompletely with EcoRI and ligated to the
corresponding cloning site of the yeast cosmid (Figure 1). Hybrids
have been isolated by transduction of E. coli, selecting for
ampicillin-resistant colonies. Since the hybrid plasmids are amplifiable to high copy numbers in the bacterial host, large amounts of
DNA can be prepared. When yeast is transformed using his3 as selective marker, the hybrids behave identically to the parent vector
YEp6, i.e., they are lost rapidly if selection pressure is relieved
(10). It is interesting to note that the hybrids, which are of high
molecular weight (40 to 50 kb), show, on a molar basis, only a 50%
reduction in transformation frequency if compared to pYcl or YEp6

Table 4

Efficiency of Transformation of Yeast Strain YHB4
by Different Vectors and Hybrids

Strain YHB4 (α, his3-532, trpl-289, gal2) was
converted to spheroplasts following a combination
of published procedures (7,9) and transformed
with samples of the DNA prepared as described in
the text.

| Vector | Efficiency* |
|--------|-------------|
| YEp6 | $10^4$ |
| pYc1 | $7 \times 10^3$ |
| pYc1 hybrids | $6 \times 10^2$ |

*Expressed as transformants/μg DNA/$10^7$ regener-
ated spheroplasts.

(Table 4).  This is in striking contrast to the strong dependence on
plasmid size of E. coli transformation frequencies (4,6,15).  Preliminary data show that the large hybrid cosmids are stable in yeast
and do not undergo detectable rearrangements.

     Figure 1 shows an alternative protocol that simplifies the
testing of hybrid cosmids (and any other hybrid plasmid) in yeast.
E. coli cells can be fused directly to yeast spheroplasts and the
hybrids can be selected in yeast by plating the whole fusion mixture
on selective yeast agar plates (manuscript in preparation).  This
technique, called yeast transfusion, is relatively inefficient
compared to yeast transformation (1% to 10% on a molar basis) but
greatly reduces the amount of labor involved in the isolation of the
hybrid DNA.


Reintroduction of Yeast Plasmid DNA into E. coli

     Due to the presence of 2μ on the yeast cosmid pYc1 and its
hybrid derivatives, autonomous replication is ensured (10) and
therefore, isolation of plasmid DNA of yeast and reintroduction into
E. coli is possible.  In the case of yeast cosmids, two possibilities
exist:  transformation and in vitro packaging.  From a series of six
different yeast transformants, DNA was isolated following lysis of
the cells with zymolyase (60,000 units per g, Kirin Brewery, Japan).
In vitro packaging and transduction into SF8 (10) and transformation
of CaCl$_2$-treated SF8 cells and subsequent selection for ampicillin-
resistant colonies were tested using all six DNA samples.  Both the

packaging and the transformation efficiencies varied from one clone
to the next, but this variation was not parallel in both procedures
(unpublished observations). This result is similar to that for re-
isolation of hybrid cosmid DNA from E. coli and can be explained by
the size dependence known for transformation frequencies, on the one
hand, and by certain structural properties of the DNA known to affect
packaging frequencies, on the other (11; for a summary of DNA molecules as substrates for DNA packaging see ref. 3, Figure 17).

As a method to recover hybrid cosmid DNA from a yeast DNA pool,
neither method is recommended because only particular subpopulations
of molecules are recovered. As a method to recover individual
clones, repackaging seems more reliable in view of its ability to
select and, in this case, to maintain large molecules.

Acknowledgments: We acknowledge receipt of bacterial, yeast and
plasmid strains donated by K. Struhl, D. Stinchcomb, A. Oppenheim, P.
Kourilsky and G. Fink. We thank T. Hohn for discussions on the
experiments and the manuscript, J. Collins for providing unpublished
information and Helgard Adelsberger for technical help.

## REFERENCES

1    Hohn, B. (1975) J. Mol. Biol. 98, 93-106.
2    Feiss, M., Fisher, R.A., Siegele, D.A., Nichols, B.P. and
     Donelson, J.E. (1979) Virology 92, 56-67.
3    Hohn, T. and Katsura, J. (1977) in Current Topics in
     Microbiology and Immunology, Vol. 78, pp. 69-110.
4    Collins, J. and Hohn, B. (1978) Proc. Nat. Acad. Sci. U.S.A. 75,
     4242-4246.
5    Feiss, M., Fisher, R.A., Crayton, M.A. and Egner, C. (1977)
     Virology 77, 281-293.
6    Kushner, S.R. (1978) in Genetic Engineering (Boyer, H.W. and
     Nicosia, S., eds.), pp. 17-23, Elsevier/North-Holland,
     Amsterdam, New York, Oxford.
7    Hinnen, A., Hicks, J.B. and Fink, G.R. (1978) Proc. Nat. Acad.
     Sci. U.S.A. 75, 1929-1933.
8    Ilgen, C., Farabaugh, P.J., Hinnen, A., Walsh, J.M. and Fink,
     G.R. (1979) in Genetic Engineering, Principles and Methods,
     Vol. 1, pp. 117-132, Plenum Press, New York and London.
9    Beggs, J.D. (1978) Nature 275, 104-109.
10   Struhl, K., Stinchcomb, D.T., Scherer, S. and Davis, R.W. (1979)
     Proc. Nat. Acad. Sci. U.S.A. 76, 1035-1039.
11   Hohn, B. and Collins, J. (submitted for publication).
12   Thomas, M., Cameron, J.R. and Davis, R.W. (1974) Proc. Nat.
     Acad. Sci. U.S.A. 71, 4579-4583.
13   Murray, N., Brammer, W.J. and Murray, K. (1977) Mol. Gen. Genet.
     150, 53-61.

14  Blattner, F.R., Williams, B.G., Blechl, A.E., Denniston-
    Thompson, K., Faber, H.E., Furlong, L.A., Grunwald, D.J.,
    Kiefer, D.O., Moore, D.D., Schumm, J.W., Sheldon, E.L. and
    Smithies, O. (1977) Science 196, 161-169.
15  Collins, J. (1977) in Current Topics in Microbiology and
    Immunology, Vol. 78, 121-170.
16  Hohn, B. (1979) Methods Enzymol. 68, 299-309.
17  Grunstein, M. and Hogness, D.S. (1975) Proc. Nat. Acad. Sci.
    U.S.A. 72, 3961-3965.
18  Hanahan, C. and Meselson, M. (submitted for publication).
19  Cami, B. and Kourilsky, P. (1978) Nucl. Acids Res. 5, 2381-2390.
20  Benton, D. and Davis, R.W. (1977) Science 196, 180-182.
21  Clarke, L. and Carbon, J. (1976) Cell 9, 91-99.
22  Bachmann, B.J., Brooks Low, K. and Taylor, A.L. (1976)
    Bacteriol. Rev. 40, 116-167.
23  Hartwell, L.H. (1974) Bacteriol. Rev. 38, 164-198.
24  Handbook of Biochemistry and Molecular Biology II (1976).
25  Galau, G.A., Klein, W.H., Davis, M.M., Wold, B.J., Britten, R.J.
    and Davidson, E.H. (1976) Cell 7, 487-505.
26  Hereford, L.M. and Rosbach, M. (1977) Cell 10, 453-462.
27  Bennet, M.D. and Smith, J.B. (1976) Phil. Trans. Royal Soc.
    London, 227-274.
28  Collins, J. and Brüning, H.J. (1978) Gene 4, 85-107.
29  Collins, J. (1979) Methods Enzymol. 68, 309.
30  Hohn, B. and Hohn, T. (1974) Proc. Nat. Acad. Sci. U.S.A. 71,
    2372-2376.
31  Hohn, B. and Murray, K. (1977) Proc. Nat. Acad. Sci. U.S.A. 74,
    3259-3263.
32  Sternberg, N., Tiemeier, D. and Enquist, L. (1977) Gene 1,
    255-280.

# DNA CLONING WITH SINGLE-STRANDED PHAGE VECTORS

W.M. Barnes

Department of Biological Chemistry
Division of Biology and Biomedical Sciences
Washington University School of Medicine
St. Louis, Missouri 63110

The filamentous single-stranded DNA coliphages M13, fd and fl are currently being developed as a new class of DNA-cloning vector having distinct advantages over other vectors, including the other two classes of E. coli-hosted cloning vectors, plasmids and phage lambda. This review describes the biology and methods of use for single-stranded phage vectors as well as rapid methods for determination of the size and orientation of inserted DNA, the advantages of a single-stranded vector particularly for DNA sequencing and in vitro site-directed mutagenesis, and the currently available single-stranded vectors, including some actual or speculative advantages of some of them over others.

The filamentous coliphage under discussion here come under three names: M13, fd and fl. They are virtually identical. Complete DNA sequences are known for fd (6408 nucleotides) (1) and M13 (6407 nucleotides) (2,3) and they are 97% homologous. The differences consist mainly of isolated nucleotides here and there, mostly affecting the redundant bases of codons with no blocks of sequence divergence. Partial information on the sequence of fl (4) indicates that it is even closer to M13 than is fd. M13 refers to these three phages as a class; the reader should not infer any justified preference for one over the others as a vector.

## BIOLOGY AND PHYSICAL METHODS OF PREPARATION

For those readers who are not as familiar with M13 as with other phages or plasmids, below is a description of its biological properties. For a more extensive review, see The Single-Stranded DNA Phages (5).

185

Table 1

Unique Restriction Sites of Available Filamentous Phage Vectors

| Phage | Phenotypic marker in inserted DNA | Unique restriction cut | | Size of integrated DNA (kb) | Integration site in phage DNA | Source of integrated DNA | Refs. |
|---|---|---|---|---|---|---|---|
| | | In pheno-typic marker | Outside phenotypic marker | | | | |
| M13+ | none | AsuI, AvaI | | 0 | | | |
| fd+ | none | AsuI | | 0 | | | |
| f1+ | | | | | | | |
| fd101 | Ap$^r$, Km$^r$ | PstI, HindIII, SmaI | | 3.6 | ∿5565 | pACYC177 | 41 |
| fd102 | Ap$^r$ | PstI | | 2.3 | | pACYC177 | 41 |
| fd103 | Cm$^r$, Ap$^r$ | PstI, EcoRI | | 3.6 | | pACYC177/18 | 41 |
| fd104 | Km$^r$ | XhoI, HindIII, SmaI | | 1.4 | | pACYC177 | 41 |
| fd105 | Cm$^r$ | EcoRI | | 1.3 | | pACYC184 | 41 |
| fd106 | Cm$^r$, Km$^r$ | EcoRI, XhoI, HindIII, SmaI | | 2.7 | | pACYC184/17 | 41 |
| fd107 | Ap$^r$ | PstI | HindIII, SalI, EcoRI | 4.4 | 5644 | pBR322 | 41 |
| fd11 | none | | EcoRI | 0.004 | 5830 | EcoRI-cleaved | 41 |
| R199(f1) | none | | EcoRI | 0.004 | 5725 | EcoRI-cleaved | 29 |
| R209(f1) | none | | EcoRI | 0.004 | 5868 | EcoRI-cleaved | 29 |
| M13mp2 | lac α-complementation | EcoRI | | 0.789 | 5868 | pMG1106 | 32 |
| M13mp5 | lac α-complementation | HindIII | | 0.789 | 5868 | | 33 |
| M13Hol76 | HisD(Hol+) | PvuII | EcoRI, SalI, PstI, KpnI(Table2) | 3.331 | 5727 | λh80dhis | 20 |

| | | | | | | |
|---|---|---|---|---|---|---|
| fdTet | Tet$^r$ | | EcoRI,HindIII,XbaI <u>Ava</u>I(<u>Xma</u>I) | | Tn10(pRT44) | 42 |
| M13Bla6 | Ap$^r$ | <u>Pst</u>I | | ~5565 | | 43 |
| M13Goril | G4 origin | | SstI,XhoI, <u>EcoRI,Kpn</u>I  2.216 | ~5565 | G4 | 44 |

M13 is male-specific, which means that it only forms plaques on
E. coli cells containing an $F^+$ or F' sex factor, since the phage
adsorbs by attachment to sex pili.  However, female (F-) cells can
be infected (and produce progeny phage in good yield) if phage DNA
is introduced by transformation (see below).  Upon natural infection,
the single-stranded virion DNA is converted to a double-stranded
replicative form (RF) which then replicates to form a steady-state
number of 50 to 100 copies per cell.  The term steady-state is
possible here, because M13 does not kill the infected cell.  Rather,
a chronic infection is set up with infected cells constantly secreting
phage particles into the medium.  M13-infected cells grow two to
three times slower than uninfected, or plasmid-containing, cells and
this is the mechanism of plaque formation.  The plaques are not areas
of phage-induced lysis, but rather foci of infection and slow growth.

The genetic engineer may consider an M13-infected cell as being
equivalent to a plasmid-containing cell (with RF as the plasmid) if
the M13 vector carries a selective marker, as many do (Table 1).  In
vitro recombination (ligation or tailing) is carried out with the
double-stranded RF DNA followed by transformation and suitable selection for infected colonies (or, alternatively, plating on a lawn of
sensitive cells to make plaques).  Infected colonies may be streak-
purified, grown up to a liter or two of saturated culture under
selective pressure, and RF can be isolated from the cell pellet by
any procedure that works for plasmids (i.e., cleared lysate followed
by ethidium bromide-CsCl banding)(6).  The culture supernatant should
be saved, however, as it contains the best part--single-stranded DNA.

M13 virion DNA is a single-stranded closed circle that may be
imagined to have been pulled from opposite sides until it is flattened.
The flattened circle is then packaged end-to-end to form a filamentous
phage particle whose length is proportional to the length of the
packaged DNA.  M13 packaging proteins do not care how long the DNA
they package is, as indicated by the propensity of wild-type to form
a small percentage of double- or longer-length particles (7).  The
maximum length of DNA that can be usefully carried by a single-
stranded cloning vector has not been established, but may well
exceed 40 kilobases.

The ease with which large-scale purification is performed depends
to a great extent on the shape of the phage particles; long skinny
particles are the easiest to precipitate with PEG.  Yamamoto et al.
(8) found that fd precipitated at only 1% PEG, ten times less than
that required to precipitate phage lambda, so that PEG precipitates
of infected-culture supernatants are already very pure preparations
of phage.  Some have found that PEG-precipitated or centrifuged
phage can be immediately phenol-extracted and ethanol-precipitated
to provide DNA of sufficient purity for DNA sequencing (9,10), although I recommend one equilibrium centrifugal banding in 30% (wt/wt)
CsCl and dialysis before the phenol extraction.  Pretreatment of
phage particles with proteinase K and SDS greatly improves the
efficiency of the phenol extraction (11).  Without this proteinase-

SDS treatment, a large interface containing trapped DNA is formed during phenol-extraction of concentrated phage suspensions.


## Transformation

M13 particles are packaged as they are secreted through the cell membrane. They do not cause lysis, or a hole in the membrane. This method of morphogenesis requires an intact bacterial membrane and cell wall structure. This makes the spheroplast method of DNA transfection unreasonable and inefficient for M13, since it would seem that only those cells in a spheroplast preparation whose cell walls have been partially removed could both take up DNA and allow phage packaging. Therefore (12), the preferred method of DNA transformation is a variation of the $CaCl_2$-shock method (13). Transformation efficiencies of $3 \times 10^6$ per µg RF DNA can be expected.


## Rapid Size Determination

The relatively small size (compared to lambda), the high concentration of phage DNA in plate or liquid cultures of M13-infected cells and the single strands make two technical tricks convenient. The first trick, the "toothpick assay" (14), rapidly determines the size of the RF (and of the single-stranded DNA) by agarose gel analysis of SDS-lysates of single colonies or small streaks on agar plates. The single-stranded DNA is even easier to visualize than RF DNA or the plasmid DNA to which the technique was first applied. The size of any inserted DNA can be inferred by comparing it with the mobility of empty vector DNA. Size differences as small as 1.5% can be detected. Culture supernatants contain enough DNA to be visible when treated directly with the SDS/agarose gel procedure used for the toothpick assay. Whole phage particles can also be analyzed on the same agarose gels (15).


## Orientation Trick (16)

A stretch of DNA can be in one of two orientations in any vector. In an M13 vector, the orientation determines which strand of cloned DNA will be packaged in the phage virions (i.e., the strand which is contiguous with the phage + strand). Knowledge of the relative orientation of the cloned DNA can be important in planning DNA sequencing or other experiments with the single-stranded DNA. Restriction analysis of RF DNA can make this determination, but that is more work than the following procedure. Unconcentrated culture supernatants contain enough phage DNA to be visible on an agarose gel (see above). If two phage-containing suspensions (0.1 ml of culture supernatant) are mixed with the SDS and buffer of the toothpick assay (above) and

then annealed by incubation overnight at 65°, they will anneal together if they contain inserts of the same DNA sequence in opposite
orientations.  The annealed molecules have a greatly altered mobility
in the agarose gel.  In this way, many small culture supernatants can
be quickly analyzed by annealing them against the same known (or unknown) standard.  This does not work with bacteria-containing cultures, perhaps because the bacterial DNA raises the viscosity of the
medium too much to allow timely annealing.  In both this technique
and the toothpick assay, the method of photography is critical for
visualization of the DNA in the agarose gels.  The photography method
described (14) allows detection of as little as 1 ng of DNA in a band.


## ADVANTAGES OF SINGLE-STRANDED VECTORS

The reader will doubtless have his or her own uses for the
single-stranded DNA that M13 can provide in large quantity (about
0.5 mg/liter of culture) and high quality (easily pure, with no
breaks or ends).  Described below are the most obvious advantages.


### Dideoxy Sequencing

Sanger's chain-termination method of DNA sequencing (17) using
thin gels (18) is the most convenient and economical way to sequence
DNA, provided a source of single-stranded DNA spanning the target
sequences is available.  M13 cloning is the best source.  The first
application of M13 cloning for this purpose was for the histidine
operon DNA sequence (19,20).
Alternative methods of obtaining single-stranded DNA, with their
comparative disadvantages, include:  a) poly U,G CsCl gradient (21)--
this is relatively expensive and has particular sequence requirements,
met however by phage lambda, a larger, more complex vector;  b) alkaline CsCl (22)--this technique also relies on particular sequence
composition of the strands being separated;  c) acrylamide gels
(23)--this technique has a low capacity, not reasonable for amounts
exceeding 10 to 50 pmoles, mostly due to the problem of reassociation of strands as they become extremely concentrated when they
enter the gel;  d) exonuclease treatment (24)--limit degradation by
exonuclease III or T7 exonuclease can turn double-stranded molecules
into single-stranded molecules.  This technique was developed especially to meet the needs of dideoxy sequencing and is being used to sequence the entire adenovirus DNA genome (25).  However the use of this
technique with adenovirus has two problems not encountered with M13:
1) more strategy and time must be expended to prepare strands for a
given target sequence, and 2) during application of the dideoxy sequencing procedure, it is often desirable to recut at the restriction
site being used (this is necessary when the primer is large, e.g.,
over 100 base pairs).  This step is not acceptable when exonuclease

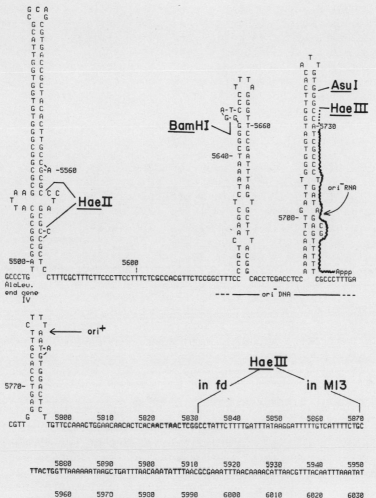

Figure 1. Intergenic region of phage fd (1) in which foreign DNA can be carried. Possible secondary structure shown is that suggested by Schaller, Beck and Takanami (4). The RNA primer for minus strand replication was determined by Geider et al. (39). The probable start point for plus strand replication (ori⁺) was determined by Schaller (40). Restriction sites indicated are those that have been used for cloning (except AsuI, a unique site in wild-type phage noted here for the first time). Figure 1 should be compared with Table 1 to deduce which vectors use which restriction sites. The M13 DNA sequence is somewhat different, so that M13 has no BamHI site in this region and one of the HaeIII sites in this region is in a different place, as indicated. The other marked sites are the same.

III-treated DNA is used as template because of the high background
of artifact bands which often result. Thus, one is restricted to
small primers when using exonuclease-treated DNA.

Once an M13 clone has been constructed, on the other hand, DNA
is easily prepared in large quantity in a ready-to-go form and the
complete versatility of the dideoxy method can then be used.

## In Vitro Mutagenesis

The same site-direction given to dideoxy sequencing by DNA
primers has also been used to create specific mutants in single-
stranded phage (26,27). In this approach, a short stretch of DNA
is synthesized in vitro with enzymatic (26) or chemical (27) methods.
The synthetic sequence is homologous to the target DNA except that
it contains the desired mutation. The synthetic sequence is annealed
to the single-stranded phage DNA and used as a primer by DNA polymerase I to copy the entire molecule into double-stranded DNA. After
transformation, replication and segregation, many of the resulting
phage DNA molecules contain the desired base change.

## CURRENTLY AVAILABLE FILAMENTOUS PHAGE VECTORS

Foreign DNA is usually carried by M13 either in the intercistronic region (I.R.) between genes IV and II, or sometimes within
the carboxy-terminal end of gene IV (28). The I.R. is about 500
nucleotides long (positions 5498 to 6005 of the DNA sequence) and
contains no genes necessary to M13 growth, except for the origins
of DNA replication of both the plus and minus strands located in
this region. Until experience proves otherwise, any restriction
site in the I.R. should be considered as a possible cloning site,
even those that appear to be located in the middle of an origin of
replication. The HaeIII site at 5726, in the ori minus site (Figure
1), has been used with great success to carry foreign DNA (20,29).

Table 1 lists filamentous phage vectors, with unique sites
(occurring only once in the genome) recommended for sites of insertion of passenger DNA. The table starts with wild-type phages
because they have been used as starting DNA for several clonings
(20,29,30). A site does not have to be unique to be useful; indeed,
the first example of M13 cloning (20,30) made use of one of 10 HaeIII
sites in the genome, two of which are in the intercistronic region
(I.R.). This use of non-unique sites begins with a partial digest
of the phage RF DNA, and since the pioneering construction of M13mp1
by Messing et al. (30), this technique has been used for all published M13 in vitro constructions (20,29,30).

Much current development of filamentous phage vectors involves insertion of foreign DNA carrying useful, unique restriction sites. Those sites presently available are listed in Table 1, or can be deduced from the computer-assisted restriction analysis shown in Table 2. This analysis has uncovered a previously unrealized, ideally located, single site for AsuI in the I.R. of wild-type phage and several similarly unrealized cloning sites in the vector M13Ho176.

It is probably best to clone into a site near the gene IV side of the I.R. (nucleotide 5500). This recommendation is based on the following observations. Most transposon insertions that create plaque-forming transducing phage map to this side of the I.R. or even within the tail-edge of gene IV (15,28,31). Several independent insertions of histidine operon DNA (20) were found in the HaeIII site on the gene IV side of the I.R., rather than in the HaeIII site on the other side of the origins of replication, which nevertheless has been shown to be capable of stably harboring lac DNA in the hybrid phage M13mp1 (30). It is difficult to clone even the histidine operon DNA into M13mp2 (position 5868) because multiple deletions of the resulting transducing phage occur spontaneously at high frequency. In contrast, histidine operon DNA is very stable (with no spontaneous deletions observed) at position 5727. Thus it is tentatively recommended that f1 vector R199 be used rather than f1 vector R209, and vectors M13Goril, M13Bla6, and M13Ho176 should be favored over M13mp2 and fd11 (see Table 1). Wild-type phage contain a single recognition site for AsuI at the recommended location. Further experience with filamentous phage vectors should establish the validity of this recommendation, which is now based on relatively little data.


## AVAILABLE SELECTIONS AND SCREENS FOR INSERTED DNA


The reader may derive many of the available phenotypic selections and screens for inserted DNA by inspection of Table 1. Perhaps the most convenient, although not the most reliable, screen is available to all of the vectors equally: plaque size is approximately inversely correlated with the size of inserted DNA (Barnes, unpublished data). The unreliability comes from the fact that some small plaques arise among wild-type phage as plating artifacts, and sometimes there is a growth-slowing effect that is specific to the DNA sequence being inserted. The possibility and range of this effect are demonstrated by an orientation effect observed among M13-histidine transducing phage (16,20). Presumably, this effect could be so serious as to make the plaques invisible. (In such a case, an infected colony selection could probably still be used.)

A more reliable screen is that built into the M13plac vectors (30,32,33). These carry the first part of the lac operon, and in their special host strain (F'lacΔM15, laci$^q$) and in the presence of

Table 2

Restriction Site Analysis of Completely Sequenced Vectors

Positions of restriction sites occurring less than three times in wild-type filamentous phage are shown, regardless of whether or not the sites are known to be useful for cloning. (Those unique sites within the intercistronic region (I.R., positions 5498 to 6005) of wild-type phage or anywhere within the inserted DNAs should be regarded as reasonable sites for cloning.) The entry "one" refers to a single site which is the reference for the DNA sequence numbering. Entries in parenthesis, i.e. (five), refer to multiple occurrences of the restriction site. The lac (45) and his (46) sequences used for this analysis are incomplete, each lacking about 50 nucleotides*. A plasmid cloning vector, pBR322 (47) (not a phage), has been included for comparison. This analysis was facilitated by computer programs modified from those of Staden (48). Not included in this table, although it qualifies as completely sequenced, is the vector M13GoriI (see Table 1) which carries a 2216-base HaeII fragment of phage G4 (49).

| | M13 wild-type(2,3) | fd wild-type(1) | M13mp1 [a] | M13Ho176 [b] | pBR322(47) |
|---|---|---|---|---|---|
| AccI GTAGAC,GTCTAC | 6090 | 6091 | 0 | 0 | 0 |
| AsuI GG-CC | 5724 | 5725 | 46,70,474 | (nine) | (fifteen) |
| AvaI CPyCGPuG | 5825 | 0 | 0 | 0 | 1424 |
| AvaII GGACC,GGTCC | 0 | 0 | 46 | 3031 | (eight) |
| AvaIII ATGCAT | 0 | 0 | 0 | 0 | 0 |
| BamHI GGATCC | 2220 | 2220,5645 | 0 | 0 | 375 |
| BalI TGGCCA | 5080 | 5081 | 0 | 0 | 1443 |
| BclI TGATCA | 0 | 0 | 0 | 0 | 0 |
| BglI GCC-----GGC | 0 | 0 | 509 | (four) | (three) |
| BglII AGATCT | 0 | 0 | 0 | 2538 | 0 |
| BstEII GGT-ACC | 0 | 0 | 0 | 2865 | 0 |
| ClaI ATCGAT | 2527,6039 | 2527,6040 | 0 | 734,1796 | 23 |
| EcoRI GAATTC | 0 | 0 | 0 | one at edge of insert | one |
| HincII GTPyPuAC | one | one | 0 | 562,2090,3073 | 650,3906 |
| HindIII AAGCTT | 0 | 0 | 0 | 0 | 29 |

| | | | | | |
|---|---|---|---|---|---|
| HgiAI | 5465 | 5466 | 0 | 118,641,945 | (five) | HgiAI GTGCTC, GAGCAC |
| HpaI | 0 | 0 | 0 | 0 | 0 | HpaI GTTAAC |
| KpnI | 0 | 0 | 0 | 61 | 0 | KpnI GGTACC |
| MstI | 0 | 0 | 0 | 3141 | (four) | MstI TGCGCA |
| PvuI | 0 | 0 | 483 | 1546 | 3734 | PvuI CGATCG |
| PvuII | 0 | 0 | 92,185,453 | 1392 | 2065 | PvuII CAGCTG |
| PstI | 0 | 0 | 0 | 3052 | 3608 | PstI CTGCAG |
| SmaI | 0 | 0 | 0 | 0 | 0 | SmaI CCCGGG |
| SstI | 0 | 0 | 0 | 0 | 0 | SstI GAGCTC |
| SstII | 0 | 0 | 0 | 0 | 0 | SstII CCGCGG |
| SalI | 0 | 0 | 0 | 3073 | 650 | SalI GTCGAC |
| XbaI | 0 | 0 | 0 | 0 | 0 | XbaI TCTAGA |
| XhoI | 0 | 0 | 0 | 0 | 0 | XhoI CTCGAG |

[a] Contains this 789-base* insert (lacPO HindII) at position 5727.

[b] Contains approximately 3331-base* insert (hisOGD) at position 5868.
Notes: 1) EcoRI site added to phage fl at position 5725 to make R199 (29); 2) EcoRI site added at position 5830 to make vector fdl1 (41); 3) EcoRI site added at position 363 to make M13mp2 (32). HindIII site added at position 363 to make M13mp5 (33); 4) the hisD gene spans 118 to 1422 and the hisG gene spans 1525 to 2424 (minus strand) (46).

isopropyl-β-D-thiogalactopyranoside (IPTG), these vectors form blue
plaques on indicator plates containing X gal agar (XG). The EcoRI or
HindIII site provided for carrying foreign DNA is within the coding
sequences of lacZ which give rise to the blue lac$^+$ phenotype, so that
the plaques carrying inserted DNA are colorless (as well as smaller).
(The lac$^+$ phenotype can be used as a carbon-source selection to
assist preparative growth of the M13plac vectors, but this selection
is not possible after more DNA has been inserted.)  It is worth
noting that the M13plac vectors, or even an M13 carrying only 203
nucleotides of the lacPO DNA (34), form blue plaques on laci$^S$ and
on wild-type male E. coli, whether or not they carry further inserts.
This is due to out-titration of the lac repressor, also observed for
lac plasmids (35).  Unfortunately, this effect cannot be used as a
carbon-source selection because laci$^S$ cells can revert to lac$^+$
too readily.

## FUTURE EXPERIMENTS

### Cosfd

The article by Hohn and Hinnen (this volume) describes the advantages of applying the phage lambda packaging system to plasmids.
The same advantages (efficient transformation, selection for large
size of inserted DNA) would apply if the cos site were added to M13.
However, the relatively large size of phage lambda (40 to 50 kilobases) can be considered to be a handicap, in terms of economy of
DNA preparation and complexity of restriction enzyme and sequence
analysis.  For applications that could utilize the efficiency of
transformation of the lambda packaging system without the need to
clone large (30 to 45 kb) amounts of DNA, the following construction is suggested.  What is needed is some temporary "baggage" DNA
in the starting vector.  This DNA would carry the cos site and the
bulk length of DNA necessary to fill up the lambda phage heads.
After infection this "baggage" DNA should reliably, efficiently,
and automatically (in vivo) excise itself and disappear.  I have
observed that the transposon Tn10 (and possibly transposon Tn5)
have the desired property of efficient and precise (or nearly
precise) excision when inserted into phage fd.  If one of these
transposons were engineered to carry enough baggage, a selective
marker and the cos site, they could simply and conveniently turn
any filamentous phage vector into a cosfd at will.

### Cloning Single-Stranded RNA Directly

Cloning mRNA sequences by first copying them into cDNA was a
major technical development in genetic engineering (36).  However,
it is technically difficult to copy long stretches (more than a

thousand nucleotides) of RNA for this purpose, and interesting RNAs,
1 to 10 kb long (such as large messages and viral RNAs), may be extremely difficult to clone intact with this approach.  It should
be possible to clone the RNA by ligating it directly into the single-
stranded form of a filamentous phage vector with T4 RNA ligase.  The
vector DNA could be opened up at the right place in one of two ways:
1) partial digestion with HaeIII, or 2) annealing of a small overlapping restriction fragment to the EcoRI region of the vector,
cleaving with RI and then denaturation of the DNA to make it completely single-stranded again.  The RNA to be cloned would need
several treatments to remove any cap or 3'-phosphate, and/or to add
a 5'-phosphate.  These last two steps are necessary to fulfill the
requirements of T4 RNA ligase, which are 3'-OH and 5'-phosphate (37).
Note that an automatic selection for full-sized RNA may operate here
since the breakdown products of RNase would ordinarily have 3'-
phosphates which would prevent such molecules from being cloned
accidentally.  The known preference of T4 ligase for short oligonucleotides (37) may be the hardest problem to overcome for this
manipulation.

## Controlled-Random DNA Sequencing

Recently R. Roberts and colleagues (personal communication),
sequencing adenovirus (strands obtained with exonuclease III) and
F. Sanger and colleagues (personal communication), sequencing
lambda (strands obtained by shotgunning lambda fragments
into M13mp2), have taken advantage of the speed of dideoxy sequencing
to sequence their target DNAs essentially at random, using little
or no restriction map information to direct each sequencing experiment.  A convenient feature in Sanger's experiments using M13 is
that the same primer, homologous to the vector DNA adjacent to
inserted DNA, is used for each sequencing experiment.  With this
approach, the first 80% to 90% of a very large sequence can be obtained, but that last 10% will require more directed effort.  This
might be called the random approach, in contrast to the following
controlled-random approach to a large sequencing project.

The entire piece of target DNA (10 kb at a time) should first
be treated with exonuclease III or T7 exonuclease for various times
so that each time-point interval corresponds to some 200 to 300
nucleotides of digestion.  Following exonuclease digestion, the DNA
should be treated with S1 nuclease and then with large fragment DNA
polymerase and triphosphates to remove the single-stranded DNA and
ensure that the undigested DNA has exactly blunt ends suitable for
T4 ligase action.  The DNA from each or several time-points of exonuclease digestion should then be ligated into the RF of a suitable
filamentous phage vector.  The toothpick assay (14) can be applied
for convenient sizing of the resulting clones.  Their size will be
a measure of the location along the original target sequence of the

edges of the inserted DNA stretches (with a two-fold ambiguity corresponding to the two possible orientations of each clone; this ambiguity can be removed by the orientation trick described above). As described for Sanger's experiments, each sequencing experiment can now use the same convenient primer adjacent to the site of insertion.  An added advantage is that the experimenter now has some systematic control over which area of the target DNA is being sequenced.

## GUIDELINES

At this writing, the NIH Guidelines for Recombinant DNA Research do not allow cloning in any cell carrying a transmissable plasmid (except for exempt experiments).  This includes the natural hosts for filamentous phages, namely male E. coli containing an $F^+$ or F' sex factor.  Two alternative hosts are officially allowed as EK1 biological containment.  1) Female ($F^-$) cells.  This has been described for the vector M13Ho176 (20) and can be used for any of the antibiotic resistance-carrying vectors in Table 1.  2) Hosts with F-factors carrying one or two tra-mutations (38).  Tra$^-$ cells may make the sex pili needed for filamentous phage adsorption, but do not mate and transfer DNA.

Acknowledgments:  I thank H. Schaller, R. Herrmann, D. Ray, J. Kaguni, G. Smith, F. Sanger and G.N. Godson for discussions and communication of preliminary results.  My research is supported by NIH grant GM-24956 and American Cancer Society grant JFRA-1.

## REFERENCES

1   Beck, E., Sommer, R., Auerswald, E.A., Kurz, C., Zink, B., Osterburg, G. and Schaller, H. (1978) Nucl. Acids Res. 5, 4495-4503.
2   van Wezenbeek, P., Hulsebos, T. and Schoenmakers, J.G.G. (personal communication).
3   van Wezenbeek, P. and Schoenmakers, J.G.G. (1979) Nucl. Acids Res. 6, 2799-2818.
4   Schaller, H., Beck, E. and Takanami, M. (1978) in The Single-Stranded DNA Phages (Denhardt, D.T., Dressler, D. and Ray, D.S., eds.), pp. 139-163, Cold Spring Harbor Laboratory, NY.
5   The Single-Stranded DNA Phages (1978) (Denhardt, D.T., Dressler, D. and Ray, D.S., eds.), Cold Spring Harbor Laboratory, NY.
6   Katz, L., Kingsbury, D.T. and Helinski, D.R. (1973) J. Bacteriol. 114, 577-591.
7   Wheeler, F.C., Benzinger, R.H. and Bujard, H. (1974) J. Virol. 14, 620-627.
8   Yamamoto, K., Alberts, B., Benzinger, R., Lawhorne, L. and Treiber, G. (1970) Virology 40, 734-744.

9   Godson, G.N. (personal communication).
10  Schreier, P.H. and Cortese, R. (1979) J. Mol. Biol. 129,
    169-172.
11  Bastia, D. (personal communication).
12  Ray, D.S. (personal communication).
13  Cohen, S.N., Chang, A.C.Y. and Hsu, C.L. (1972) Proc. Nat. Acad.
    Sci. U.S.A. 69, 2110-2114.
14  Barnes, W.M. (1977) Science 195, 393-394.
15  Herrmann, R., Neugebauer, K., Zentgraf, H. and Schaller, H.
    (1978) Mol. Gen. Genet. 159, 171-178.
16  Barnes, W.M. (unpublished data).
17  Sanger, F., Nicklen, S. and Coulson, A.R. (1977) Proc. Nat. Acad.
    Sci. U.S.A. 74, 5463-5467.
18  Sanger, F. and Coulson, A.R. (1978) FEBS Lett. 87, 107-110.
19  Barnes, W.M. (1978) Proc. Nat. Acad. Sci. U.S.A. 75, 4281-4285.
20  Barnes, W.M. (1979) Gene 5, 127-139.
21  Szybalski, W., Kubinski, H., Hradecna, Z. and Summers, W.C.
    (1971) in Methods in Enzymology (Grossman, L. and Moldave, K.,
    eds.), Vol. XXID, pp. 383-413, Academic Press, New York, NY.
22  Vinograd, J., Morris, J., Davidson, N. and Dove, W.F. (1963)
    Proc. Nat. Acad. Sci. U.S.A. 49, 12-17.
23  Maxam, A.N. and Gilbert, W. (1977) Proc. Nat. Acad. Sci. U.S.A.
    74, 560-564.
24  Smith, A.J.H. (1979) Nucl. Acids Res. 6, 831-848.
25  Roberts, R.J. et al. (personal communication).
26  Hutchison, C.A., Phillips, S., Edgell, M.H., Gillam, S., Jahnke,
    P. and Smith, M. (1978) J. Biol. Chem. 253, 6551-6560.
27  Razin, A., Hirose, T., Itakura, K. and Riggs, A.D. (1978)
    Proc. Nat. Acad. Sci. U.S.A. 75, 4268-4270.
28  Ravetch, J.V., Ohsumi, M., Model, P., Vovis, G.F., Fischoff, D.
    and Zinder, N.D. (1979) Proc. Nat. Acad. Sci. U.S.A. 76,
    2195-2198.
29  Boeke, J.D., Vovis, G.F. and Zinder, N.D. (1979) Proc. Nat.
    Acad. Sci. U.S.A. 76, 2699-2702.
30  Messing, J., Gronenborn, B., Mueller-Hill, B. and Hofschneider,
    P.H. (1977) Proc. Nat. Acad. Sci. U.S.A. 74, 3642-3646.
31  Ray, D.S. and Kook, K. (1978) Gene 4, 109-119.
32  Gronenborn, B. and Messing, J. (1978) Nature 272, 375-376.
33  Messing, J. (1979) Recombinant DNA Technical Bulletin 2, 43-48.
34  Boguski, M. and Barnes, W. (unpublished data).
35  Backman, K., Ptashne, M. and Gilbert, W. (1976) Proc. Nat. Acad.
    Sci. U.S.A. 73, 4174-4178.
36  Maniatis, T., Kee, S.G., Efstratiadis, A. and Kafatos, F.C.
    (1976) Cell 8, 163-182.
37  Sugino, A., Snopek, T.J. and Cozzarelli, N.R. (1977) J. Biol.
    Chem. 252, 1732-1738.
38  Achtman, M.N., Willets, N. and Clark, A.J. (1971) J. Bacteriol.
    106, 529-538.
39  Geider, K., Beck, E. and Schaller, H. (1978) Proc. Nat. Acad.
    Sci. U.S.A. 75, 645-649.

40   Schaller, H. (1978) Cold Spring Harbor Symp. Quant. Biol. 43,
     401–408.
41   Herrmann, R., Neugebauer, K., Zentgraf, H. and Schaller, H.
     (personal communication).
42   Zacher, A.N., Stock, C.A., Golden, J.W. and Smith, G.P. (per-
     sonal communication).
43   Hines, J. and Ray, D.S. (personal communication).
44   Kaguni, J. and Ray, D.S. (personal communication).
45   Maxam, A.N., Gilbert, W., Chapman, N., Copenhaver, G., Donis-
     Keller, H., Rosenthal, N. and Herr, W. (personal communication).
46   Barnes, W.M. and Husson, R.N. (unpublished data).
47   Sutcliff J.G. (1978) Cold Spring Harbor Symp. Quant. Biol.
     43, 77–90.
48   Staden, R. (1977) Nucl. Acids Res. 4, 4037–4051.
49   Godson, G.N., Barrell, B.G., Staden, R. and Fiddes, J.C. (1978)
     Nature 276, 236–247.

# BACTERIOPHAGE LAMBDA VECTORS FOR DNA CLONING

Bill G. Williams[1]

Cetus Corporation
600 Bancroft Way
Berkeley, California  94710

and

Frederick R. Blattner[2]

Department of Genetics
University of Wisconsin
Madison, Wisconsin  53706

The bacteriophage lambda, its family of lambdoid phages and their hybrids have provided a wealth of information about several fundamental areas of molecular biology over the last 27 years.  This is at least in part due to having been the first well-studied pieces of naturally cloned DNA in Escherichia coli.  The naturally lysogenic or integrated phase of the lambda life cycle allowed investigators to introduce and explore mutations in any of the organism's essential genes, and they observe the effects of these mutations on viral physiology.  It should not go unnoticed that the "cloned" state allows one to proceed directly to the study of mutations in essential genes, a task only otherwise approachable with conditionally lethal mutations or more complex diploid organisms.  And it is also no coincidence that the lactose operon, the most well understood operon at this time of writing, exists as a fragment of DNA cloned into the lambda genome by traditional in vivo genetic techniques (1).

The genomes of the entire family of lambdoid phages are similarly organized (2,3) (for the convenience of traditional geneticists as well as contemporary DNA cloners) so that the central one-third, shown as the replaceable region in Figure 1, contains genes that are

---

[1]Contributor of the text and drawings.
[2]Contributor of the tabulated restriction sites and cloning capacities.

entirely dispensable for lytic growth.  As a consequence of the
ability of the virus to carry foreign DNA replacing chunks of its
dispensable regions, specialized transducing phages of lambda had
been used for many years prior to the development of in vitro DNA
joining techniques to propagate foreign DNA segments of E. coli.
These transducing phages have provided a rich background of techniques for studying the selection, organization and expression of DNA
inserted into lambda.  With the emergence of recombinant DNA technology, it was a natural extension for students of lambda to adapt the
phage so that DNA from any source could be introduced into the
replaceable region of its genome.  Unfortunately, the wealth of
sophisticated detail that makes lambda so versatile as a DNA cloning
vector has often shrouded its utility from the uninitiated.  This
chapter is being written in part to provide a concise explanation of
the simple ways in which lambda can and is being used to clone DNA,
to provide an up-to-date catalog of the various genetic configura-
tions of lambda vectors now available from several laboratories
(4-18) and to present restriction maps of all the published vectors
with respect to all the restriction sites commonly used for DNA
cloning.  The maps of the EK2 certified lambda vectors are presented
in Figure 3* and all the remaining published structures are shown in
Figure 4.  The detailed restriction site positions are tabulated in
base pairs from the left end and are presented in Table 1.  The com-
puted cloning capacities of these vectors with each relevant enzyme
and enzyme combination are presented in Table 2.  The listings of
specific restriction sites have been made possible by the development
of the least-squares fit computer program of Schroeder and Blattner
(86) for restriction maps, and the application of that program to
provide a comprehensive map of lambda in Daniels et al.(87) and of
the first 21 of the Charon phage series in deWet et al.(88).  The
recent comprehensive map of lambda by Szybalski and Szybalski (76)
has also been very useful in the preparation of these figures, especially for defining genetic end points, and is highly recommended for
general reference.


THE ORGANIZATION OF THE LAMBDA GENOME: A CLONE'S EYE VIEW

      For the purpose of DNA cloning, the genetic map is most simply
considered as the circular form of Figure 1.  The promoter $p_R$ is a
benchmark, since it initiates the essential transcription to the
right that leads to expression (in regulated cascades) of DNA replication functions, the genes for cell lysis, then proceeds across the
cohesive end to express the genes for cutting and packaging the
mature phage DNA.  Transcription from $p_R$ is normally terminated at
$t_{R2}$ until gene N product accumulates to "antiterminate" and allow
transcription to proceed to the right.  The deletion nin5 (N indepen-
dence, 19) removes $t_{R2}$ and thereby enlarges the replaceable region

_____
*Figures 3 and 4 and Tables 1 and 2 will be found on pages 231ff.

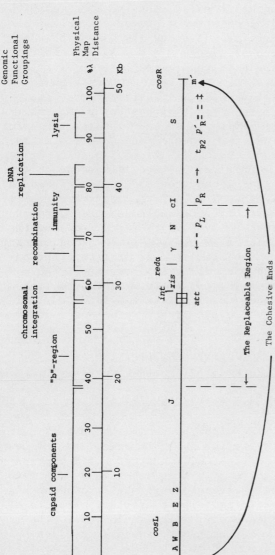

Figure 1. A simplified depiction of phage lambda including the genetic features necessary to con-sider its behavior as a cloning vehicle. The individual genes listed in the capsid components re-gion are essential genes that contain amber mutations in the different HV2 certified vectors shown in Figure 3. The replaceable region includes genetic material which is entirely dispensable for lytic phage growth (the essential gene N becomes dispensable when the transcription terminator $t_{R2}$ is deleted). This replaceable region is defined by the end of the last essential gene on the left arm, gene J, and by the promoter $P_R$ on the right arm, which initiates the rightward transcriptional cascade across all the essential gene regions. Late transcription is initiated at $P_R'$ and proceeds across the cos site into the capsid component genes. The cohesive ends m and m' are complementary 12 base-pair single strand tails in the mature phage which hybridize to circularize the genome upon infection, and can be efficiently annealed in vitro. m and m' are formed by the restriction enzyme-like cleavage of the replicated DNA at the recognition sequences cosL and cosR. The relevant uses for cloning of the various genes and DNA sites within the replaceable region are discussed in the text. (See ref. 76 for a comprehensive map of lambda correlating the physical, genetic and restric-tion maps of bacteriophage lambda.)

to include gene $\underline{N}$.  The promoter $p_R$ defines one end of the replaceable region of the genome since any piece of DNA inserted to the
right of $p_R$ could interrupt the pattern of transcription necessary
to replicate and package the vector.  The other end of the replaceable region is defined by the end of gene J, the last essential gene
in the "heads and tails" operon, known to be close to, but less than,
39.5% of the wild-type genome length to the right of the cohesive
end.  The replaceable region therefore includes the DNA to the left
of $p_R$ involved with the formation of lysogens and phage recombination.  Gene $\underline{cI}$ codes for the repressor that keeps both major phage
promoters $p_R$ and $p_L$ shut off when the phage DNA is integrated in
the $\underline{E. \ coli}$ chromosome.  Promoter $p_L$ drives transcription to the
left across gene $\underline{N}$ and the phage recombination systems gamma and $\underline{red}$
(which are dispensable for growth if the host supplies the recA pro-
duct), then across the phage genes for integration ($\underline{int}$) and excision
($\underline{xis}$) of the phage genome into and out of its $\underline{E. \ coli}$ host.  The attachment site ($\underline{att}$) of the phage provides the specific sequence to
direct the recombination event that integrates the lambda genome into
the $\underline{E. \ coli}$ chromosome at a corresponding bacterial attachment site.
The DNA to the left of the $\underline{att}$ site is referred to as the b2 region
(after the name of a deletion that removes it) and need only be of
concern to the extent that it does carry several promoters directing
transcription in both directions, and terminators of transcription
(E. Rosenvald, personal communication, 73).  In short then, the
working part of a lambda vector is the fusion of a right end that
must include $p_R$ and a left end that must include the head and tail
genes A-J, fused through the cohesive end ($\underline{cos}$).

## PHAGE PACKAGING AND DNA CLONING APPLICATIONS

### A Requirement for 38.5 to 52.0 kb of DNA

One clear advantage to working with phages is that DNA is deliv-
ered to the user in packages easy to purify and handle, and may be
concentrated to a high level.  Routine CsCl banding of phage preparations yields DNA concentrations of 0.5 to 4 mg/ml.  These have been
shown to be remarkably stable over several years when stored as the
CsCl band at refrigerator temperatures.  There are three aspects of
packaging relevant to cloning technology, which include: 1) the
so-called "size selection" for ensuring that plaques in a given
experiment result from the inclusion of foreign DNA, 2) the availability of an $\underline{in \ vitro}$ packaging system (78-80), and 3) the emergence
of hybrid plasmids, called cosmids by Collins and Hohn (20), containing the lambda cohesive end, which allows them to be packaged inside
phage particles if they meet the appropriate length requirements.
      The requirements for a DNA molecule to be efficiently packaged
$\underline{in \ vivo}$ are to contain both the left and right cohesive ends of

lambda (shared by all the members of the lambdoid phages) separated by from 78% to 105% of the wild-type genome length.

There is a precipitous decrease in viability in phages containing less than 78% of the genome.  A recombinant phage can be forced by selection to be 76.5% of the genome, but will grow very poorly and will soon be overgrown by phages containing random duplications that supply the additional DNA length necessary for efficient packaging. Similarly, one will rarely find recombinant phages with genomes larger than 105% lambda wild-type length.  The fact that there is both an upper and lower limit on the size of a phage genome means that either different vector designs or different vector-enzyme combinations must be used to clone the smallest and the largest possible DNA fragments.  Given that about 60% of the phage genome is necessary for lytic growth, then a cloning vector designed to accommodate the largest possible DNA fragment would have restriction cuts at the extreme left and right ends of the replaceable region.  The fused left and right end vector fragments then would contain 60% of the genome length (29.6 kb as a theoretical minimum) and would have a requirement for an included fragment of at least 8.9 kb, which would be up to 22.4 kb.  On the other hand, a vector used for cloning a very short DNA segment must produce a left and right end fusion fragment of at least 38.5 kb after digestion with the restriction enzyme of choice.


## The Size Selection Concept

When the left and right end fusion fragment of a vector is itself too short to package after digestion with restriction endonuclease, there is an absolute requirement for the incorporation of additional DNA to produce viable phage and plaques.  But the digestion mix will also contain the extraneous fragment or fragments that were cut out of the vector.  These internal stuffer fragments must have been retained to have propagated the vector.  They will be present in equimolar amounts to the vector, and in some cases it will not be worthwhile to try to remove them by biochemical separation techniques.  Even in those cases where the stuffer fragments are removed physically, the level of residual contamination will likely be on the order of 1%.  This limits the power to select for rare cloned inserts because of the background of reconstituted vectors.  (An enhancement of $10^2$ will be very helpful in some experiments, but compare this with the efficiency of a routine genetic selection in lambda where a recombinant may easily be selected from among $10^6$ to $10^8$ parental types.)  Two principles emerge from consideration of the problems of vectors reincorporating their stuffer fragments instead of target DNA fragments to form plaques.  First, stuffer fragments that can be genetically selected against, or that can be quickly and easily screened for, simplify the task of finding a rare cloned insert. Second, the number of stuffer fragments produced by restriction enzyme digestion should be minimized (optimally to one) since they

compete with target DNA for the vector in the ligation step.
Furthermore, the number of different stuffer fragments produced on
restriction of the vector DNA determines the number of tests necessary positively to identify a phage carrying a cloned insert by
genetic criteria.  With those vectors for which the "size selection"
applies, the presence of a plaque means that the vector contains DNA
included between its flanking restriction sites.  If there had been
only one kind of stuffer fragment removed by restriction (for example, a vector like Charon 10 containing the lac operator), then a
single test to show that phage from a single plaque of a Charon 10
EcoRI cloning experiment did not contain any lac operator would be
sufficient positively to identify that phage as containing cloned
foreign DNA.  By contrast, an EcoRI cloning experiment using Charon 9
produces three different stuffer fragments, requiring three different
genetic tests to eliminate the possibility that any plaque is the
result of the reincorporation of vector DNA.

     If a great reduction in the efficiency of the cloning process
can be tolerated in order to get the desired clone, then the fore-
going problem of vector stuffer fragment reincorporation can be dealt
with by flooding the ligation reaction with target DNA.  This will
drive most molecules into oligomers of target DNA.  Although most
vector fusion fragments will initially undergo reaction with target
fragments, those dimers will principally be consumed in the formation
of target oligomers favored by high target DNA concentration.  The
problems that may result from ligation with high target DNA concentration include lower efficiency and complex clones containing multiple cloned fragments and deletions of either vector or target DNA.

     In circumstances where vector-stuffer fragments reconstitution
does pose a serious problem, the technique of treating the restricted
vector DNA with phosphatase previous to the ligation step (21) ought
to provide a substantial decrease in reconstituted vectors.  The
removal of 5' phosphate from both vector and stuffer fragments blocks
their ligation and therefore effectively selects them out of the
transfecting population.  This helps to select the desired product
against a lowered background of reconstituted vectors.  It may be
possible to increase the efficiency of constructing target-vector
combinations by this method, too.  If a vector stuffer fragment hybridizes its restricted sticky end to the vector sticky end and this
association is stable throughout the ligation reaction, then those
vector molecules will have been competed away from target fragments.
If the ligation reaction is carried out at a temperature far above
the Tm of the sticky end, however, for example 15°C for EcoRI, then
the DNA ligase would be expected to fix the desired transient vector-
target fragment hybrids irreversibly.  Whether or not a protocol
using a constant high-temperature ligation would be more effective
than occasionally raising the temperature to melt out the unwanted
hybrids and returning the ligation mix to a low temperature has not
yet been determined.  Whenever the bacterial alkaline phosphatase
(BAP) procedure is used, it must be recalled that BAP is extra-

ordinarily resistant to the diethyl pyrocarbonate used by Blattner et
al. (ref. 16, footnote 17), to inactivate other nucleic acid enzymes
used sequentially (so much so that this resistance is used in the BAP
purification cited by Efstradiatis et al. (21)).

CONSIDERATIONS OF MONOMOLECULAR CYCLIZATION:<br>A CONCEPTUAL BASIS FOR PROTOCOL DESIGN

In those cases where the sticky ends generated by a single restriction enzyme are used to seal target DNA fragments into vectors,
the possibility exists for both the target fragments and the vector
to undergo a monomolecular cyclization. This cyclization removes
them from the pool of molecules participating in the bimolecular
cloning reaction, and can in some cases be crucially important. The
rate of molecular self-closure is independent of DNA concentration,
and depends on molecular length. As long as the DNA molecules are
behaving like freely-jointed polymers, the ends of short molecules
tend to find each other much more rapidly than the ends of long molecules, since they are constrained by molecular length to occupy a
much smaller sphere of volume. In order to ensure that an appreci-
able fraction of DNA molecules participate in the desired bimolecular
reaction, one must ensure that a given molecular end collides with
the ends of different molecules more frequently than it collides with
its own other end. This problem can be approached using the concept
of the apparent concentration of one molecular end in the neighborhood of its own other end. The probability density, j, for one end
of a random coil to be lying in the neighborhood of the other is
expressed by a simplification of the well-known (23) Jacobson-
Stockmayer expression (24) as $j = [h/L]^{3/2}$, where h is a constant
and L is molecular length. Wang and Davidson (25) measured j using a
well-characterized lambda DNA, which allows the determination of the
constant in the above formula and the evaluation of j for any L.
Here, the quantity j is used as a concentration term (simply discussed in Davidson and Szybalski (23)) and the "j concentration" refers
to that concentration of a given DNA fragment at which the theory
predicts that the initial rate of reaction is equal for bimolecular
joining and monomolecular cyclization.

This formulation leads to a very simple approach to determining
the optimum DNA concentrations for the ligation of target DNA fragments into vectors. From the curve of Figure 2, one can determine
the mass concentration of vector DNA needed to compete effectively
with the more rapid cyclization of shorter target fragments. Then one
can calculate the mass concentration of target DNA to aim for in the
ligation step by applying its fractional molecular weight of the
vector to the vector concentration called for by the figure, i.e.,

[Target] μg/ml = (MW Target/MW Vector) [Vector] μg/ml .

These values arise from the principles that the <u>length of the short-
est target fragment must determine the molar concentration of vector</u>

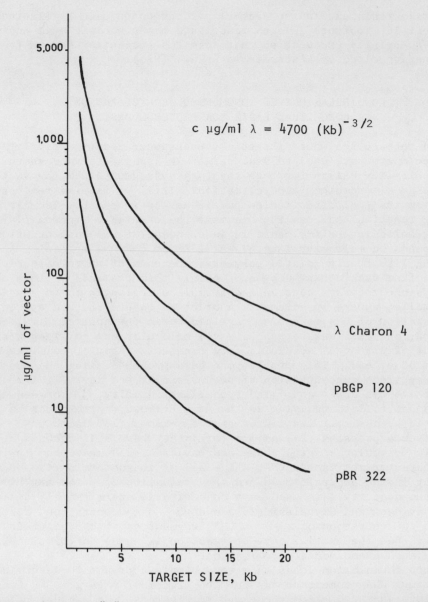

Figure 2. The "j" concentration of vector DNA versus target fragment size. This is the minimum vector DNA concentration in a ligation mix necessary for the initial rate of joining of target and vector to compete one-to-one with the rate of monomolecular cyclization of a target fragment of given length. The concentration of lambda DNA in µg/ml has been calculated by substituting the measured value of j determined by Wang and Davidson (25) into the Jacobson-Stockmayer expression (24) $j = (3/2\pi Lb)^{3/2}$. The mass concentrations of pBGP120 and pBR322 were calculated to yield the same molar j concentration as lambda DNA for a given target fragment length.

ends necessary to compete with the loss of target due to monomolecular cyclization, and that the most efficient cloning reaction will
result from molar ratios of about one-to-one target-to-vector fragments.  Figure 2 relates the target fragment length to lambda vector
concentration in $\mu$g/ml.  This relationship holds for any lytic lambda
vector, and is simply extended to other plasmid vectors.  Since the
important concentration is the molar concentration of vector ends for
a given target fragment length, the mass concentration of any other
vector can be determined by applying its fractional molecular weight
of lambda ($31 \times 10^6$) (23) to the mass concentration given on the
ordinate of Figure 2.  Vectors pBR322 (MW = $2.6 \times 10^6$) and pBGP120
(MW = $11.5 \times 10^6$) have been included in Figure 2 as examples.

The model underlying the theoretical curve of Figure 2 breaks
down in the region of very short DNA fragments.  In the lower limit,
the length of the DNA fragment will not be long enough to allow the
bending necessary for one end to join the other.  In the most comprehensive documentation published to date, Dugaiczyk et al. (28) note
the inability to form circular monomers of a 255 base-pair fragment
using E. coli DNA ligase.  However, R. Davis (personal communication)
reports having observed circles of various sizes down to the limits
of resolution in the electron microscope; with T4 DNA ligase at high
concentration (blunt-end joining conditions), we have observed the
circularization of a 123 base-pair fragment (unpublished data) in the
electron microscope and C. Bahl (personal communication) has observed
apparent circularization of dimers and higher oligomers of a synthetic 21-mer, as judged by its altered gel mobility and the inability
of BAP to remove $^{32}$P introduced as 5' terminal label of the linear
fragment.  From the data (28), it is clear that the kinetics of circularization do indeed fall off as one approaches short DNA lengths.
The take-home lesson is that fragments up to about 200 base pairs,
and those larger than 1 kb, are quite reasonable to clone with a
single enzyme and competing out-target circularization with high
vector concentration.  Although no systematic study has been done,
the experience of several laboratories has confirmed that the predic-
tion of the Jacobson-Stockmayer relationship results in acceptable
molecular cloning efficiencies for target fragments in the larger
than 1 kb size range (F. Blattner and B. Williams, unpublished data;
R. Davis, personal communication; 29).  The optimal conditions
reported (29) correspond to phage vector ends at 2 x "j", and target
fragment at 1.7 x the molar concentration of vector.

Lambda vector DNA preparations can be made directly from phage
(51) to yield DNA at 4 mg/ml (although 0.5 to 1.0 mg/ml is more
routine), which defines an upper limit to the approach of high vector
concentration.  Several other options present themselves for the
cloning of troublesomely small DNA fragments.  The well-established
tailing procedure (77) using terminal transferase avoids the compli-
cation of cyclization by extending target and vector fragments with
different complementary homopolymers.  The problem here is the necessity to provide vector DNA free of nicks that are templates for
branch formation.  This problem is accentuated in phage cloning since

the vectors tend to be much longer than plasmids, with proportionally greater chances to contain nicked genomes. It is likely that, on the average, a single branch per phage genome is lethal (H. Faber, personal communication). This is not an insurmountable problem but does require that the vector DNA be examined for integrity on a denaturing gel previous to the tailing procedure. It has been found that withdrawing phage bands from either the top or the side of one step gradient and two subsequent CsCl equilibrium gradients vastly improves the single-stranded integrity of the resulting DNA over dripping the phage bands and running only two gradients. The other major drawback to the use of molecular tailing is the difficulty in recovering the cloned fragment cleanly from the vector. The procedure of Goff and Berg (30) demonstrated the ability to excise a poly(dA)-poly(dT) cloned fragment from the plasmid pMB9 and reinsert it into a second subsequent vector using the original conserved poly(dT) tails. This procedure relies on the production of single-stranded snap-back structures produced by the intramolecular pairing of the poly(dA)-poly(dT) tracts flanking the insert that protect the insert from subsequent digestion by ExoVII. It is likely that this method will prove useful in work with lambda phage vectors as well as in poly(dG)-poly(dC) tailing experiments.

A second option for avoiding circularization is to prepare both target and vector with two different restriction enzymes to produce heterologous ends that cannot join. This approach immediately eliminates the possibility of using any selections or screens based on insertional inactivation. It must also be kept in mind that vector digestion with two enzymes will always produce at least one internal vector fragment that will be competing with target for vector joining. Since the simplest approach to this problem is to add sufficient target fragments to outcompete vector fragments, it must also be recalled that too high a target fragment concentration will complicate the results.

A third option is to use alkaline phosphatase treatment of the target fragment. The choice of whether to treat the target or vector with phosphatase depends on where the experimental difficulty lies. Treating the target will increase the efficiency of ligating target and vector, whereas treating the vector will provide an efficient selection for what may be a very rare ligation event.

## SELECTIONS AND SCREENS AVAILABLE IN LAMBDA VECTORS

A relatively few genes in lambda have been utilized as the basis for selections or screens in molecular cloning experiments. The term "selection" will be reserved for those manipulations in which only the desired products, i.e. vectors carrying cloned fragments, form plaques. The size selection alluded to earlier could provide a strong selection for the replacement of vector stuffer fragments by large target fragments if, for example, the digested vector DNA were phosphatase treated prior to ligation with restricted target. A

vector of substantially increased size can also be selected on a pel⁻
host (31,34,35).

This opens the possibility of inserting (versus substituting) a
moderate to large target fragment into a vector itself too short to
plate on pel⁻. This system was used to good advantage by Cameron et
al. (32), to select populations of λgt1 or λgt2 containing EcoRI-
cleaved fragments of yeast DNA in different size classes. The size
class of inserts that can be selected using this technique can be
adjusted according to the genome size of the vector used. Philippsen
et al. (32) point out that one can avoid having to amplify the
ligated DNA population through a rk⁻ mk⁺ host by using the rk⁻ mk⁻
pel⁻ strains of Scandella and Arber (34,35).

Resistance to chelating agents like EDTA, citrate and pyrophosphate (36) provides a complementary selection for vectors that have
undergone a net deletion by replacing a large stuffer fragment with a
smaller cloned piece (12). This will be most effective when the
vector fusion fragment alone is too short to plate. In the case that
the vector contains a single stuffer fragment and falls near either
the minimum or maximum of the included DNA size range, fractionation
on a cesium chloride equilibrium gradient may be used either to
detect the size distribution of clones or select vectors containing
cloned fragments of different length than the stuffer fragment. (The
density of the phage particle containing the full-length genome in
CsCl is 1.5080, while that of a minimum-length phage deleted for
∿10.9 kb would be 1.4774.) The resolving power of the CsCl gradient
used as a physical selection was demonstrated by Philippsen et al.
(33) to provide a 30-fold enrichment of vectors containing a large
cloned fragment from a pool of vectors ligated with HindIII fragments
of yeast DNA (see ref. 37). Sternberg et al. (66) also show that an
equilibrium CsCl gradient readily fractionates a population of phages
composed of greater than 85% hybrids, selected by the bias of their
in vitro packaging system for full-length genomes, on the basis of
the size of the cloned insert.


## THE Spi PHENOTYPE AS A POSITIVE SELECTION FOR CLONED INSERTS

In contrast to selection based on size, a unique positive genetic selection available in lambda is based on the "Spi" phenotype
(sensitivity to ϕP2 inhibition (38)). Wild-type lambda cannot plate
on a lysogen of ϕP2 if it contains either red and gam functions
located between 65 and 68% on the physical lambda map. Therefore,
derivatives that have replaced the DNA in this region with a foreign
cloned fragment will plate. This selection works for lambda vectors
with hybrid immunity regions from ϕ434 and ϕ21, but not with immunity
of ϕ80. It has been observed by several workers that many red⁻
phages grow poorly (33,39). This must be kept in mind if any amplification step on the population of ligated molecules is anticipated.
Since reconstituted vectors might have a growth advantage over vectors containing cloned fragments, then clones could be underrepre-

sented or lost from a replicating pool. This should not pose a
problem when picking or screening individual plaques formed by
plating either transfected spheroplasts or in vitro packaged molecules directly on a lawn of φP2 lysogens.

   It should be noted that not all red⁻ phages grow poorly. All
the bio phages of the Charon series grow to usable titers, even
though they are red⁻, whether or not they are gam⁻. Even the most
fragile grower of the series, Charon 4A, routinely produces titers of
about $10^{10}$ when propagated in liquid culture by the preadsorb dilute
shake (PDS) method (16) and has occasionally grown to $10^{11}$. The fact
that EcoRI clones replacing the lac and bio containing stuffer fragments of Charon 4 have themselves been observed to grow at least as
well, and often better, than the parent vector, means that whichever
mutations have overcome the red⁻ growth disadvantage (40,41) are not
confined to stuffer fragments. The fact that vectors like Charons 4
and 4A are not themselves powerful growers and that derivations containing cloned fragments grow as well as or better than the parent
vector, also make them good candidates for vectors in shotgun experiments where it is desirable to amplify the library of clones provided
from the original packaging or transfection. The addition of a known
chi mutation in the essential region of a vector has demonstrated the
feasibility to using this selective phenotype in DNA cloning on a
large replacement vector (81). A unique selection for inserts has
become available with the vector λgtWES·T5622 constructed by Davison
et al. (89). By replacing the λB stuffer fragment of the λgt vector
with two identical 1.8 kb fragments from phage T5, the vector cannot
be propagated on an E. coli host carrying plasmid ColIb. Such a host
selects against the T5 gene A3 carried on each stuffer fragment and
therefore the size selection ensures that virtually all plaques produced from ligations of this restricted vector and target DNA will
result from vectors carrying cloned inserts.


                SCREENS TO INDICATE SUCCESSFUL CLONING


                        Visual Screens

   The genetic screens available as indicators of successful cloning fall into two classes. The visual screens are those that allow
one to tell by direct inspection of the experimental plates whether
or not the plaques are made from vectors containing cloned fragments. These now include the blue versus colorless lac screen, the
turbid versus clear cI insertion screen, and the bio and ara complementation screens. The other category of screens involves the genetic testing of individual plaques, usually by picking and testing for
the ability to replicate on subsequent hosts. When one is looking
for indications of success at cloning any target versus looking for
rare targets, these genetic screens can be quite an adequate and
trivially easy way to identify clones. The simplest of the visual

screens is the clear versus turbid plaque morphology that was shown
by N. Murray (12) to discriminate reconstituted or uncleaved vectors
(turbid plaques) from those carrying cloned fragments inserted into
either of the single EcoRI or HindIII sites (clear plaques) in the
immunity region substitution of φ434.  Several vectors among the
Murray series and the Charon vectors offer this screen and Charon 7
has the additional capability to accommodate heterologously
terminated EcoRI/HindIII fragments (to avoid the problem of monomolecular cyclization).  The vectors carrying cloned fragments in the
repressor gene also enjoy the substantial growth advantage over the
native vectors usually seen in clear plaque-forming mutants.  Other
relevant features of vectors carrying fragments cloned into the
repressor gene are that the vector would provide transcription of the
cloned fragment from the right and that, in the absence of a cloned
terminator, one would expect a polypeptide coded for by the cloned
fragment to be fused with the repressor itself.  The DNA sequence to
the right of the φ434 cI EcoRI site has been published by Grosschedl
and Schwarz (75) defining the frame and amino acid sequence of the cI
repressor up to the point of insertion with EcoRI.  The absence of
functional repressor need not eliminate the possibility of lysogenizing the vector into an E. coli host, since one could infect (or
transfect) directly into a preformed lysogen containing a prophage of
φ434 immunity.  Insertion of a DNA sequence into the cI gene has the
additional feature that infection of UV-treated homoimmune lysogens
will produce the hybrid gene product virtually exclusively.  This
provides a very convenient assay for fusion proteins by polyacrylamide gel electrophoresis of the labeled products (52).

The visual screen provided by the lac5 substitution really has
two components.  The insertion of a segment of foreign DNA into
either the EcoRI SstI site within the coding sequence of β-galactosidase (lacZ) eliminates the enzyme's ability to convert colorless
5-bromo 4-chloro 3-indolyl-β-D-galactoside (XG, Bachem Inc.) into a
deep blue dye.  Thus, a vector containing the intact lac5 substitution produces blue plaques with a large blue halo on a host strain
that is either proficient for or entirely deleted for the lac operon,
while the same vector carrying an inserted DNA sequence in either the
EcoRI or SstI lac site will produce a colorless plaque when plated on
a fully lac deleted host.  The physical removal of the lac5 DNA,
which includes the lac operator, also results in a colorless plaque
when the phage is plated on either a Lac$^+$ or Lac$^-$ host.  However, the
presence of the functional lac operator in a vector that contains a
cloned fragment in the lac5 EcoRI or Sst site is sufficient to bind a
Lac$^+$ host cell's lac repressor and induce synthesis of the host's own
β-galactosidase (42–44).  This produces a distinct plaque appearance
on plates containing XG, which is a blue-colored plaque without the
strong blue halo generated by a lacZ$^+$ phage.  Since transfection
plates tend to produce plaques of very different sizes and, therefore
different intensities of blueness, the discrimination between vectors
with and without insertions in lac5 can be enhanced by the use of a
lac deleted host, thereby producing only blue (intact uninterrupted

vector) or colorless (vectors carrying cloned fragments) plaques.
Using the blue color on XG plates can also be useful visually to
identify vectors containing cloned fragments that have replaced a
single stuffer fragment carrying the lac operator as opposed to
inserting into a single site in lac. Since a vector like Charon 10
contains only a single such EcoRI stuffer fragment and is too short
to form plaques without the inclusion of DNA between its EcoRI sites,
then a colorless plaque from an EcoRI cloning experiment plated on a
Lac[+] host would necessarily contain a cloned fragment of foreign DNA.
Several similar vectors (12) employ the lac color test by including
suppressor genes supE or supF in the vector as single stuffer fragments, which function to suppress host lac amber mutations. So the
take-home lesson for the use of the lac-XG color screen is to plate
vectors using EcoRI or SstI insertion into lac5 on lac deleted hosts
and plate those that will replace their lac operator-containing fragment by cloned DNA on Lac[+] host cells.

Two other advantages that accrue to using insertion in lac5 are
that the cloned fragment will be under the transcriptional control of
the strong lac promoter, which can be sensitively controlled to give
high levels of transcription from the right, and the translated polypeptide would be protected from intracellular degradation by fusion
with β-galactosidase. These features have been best demonstrated by
cloning the λplac5 EcoRI fragment, containing all of the lac5 substitution upstream of the EcoRI site in lacZ, into the plasmid pBG120
(45). It was virtually this same plasmid design that was used to
achieve both expression and protection of the cloned synthetic gene
for somatostatin (46).

Another visual screen that is available with several Charon
vectors is the Bio test (82). Both the Bio substitutions bio1 and
bio256 can complement a host bioA deficiency. Therefore, plating
these vectors with bioA hosts on plates limiting in biotin produces a
heavy ring of cell growth around plaques formed by vectors retaining
the Bio DNA and no additional growth around those vectors in which
the Bio DNA has been replaced by foreign cloned DNA (47). The bio256
EcoRI fragment contains all the genes of the operon and therefore
produces biotin on hosts completely deleted for the operon. This
expression is independent of the EcoRI fragment orientation in Charon
10 and would also be expected to be independent in Charon 4 since the
Bio promoters are internal to the bio1 and bio256 substitutions
(48-50). An advantage to cloning into the EcoRI site within bio256
(Charons 4, 4A, 10 and 11) is that transcription will be provided
from the right by the strong lambda promoter $p_L$ (51,52).

A 5.7 kb EcoRI fragment carrying the araB, araA and araD genes
has been utilized as a single stuffer fragment in several vectors
from the laboratory of R. Davis (17; Figure 4). These allow
reconstituted vectors to be discriminated from those carrying foreign
cloned fragments by a color test using Ara[−] host cells on MacConkey
plates supplemented with arabinose. In the deleted vectors λgt6, 7
and 8, the size selection also applies for EcoRI cloning so

that any Ara⁻ plaque is positively identified as carrying a cloned
fragment.

## Replica Screens

The fact that insertion into sRIλ3 has been shown to inactivate
the red recombination function has been used as the basis for a
genetic screen since red⁻ phages fail to plate on polA hosts (4).
This test requires that plaques be individually picked and tested for
growth on polA⁺ and polA⁻ hosts but the level of resolution of such a
screen is easily within $10^{-3}$.
There are mutations available over most regions of the genome
that can be used as the basis for qualitative complementation or
marker rescue tests.  If the size selection applied to the particular
restriction enzyme and vector combination used, then the genetic
demonstration by complementation tests that vector internal fragments
are missing indicate that the vector must contain foreign cloned DNA.
The red⁻ phenotype of failure to plate on polA can therefore also be
used to demonstrate the replacement of a vector fragment carrying a
redα gene fragment by cloned DNA.  The red (here a color, not a
gene!) plaque assay of Enquist and Weisberg (53) depends on expression of the lambda genes int and xis and has been shown to identify
vectors carrying substituted cloned DNA for the λC fragment (6).
These clones lacked int and xis and therefore could not help out a
cryptic prophage inserted into galT.  Prophage excision produces red
plaques due to reverted Gal⁺ lysogenic host cells.  The EMBO test
employed in Thomas et al. (6) was yet another color test that is
sensitive to the presence of the λC fragment in either orientation
and also tests plaques for the presence of lysogens within them.  The
EMBO test utilizes eosin and methylene blue plates with no sugar
overlaid with $10^9$ λb2cI⁻ phage, spotted with cells from each plaque
to be tested and incubated at 32°C (83).

## SCREENING BY RESTRICTION ANALYSIS

It is possible to screen moderate numbers of isolated candidates
by restriction enzyme digestion and gel electrophoresis of DNA from
as little as 2 ml of lysates grown from single plaques.  This method
is detailed in footnote 29 of Blattner et al. (54) and the phage from
a single plaque can be used to produce volumes of lysate from 100 μl
to one liter by the PDS lytic growth method (51).  The method of DNA
preparation directly from liquid lysates was modified from the original method by Cameron et al (32), which used plate lysates.  Both
methods take advantage of the substantial levels of unpackaged phage
DNA in a lysate.  Treatment of these samples by the addition of
diethyl pyrocarbaonate, or the more stable dimethyl dicarbonate,
immediately after isolation helps to relieve the problem of DNA
degradation during the restriction digests, since such digests are

often contaminated with nucleases.  Individual isolates from shotgun
cloning experiments have been sensitively screened by cleavage with
enzymes that cut conveniently on both sides of the points of cloned
DNA insertion so that any change in the mobility of that fragment
indicates the presence of cloned DNA.  This approach has been used
with Charon 3A (16,54).  It has the advantage of being able to detect
small inserts and does not require the integrity of the original
enzyme sites used in cloning.  Shotgun cloning techniques that do
preserve the sites used for cloning allow the entire population of
phage produced to be screened in a single digest and directly demonstrate the size distribution of the cloned fragments (55,63).

## SELECTING COMPLEMENTING GENETIC ACTIVITIES

     It was the history of lambda as an established transducing phage
that originally made it so attractive as a vehicle for recombinant
DNA cloning.  Several groups have used the wealth of knowledge of E.
coli genetics to select vectors carrying DNA fragments that could
complement the growth of hosts carrying defects in genes of interest.
Cameron et al.(55) used the inability of $\lambda red^-$ to grow on E. coli
ligts7 unless it carries a gene that supplies either DNA ligase or a
recombination function to select for the E. coli EcoRI fragment
carrying the overproducing lop-11 $lig^+$.  Borck et al. (10) used
lambda vectors for EcoRI and HindIII shotgun cloning of E. coli DNA.
Plaque-forming phages carrying a wild-type allele for several auxotrophic markers were selected by looking for halos of bacterial
growth around what would have otherwise been invisible plaques of the
appropriate auxotroph spread on plates limiting in the required
nutrient (56).  This technique also allowed the isolation of phages
carrying supD, supE and supF by demanding suppression of lacZ or trpE
amber mutations with the appropriate hosts and minimal plates.  This
set of selections additionally showed that a catabolic enzyme activity (tna) can complement a synthetic defect (trpB) when glycerol is
used as a carbon source to avoid catabolite repression.  A selection
for E. coli K modification genes hsdS and hsdM was provided by alter-
nately growing phages from the EcoRI shotgun bank on a nonmodifying
strain and a restricting strain.  The isolates from this HindIII bank
gave the first direct genetic evidence that integration into the
sHindIII3 site produces phages proficient for lysogeny.  These phages
could form lysogens efficiently upon infection, to provide further
characterization of the cloned gene product.  An alternative strategy
for lysogenization included infection of lytic phages into preformed
lysogens of like immunity (efficiency about 1% (9)).  This approach
will be generally useful for clones made in $cI^-$ and/or $att^-$, $int^-$
vectors.  Any clones made in the EK2 phage vectors, except $\lambda vir$,
which could be handled in EK1 conditions, could be efficiently
lysogenized with this technique.
     A powerful additional advantage to lysogenization is the option
to utilize the homology provided by the cloned fraagment to direct

the integration.  The only prerequisites here are that the vector or host lack the att site and that the vector provide lysogenization machinery cI, int and xis (57,58).  This should allow the isolation of transducing phages carrying sequences flanking the cloned fragment due to aberrant excision events.  The clones obtained through the above selections also established that fragments inserted in sHindIII3 are under the transcriptional control of $\lambda_{P_L}$ and that DNA from other prokaryotes, including Gram-positive bacteria, can be used with these techniques.

## SELECTING EUKARYOTIC GENE EXPRESSION IN E. coli

Using $\lambda$gt and unfractionated yeast DNA, Struhl et al. (9) selected colonies that could grow in the absence of histidine by coinfecting a nonreverting His$^-$ E. coli host with phages from the pool of $\lambda$gt-yeast hybrids and an integration helper phage.  Subsequently, Struhl and Davis (17) demonstrated that the cloned fragment did not simply suppress the original hisB463 lesion used to select it, but coded for an active imidazole glycerolphosphate (IGP) dehydratase in an E. coli host deleted for the entire his operon.  Furthermore, the expression of IGP dehydratase from the cloned fragment was shown to depend on a functional his3 locus in the yeast DNA donor.  Homologous fragments of yeast DNA from two haploid strains isogenic to the original donor strain, except carrying separate lesions in his3, were cloned and selected by the plaque filter hybridization technique of Benton and Davis (59).  The fragments, each carrying a separate mutant allele, were shown to be indistinguishable from the original donor fragment by restriction enzyme and DNA heteroduplex analysis.  Upon testing, neither cloned fragment from either of the his3$^-$ yeast strains could complement the E. coli hisB463.  A three-factor phage cross between the two his$^-$ clones produced a phage pool containing recombinants capable of lysogenizing the His$^-$ E. coli strain to His$^+$.  Analysis of the selected recombinants also allowed the order of the two his3 yeast lesions to be mapped.  The "yeast-in-coli" IGP dehydratase activity showed a pH optimum clearly distinguishable from the E. coli activity and parallel to the yeast enzyme, and demonstrated other enzymological parameters similar but not identical to the yeast enzyme.  Since the expression of the cloned fragment was independent of the orientation with which it was inserted into the phage vector and expression occurred in the absence of known lambda transcription, it is likely that the cloned fragment carried its own promoter.  This is further substantiated by the fact that the complementation also occurred when the cloned fragment was transferred into three different plasmid vectors.  At least in one case, the IGP dehydratase activity was not directly selected for, but selected and maintained by the tetracycline resistance encoded by the plasmid vector.  Even in this case, the level of enzyme activity found in the extract of the his deleted E. coli host carrying the cloned fragment was comparable to the level

found in His$^+$ $\underline{E}$. $\underline{coli}$.  This demonstrated that the transcription and
translation of the eukaryotic DNA fragment in $\underline{E}$. $\underline{coli}$ is occurring at
reasonably normal efficiencies and we are not seeing the result of
having selected for highly inefficient and unusual activities of
either the transcriptional or translational apparatus.

## LEVELS OF EXPRESSION OF CLONED DNA

It is possible to obtain moderate to massive levels of expression of cloned gene products using lambda cloning vectors, independent of whether or not the cloned gene carries its own promoter.
Lambda cloning vectors available to date offer control and amplification of cloned fragment expression both by supplying transcription
from phage promoters and by increasing gene copy number when the
vector carrying the cloned fragment is replicated.  By following
protein synthesis after the infection of U.V.-irradiated host cells
with different clones, Jaskunas et al. (52) were able to demonstrate
the expression of the ribosomal proteins encoded by the cloned DNA
fragments.  Having identified the presence of the individual protein
genes on different phages, each phage was assayed for expression of
the protein when its transducing phage was infected into a U.V.-
irradiated lysogen.  This shut down transcription from the phage
promoters $p_L$ and $p_R$ and revealed whether or not the cloned gene
was being expressed by its own or a phage promoter.  Several of these
ribosomal protein genes cloned without their own promoters into the
Charon vectors 3 and 4 were shown to be expressed under the exclusive
transcriptional control of either $p_L$ or $p_R$ (51,52).  In order to
assess the quantitative levels of expression possible for genes
cloned into lambda vectors, Moir and Brammar (60) investigated the
expression of the genes for tryptophan synthesis present on several
specialized transducing phages.  These workers investigated expression from infection of host cells rather than induction of lysogens.
In the case where DNA replication was blocked and the infected cells
were incubated in the presence of low levels of L-tryptophan, it was
shown that anthranilate synthetase activity increased steadily for at
least 20 min to moderate levels (0.14 units/mg protein).  This was a
model for the levels one might expect in an integrated prophage,
since it was shown to be independent of both replication and transcription from $p_L$.  To take maximum advantage of the increased gene
dosage available when phage DNA replicates, phage mutations in gene
$\underline{Q}$, a positive regulatory gene for lambda late genes, and genes $\underline{S}$ and
$\underline{R}$ for cell lysis, were tested singly and in combination.  The spec-
tacular results with gene $\underline{Q}$ amber mutants, singly or in combination
with $\underline{S}$ amber7, were that they produce >60 units/mg protein when
assayed 2 hr post infection.  SDS gel analyses of crude extracts
of these cells confirm that approximately 50% of the soluble
protein in the crude extracts, and an even greater percentage of
the protein synthesized during the second hour of infection,
was due to the five cloned $\underline{trp}$ genes.  This is nearly 500 times

the level observed in the absence of replication. In addition, this
phage was Red$^-$ and therefore not replicating maximally. Similar high
level results were obtained using N$^-$ (nin$^+$) phages, which are somewhat simpler to make and grow when the block in N can be a double
amber mutation.  The principle also emerges here that any negatively
controlled gene is susceptible to this kind of amplification, while
expression of positively controlled genes will be limited by the
availability of the positive effector.

        Extensive efforts by Moir and Brammar to utilize the powerful
phage promoter p$_L$ were less astounding due to the interactions of
lambda genes N, cI and cro.  The cro$^-$ mutants necessary for high-
level transcription from p$_L$ did not provide the high level of
replication that is the cornerstone of this very powerful amplification.  However, simply increasing the multiplicity of phage infection
per cell from 1 to 5, with accompanying mutations in cI, cro, Q and
S, brought the level of expression to the point that the five trp
gene products would have constituted 16% of the cellular protein.

        Further substantiation that these high levels of expression are
obtainable comes from the different approach of Panasenko et al.
(61).  Here, an overproducing mutant of E. coli DNA ligase was transferred from the original λgt vector into which it had been cloned
onto the new λgt4 and subsequently lysogenized into E. coli.  Since
the λgt4 carried the temperature-sensitive repressor mutation cI857,
shifting the temperature transiently to 43°C induces prophage development and expression of the cloned gene.  The measured specific
activity of the DNA ligase was found to be enhanced by about 100-
fold.  When the mutation S amber7 was introduced into the vector to
delay the lysis of the infected cells after induction, the level of
DNA ligase rose to nearly 500 times the specific activity of the
uninfected cells.  This was estimated to be 5% of the total cellular
protein.  It is likely that additional mutations introduced in the
vector to block the concomitant synthesis of phage proteins upon
induction would further increase this already impressive level of
specific activity.

        A point to be kept in mind when considering the transcription of
a cloned fragment comes from the recent confirmation by Ward and
Murray (74) that convergent transcription can totally block gene
expression.  With a trp DNA fragment carrying its own promoter, they
were able to show that anthranilate synthetase activity accumulated
only in the absence of opposing transcription from λp$_L$.  In the
case where the DNA fragment was inserted in the opposite orientation,
the rate of anthranilate synthetase production was insensitive to
transcription from λp$_L$.

        Two fundamental advantages exist in using lambda vectors over
their plasmid counterparts to achieve expression of cloned genes.
With either the induction of an integrated prophage or lytic infection as the route to expression, there is no further requirement for
cell survival after either induction or infection.  Therefore, phage
cloning intrinsically accommodates either the cloning of DNA sequences, which themselves might be lethal to the host in multiple

copy, or the cloning of genes specifying products that could be
lethal.  One could additionally expect that a cloned gene would be
most stable in an integrated prophage state since it is not subject
to loss by dilution and it is not subject to sequence variation due
to recombination among replicating forms.

   The report by Chang et al. (62) of the phenotypic expression of
mouse dihydrofolate reductase in E. coli removes the last lingering
doubt that there may generally be fundamental barriers to the expression of genes of higher eukaryotes in prokaryotic hosts.  This opti-
mism is further encouraged by the fact that Chang et al. (62) were
able to obtain a substantial fraction of cells expressing the cloned
gene product by merely screening among a population shown by hybridization analysis to contain the desired DNA sequences.  This assembly
of clones will provide the basis for a wealth of additional information on expression of cloned genes, both in general and in the PstI
site in the β-lactamase gene of the plasmid vector pBR322.


## SHOTGUN CLONING IN PHAGE LAMBDA VECTORS

   The cloning of random or restriction fragments from unfractionated complex genomes and the subsequent isolation of specific clones
from those libraries have been demonstrated by Maniatis et al. (63)
and Blattner et al. (54).  The success of this technology has had to
await three factors:  1) development and certification of phage EK2
vectors (Figure 3); 2) screening procedures that would allow one to
detect the desired recombinant among the entire library of shotgun
clones, independent of the expression of cloned gene products (plaque
hybridization in situ, 59), and 3) an in vitro packaging system
efficient enough to reduce the volume of cloned material necessary to
plate to produce a library of clones including the entire genome to a
manageable size (64-66).  Figure 3 shows the structure of the
certified EK2 vectors now available, including the recently certified
Charon 21A usable with HindIII, XhoI and SalI as well as EcoRI.
Maniatis et al. (63) detail the construction of cloned genomic DNA
libraries of Drosophila and silkmoth in Charon 4 and rabbit genomic
DNA in Charon 4A.  Blattner et al. (54) used Charon 3A to produce
unfractionated shotgun collections of human and mouse genomic DNA
that were shown to contain human fetal gammaglobin and mouse α-type
globin cloned sequences.  Lawn et al (84) describe a human library in
Charon 4A from which clones were obtained containing globin genes and
Tucker et al. (85) document the isolation of an immunoglobulin heavy
chain gene from a library of partial EcoRI digested mouse DNA in
Charon 4A.  Several immunologically interesting strains of mice
have recently been used to prepare BamHI fragmented DNA for the
production of shotgun collections in Charon 28.  That vector contains a single BamHI stuffer fragment that was purified away from
the vector fusion fragment by 5 to 20% NaCl gradients.  The ligated mixtures were in vitro packaged and the size selected hybrids

were verified to contain eukaryotic DNA by their relative efficiency of plating on restricting versus nonrestricting strains of E. coli K.

The argument that a phage DNA packaging system is necessary to produce a representative library of shotgun clones of complex genomes is stressed by both Maniatis et al. (63) and Blattner et al. (54). The problems of scale that accrue to the use of a less efficient system for introducing restricted and ligated DNA into cells become prohibitive in consideration of the numbers of individual isolates necessary to ensure that the cloned genome is completely represented in the library. The three different procedures used for introducing DNA into cells are calcium-shocked cells (69), EDTA and lysozyme treated spheroplasts (70) and in vitro packaging (64-66). The typical efficiencies with which these systems convert native phage DNA molecules into plaques is on the order of $10^{-5}$ for both calcium shock and lysozyme-EDTA spheroplasts, while in vitro packaging is routinely $2 \times 10^{-4}$ with occasional incidences of $10^{-2}$. Lysozyme-EDTA spheroplasts have the advantage that they keep well over weeks, so a batch can be made and assayed prior to the commitment of one's precious DNA sample. Its results are not particularly reliable from batch to batch and different strains have markedly different optimal points in their growth curves to harvest. We have recently found strain QR48, recAsupE, to give reproducibly better efficiencies than standard strains, up to $10^{-4}$. However, one cannot exceed more than about 500 plaques per plate using the optimal transfection conditions. This means that the generation of a complete mammalian library would probably require about 100 μg of vector DNA in the ligation mix plated on about 2000 transfection plates. If the cloning system is such that only 10% of the vectors have inserts (whether they plate or not) those numbers go up proportionally. And even if the experimenter were willing to plate 2000 to 20,000 plates to produce the library, the yield of plate lysates made from plates of such low plaque densities are low, while the task of picking $10^6$ individual plaques is monstrous. So even above efficiency, the in vitro packaging method has the advantage that the packaged reaction can be plated at an optimal density of about $10^4$ phage per plate. This allows the convenient preparation of a complete library by plate lysates, which can be easily processed to yield a cesium chloride band of purified phage in one day.

Sternberg et al. (66) developed an in vitro packaging system that provided a substantial selection for full-length lambda genomes. When they compared their $\lambda D^- sR13^+$ insertion vector, which is 78% of the genome length, to wild-type DNA, they found a 200-fold increase in the efficiency of packaging the full-length genome. There was no such size selection shown by transfection of the same DNA samples. The power of such an advantage is seen by comparing the fraction of the plaques containing inserts in their shotgun pools. While 5 to 10% of the transfected ligation mixes contained inserts, 84 to 91% of the plaques from the in vitro packaged ligations were hybrids. Blattner (personal communication) reports that this size bias in the packaging reaction can be eliminated by the use of a 50 mM spermidine

and 100 mM putrescine mixture in the buffer B of Sternberg et al.
(66).

The use of a Damber short vector in this system allows convenient characterization of the size of the insert by two criteria:
large inserts are not viable on Su° hosts and are also exquisitely
sensitive to chelating agents after growth on Su⁺ hosts. Combining
these features with a single EcoRI insertion site in the gene redα
allows the detection of inserts by the inability of the phage to grow
on ligts7, and the subsequent genetic characterization of the inserts
as large or small.

A potentially severe problem encountered with the use of at
least one in vitro packaging protocol was the unexpected observation
that the packaging size limits were more restrictive than they are in
the EDTA and lysozyme transfection system. Blattner et al.(54) show
the distribution of Charon 3A phage shotgun collections of mouse and
human genomes in CsCl equilibrium density gradients. Both the human
and mouse genomes showed substantial fractions of their EcoRI cleaved
fragments to be greater than 7 kb by agarose gel analysis. Yet
virtually none of the vectors containing inserts, in either shotgun
collection, have included fragments larger than 7 kb. All the shotgun collection is smaller than the original vector containing the
lac5 EcoRI fragment. Since several clones of E. coli DNA in the
Charon 3 phage vector have been demonstrated to be stable and grow
well with inserts up to 9.6 kb (51), there appears to be a more
restrictive upper size limit on in vitro packaged clones in Charon
3A. Whether this is due to the difference between Charons 3 and 3A
or transfection and packaging is not yet clear.

Maniatis et al. (63) approached the problem of adequate representation of cloned sequences in shotgun libraries in two ways. They
prepared blunt-ended random fragments both by shearing followed by S1
digestion and by partial digestion with restriction enzymes that make
frequent cuts and leave blunt ends. In both cases, they purified
target fragments of about 20 kb in order to reduce the number of
cloned inserts necessary to represent the entire genome. These 20 kb
fragments were then blunt-end ligated to EcoRI linkers, treated with
EcoRI and ligated into either Charon 4 or Charon 4A. The ligation
reactions were carried out at high DNA concentrations to favor
concatenates that are the preferred substrate for the in vitro
packaging system. The authors assessed the efficiency of cloning,
measured in plaque-forming units (pfu) per μg of eukaryotic DNA.
Analysis of these efficiencies shows that nonlimit digestion produces
fragments that are 3 to 15 times more efficiently cloned (rabbit)
than those prepared by shear and S1 (Drosophila and silkmoth).
Additionally, it is observed that the EK2 vector Charon 4A is a less
efficient packaging substrate (0.4 to 5 x $10^7$ pfu/μg intact Charon 4A
DNA) than the EK1 vector Charon 4 (2 to 20 x $10^7$ pfu/μg intact Charon
4 DNA).

After packaging, the phage libraries were amplified about
$10^6$-fold by the preparation of low density plate lysates. Analysis
of the buoyant density of the resulting phage population reflects the

size range of the inserted fragments. In marked contrast to the
stringent upper size limit imposed by the in vitro packaging procedure on Charon 3A of Blattner et al. (54), the observed distribution of the rabbit library produced from Charon 4A was broader and
reflected substantially larger cloned fragments than those carried by
the parental vector. In fact, the curve shown by Maniatis et al.
(63) nicely approximates the size distribution one would expect from
the sucrose gradient fractionation of the target fragments purified
for cloning. Whether this lack of size bias observed by Maniatis et
al. is the result of differences in the target fragment population,
the in vitro packaging protocols or the vectors themselves used by
Blattner et al., is not yet apparent. The libraries of Maniatis et
al. are also analyzed by agarose gel electrophoresis of EcoRI digests
to substantiate the size distribution of the inserts, demonstrating
that, at most, the libraries contain a few percent of the vector
stuffer fragments among the population of inserts. Since the size
selection applies to Charons 4 and 4A, that means all but a few percent of the phages contain cloned eukaryotic DNA. When these amplified libraries were compared to embryo DNA by Cot analysis with
single copy $^3$H labeled tracer DNA, the resultant curves were super-
imposable. This demonstrates that the amplification of the library
does not result in a gross loss of complexity. Furthermore, all
three libraries have been successfully screened with gene-specific
hybridization probes. Using cDNA from total chorion mRNA of the
silkmoth as a probe, Maniatis and co-workers calculate an expected
one positive signal per 530 plaques of the silkmoth library if the
genes for the many chorion proteins are unlinked. Screening of
350,000 plaques has yielded 350 independent isolates. Similar calculations for adult rabbit β-globin predict four to five clones per
750,000 plaques and a total of four independent isolates were obtained by screening 750,000 plaques with labeled DNA from rabbit
β-globin cDNA plasmid. These results beautifully validate the power
of the detection procedures as well as the generality of the cloning
procedure and the fidelity of the amplification step available with a
phage cloning system.

Several technical details for the screening of phage libraries
by the in situ plaque hybridization technique of Benton and Davis
(59) are presented in detailed protocols both in Blattner et al. (54)
and Maniatis et al. (63) (both works are highly recommended for their
extraordinarily complete and useful protocols). Blattner et al (54)
present the use of cafeteria trays as the basis for a megaplate technology, which allows the plating of ∿5 x 10$^5$ plaques per plate and
is conveniently fitted to a single sheet of X-ray film. An ingenious
and simple device, the "Tucker Box", is described which allows the
convenient alignment of the visual image of the exposed X-ray film
and the megaplate, to facilitate recovery of the plaque responsible
for a positive hybridization spot on the film. It was found to be
essential to include poly(rA) and denatured sonicated E. coli DNA in
the hybridization cocktail as well as to include very stringent
washes to reduce the level of nonspecific hybridization to acceptable

levels.  Maniatis et al.(63) also adsorb DNA onto duplicate filters,
applied sequentially (not stacked), in order to increase their signal
reliability from 88 to 100%.


## RESTRICTION MAPS OF CURRENTLY AVAILABLE PHAGE VECTORS

The variety of lambda phage vector genomes presented in Figures
3 and 4 reveals the wealth of diversity available to users of the
lambda phages for DNA cloning.  They are presented in the hope that
the brief introduction given here in the ways lambda can be used will
render this compendium more useful than intimidating.  The recent
expansion of vector capability is the possibility of using BamHI in
the vectors λNM570-BV1 and λNM570-BV2 (72), λL47.1 (81) and Charons
27 and 28 (F. Blattner, unpublished data).  Charon 22 is the only
currently available lambda vector usable with XmaI.


## A COMPARISON WITH PLASMIDS

A point in apology for what may look like hopeless complexity of
these phages, when compared to the diagrammatic simplicity of available plasmids, is that it is the current state of (diminishing) ig-
norance about plasmid relatedness, genetics and physical mapping that
keeps their cartoon representations so simple.  Certainly when the
promoters become closely mapped, their positions will crop up on most
drawings.  And, as incompatibility functions become defined to a
specific length of DNA, it would be surprising if those using more
than one incompatibility type did not look to the different phage
immunity representations for models of drawings.  As regions of
heteroduplex homology are found in common and as mutant activities
are found to be complementary between different plasmids, even
further complexity will emerge in the diagrams that once were represented as a simple circle with an EcoRI cut positioned at the top.
This is a suggestion that the lambda phage system may not be any more
complex than the plasmids usually represented so simply and that the
increased detailed knowledge about lambda should certainly put substantially more capability into the hands of those who wish to use it
for DNA cloning.  The point that plasmids offer some unique advantages over lambda probably needs no emphasis but those advantages are
principally that plasmids have no constraints on size and, therefore,
can be used to propagate DNA fragments as large as one wishes and
that a small DNA fragment is much more efficiently propagated for
preparative recovery in a small plasmid than in a phage.  A 1 kb
fragment propagated in a lambda phage would constitute only 2% of the
isolated DNA, while that same fragment propagated in a small plasmid
like pBR322 would constitute 20%.  This latter advantage is probably
the more substantial since very large plasmids themselves present a
difficulty to efficient isolation.  Even here, however, the advantage
is not absolute.  A pBR322 carrying a cloned eukaryotic fragment

recently proved difficult to propagate in X1776. When prepared by
propagating the entire plasmid in the EcoRI site of Charon 16A grown
in DP50 supF, very respectable recoveries were obtained in our
laboratory.

## SECOND GENERATION VECTORS

The development of hybrid cloning vectors containing a plasmid
origin of replication and the lambda cohesive end site, the cosmids
of Collins and Hohn (20), demonstrate a clever application of both
the size selection and in vitro packaging advantages of lambda into
plasmid cloning, which will fill a useful niche in the cloning of
large DNA fragments on the order of 16 to 30 megadaltons. Phage
vectors carrying segments of DNA homology may be used to direct the
integration of cloned fragments into the chromosomes of any host for
which a DNA transformation system is available. In addition, there
is a report from R. Davis that suggests that a cloned fragment of
yeast carrying a trp1 allele in a lambda vector can direct the replication of that vector in yeast providing a system whereby cloned
DNA fragments may be subject to all the established manipulations in
E. coli and also manipulated for genetic studies and expression back
into yeast (R. Davis, 9th International Conference on Yeast Genetics
and Molecular Biology, June 1978; see also this volume, pp. 169-183).
Clearly, the application and extension of the prokaryotic DNA vectors
such as those enumerated here are only the beginning.

Acknowledgments: Our thanks to colleagues who made unpublished
information and vectors available and to the laboratories who produced this massive variety of vectors. Substantial contributions to
the mapping were provided by J. Schroeder in helping to develop the
computerology, by Donna Daniels and Jeff deWet for putting it all
together and by David Moore and Frances Lawyer for helping to fill in
the cracks. Bill Williams acknowledges a great debt to Jane Setlow,
who has been more than an editor should ever have to be, and to his
wife, Barbara Jean, for her support during this grumpy production.

## REFERENCES

1  Shapiro, J., MacHattie, L., Eron, L., Ihler, G., Ippen, K. and
   Beckwith, J. (1969) Nature (London) 224, 768-774.
2  Simon, M., Davis, R. and Davidson, N. (1971) in The Bacteriophage Lambda (Hershey, A.D., ed.), pp. 313-328, Cold Spring
   Harbor Laboratory, Cold Spring Harbor, NY.
3  Fiandt, M., Hradecna, Z., Lozeron, H. and Szybalski, W. (1971)
   in The Bacteriophage Lambda (Hershey, A.D., ed.), pp. 329-354,
   Cold Spring Harbor Laboratory, Cold Spring Harbor, NY.
4  Murray, N.E. and Murray, K. (1974) Nature (London) 251, 476-481.
5  Rambach, A. and Tiollais, P. (1974) Proc. Nat. Acad. Sci.
   U.S.A., 71, 3927-2930.

6    Thomas, M., Cameron, J.R. and Davis, R.W. (1974) Proc. Nat.
     Acad. Sci. U.S.A. 71, 4579-4583.
7    Murray, K. and Murray, N.E. (1975) J. Mol. Biol. 98, 551-564.
8    Enquist, L., Tiemeier, D., Leder, P., Weisberg, R. and
     Sternberg, N. (1976) Nature (London) 259, 596-598.
9    Struhl, K., Cameron, J.R. and Davis, R.W. (1976) Proc. Nat.
     Acad. Sci. U.S.A. 73, 1471-1475.
10   Borck, K., Beggs, J.D., Brammar, W.J., Hopkins, A.S. and Murray,
     N.E. (1976) Mol. Gen. Genet. 146, 199-207.
11   Tiemeier, D., Enquist, L. and Leder, P. (1976) Nature (London)
     263, 526-527.
12   Murray, N.E., Brammar, W.J. and Murray, K. (1977) Mol. Gen.
     Genet. 150, 53-61.
13   Donoghue, D.J. and Sharp, P.A. (1977) Gene 1, 209-227.
14   Pourcel, C. and Tiollais, P. (1977) Gene 1, 281-286.
15   Panasenko, S.M., Cameron, J.R., Davis, R.W. and Lehman, I.R.
     (1977) Science 196, 188-189.
16   Blattner, F.R., Williams, B.G., Blechl, A.E., Denniston-
     Thompson, K., Faber, H.E., Furlong, L.A., Grunwald, D.J.,
     Kiefer, D.O., Moore, D.D., Schumm, J.W., Sheldon, E.L. and
     Smithies, O. (1977) Science 196, 161-169.
17   Struhl, K. and Davis, R.W. (1977) Proc. Nat. Acad. Sci. U.S.A.
     74, 5255-5259.
18   Philippsen, P., Kramer, R.A. and Davis R.W. (1978) J. Mol. Biol.
     123, 371-386.
19   Court, D. and Sato, K. (1969) Virology 39, 348-352.
20   Collins, J. and Hohn, B. (1978) Proc. Nat. Acad. Sci. U.S.A. 75,
     4242-4246.
21   Ullrich, A., Shine, J., Chirgwin, J., Pictet, R., Tischer, E.,
     Rutter, W.J. and Goodman, H.M. (1977) Science 196, 1313-1319.
22   Efstratiadis, A., Vournakis, J.N., Donis-Keller, H., Chaconas,
     G., Dougal, D.K. and Kafatos, F.C. (1977) Nucl. Acids Res. 4,
     4165-4174.
23   Davidson, N. and Szybalski, W. (1971) in The Bacteriophage
     Lambda (Hershey, A.D., ed.), pp. 45-82, Cold Spring Harbor
     Laboratory, Cold Spring Harbor, NY.
24   Jacobson, H. and Stockmayer, W.H. (1950) J. Chem. Phys. 18,
     1600-1606.
25   Wang, J.C. and Davidson, N. (1966) J. Mol. Biol. 19, 469-482.
26   Bolivar, F., Rodriguez, R.L., Green, P.J., Betlach, M.C.,
     Heyneker, H.L., Boyer, H.W., Crosa, J.H. and Falkow, S. (1977)
     Gene 2, 95-113.
27   Polisky, B., Bishop, R.J. and Gelfand, D.H. (1976) Proc. Nat.
     Acad. Sci. U.S.A. 73, 3900-3904.
28   Dugaiczyk, A., Boyer, H.W. and Goodman, H.W. (1975) J. Mol.
     Biol. 96, 171-184.
29   Tiemeier, D.C., Tilghman, S.M. and Leder, P. (1977) Gene 2,
     173-191.
30   Goff, S.P. and Berg, P. (1978) Proc. Nat. Acad. Sci. U.S.A. 75,
     1763-1767.

31   Emmons, S.W., MacCosham, V. and Baldwin, R.L. (1975) J. Mol.
     Biol. 91, 133-146.
32   Cameron, J.R., Philippsen, P. and Davis, R.W. (1977) Nucl.
     Acids, Res. 4, 1429-1448.
33   Philippsen, P., Kramer, R.A. and Davis, R.W. (1978) J. Mol.
     Biol. 123, 371-386.
34   Scandella, D. and Arber, W. (1974) Virology 58, 504-513.
35   Scandella, S. and Arber, W. (1976) Virology 69, 206-215.
36   Parkinson, J.S. and Huskey, R.J. (1971) J. Mol. Biol. 56,
     369-384.
37   Szybalski, E. and Szybalski, W. (1971) Proc. Nucl. Acids Res. 2,
     311-354.
38   Zissler, J., Signer, E. and Schaefer, F. (1971) in The Bacterio-
     phage Lambda (Hershey, A.D., ed.), pp. 469-476, Cold Spring
     Harbor Laboratory, Cold Spring Harbor, NY.
39   Cameron, J.R. and Davis, R.W. (1977) Science 196, 212-215.
40   Henderson, D.A. and Weil, J. (1975) Genetics 79, 143-174.
41   Lam, S.T., Stahl, M.M., McMilin, K.D. and Stahl, F.W. (1974)
     Genetics 77, 425-433.
42   Bahl, C.P., Marians, K.J. and Wu, R. (1976) Gene 1, 81-92.
43   Marians, K.J., Wu, R., Staminsky, J., Hozumi, T. and Narang,
     S.A. (1976) Nature 263, 744-748.
44   Heyneker, H.L., Shine, J., Goodman, H.M., Boyer, H., Rosenberg,
     J., Dickerson, R.E., Narang, S.A., Itakura, K., Liu, S. and
     Riggs, A.D. (1976) Nature 263, 748-752.
45   Polisky, B., Bishop, R.J. and Gelfand, D.H. (1976) Proc. Nat.
     Acad. Sci. U.S.A. 73, 3900-3904.
46   Itakura, K., Hirose, T., Crea, R., Riggs, A.D., Heyneker, H.L.,
     Bolivar, F. and Boyer, H.W. (1977) Science 198, 1056-1063.
47   Kayajanian, G. (1968) Virology 36, 30-41.
48   Cleary, P., Campbell, A. and Chang, R. (1972) Proc. Nat. Acad.
     Sci. U.S.A. 69, 2219-2223.
49   Ketner, G. and Campbell, A. (1975) J. Mol. Biol. 96, 13-29.
50   DasGupta, C.K. and Guha, A. (1978) Gene 3, 233-246.
51   Williams, B.G., Blattner, F.R., Jaskunas, S.R. and Nomura, M.
     (1977) J. Biol. Chem. 252, 7344-7354.
52   Jaskunas, S.R., Fallon, A.M., Nomura, M., Williams, B.G. and
     Blattner, F.R. (1977) J. Biol. Chem. 252, 7355-7364.
53   Enquist, L.W. and Weisberg, R.A. (1976) Virology 72, 147-153.
54   Blattner, F.R., Blechl, A.E., Denniston-Thompson, K., Faber,
     H.E., Richards, J.E., Slightom, J.L., Tucker, P.W. and Smithies,
     O. (1978) Science 202, 1279-1284.
55   Cameron, J.R., Panasenko, S.M., Lehman, I.R. and Davis, R.W.
     (1975) Proc. Nat. Acad. Sci. U.S.A. 72, 3416-3420.
56   Franklin, N.C. (1971) in The Bacteriophage Lambda (Hershey,
     A.D., ed.), pp. 621-638, Cold Spring Harbor Laboratory, Cold
     Spring Harbor, NY.
57   Shimada, K., Weisberg, R.A. and Gottesman, M.E. (1972) J. Mol.
     Biol. 63, 483-503.

58   Schrenk, W.J. and Weisberg, R.A. (1975) Mol. Gen. Genet. 137,
     101-107.
59   Benton, W.D. and Davis, R.W. (1977) Science 196, 180-182.
60   Moir, A. and Brammar, W.J. (1976) Mol. Gen. Genet. 149, 87-99.
61   Davis, R. (personal communication).
62   Chang, A.C.Y., Nunberg, J.H., Kaufman, R.J., Erlich, H.A.,
     Schimke, R.T. and Cohen, S.N. (1978) Nature 275, 617-624.
63   Maniatis, T., Hardison, R.C., Lacy, E., Lamer, J., O'Connell,
     C., Quon, D., Sim, G.K. and Efstratiadis, A. (1978) Cell 15,
     687-701.
64   Becker, A. and Gold, M. (1975) Proc. Nat. Acad. Sci. U.S.A. 72,
     581-585.
65   Hohn, G. and Murray, K. (1977) Proc. Nat. Acad. Sci. U.S.A. 74,
     3259-3263.
66   Sternberg, N., Tiemeier, D. and Enquist, L. (1977) Gene 1,
     255-280.
67   Certified EK2 Systems (1977) Recombinant DNA Technical Bulletin
     2(2), 92-93.
68   Blattner, F.R., Kiefer, D., Moore, D., deWet, J. and Williams,
     B.G. (1978) Application for EK2 Certification of a Host Vector'
     System for DNA Cloning, Supplement IX: Data on Charon 21A, April
     7, 1978.
69   Mandel, W. and Higa, A. (1970) J. Mol. Biol. 53, 159-162.
70   Henner, W.D., Kleber, I. and Benzinger, R. (1973) J. Virol. 12,
     741-747.
71   Wilson, G.G. and Murray, N.E. (1979) J. Mol. Biol. 132, 471-491.
72   Klein, B. and Murray, K. (1979) J. Mol. Biol. 133, 289-294.
73   Kravchenko, V.V., Vassilenko, S.K. and Gracher, M.A. (1979) Gene
     7, 181-195.
74   Ward, D.F. and Murray, N.E. (1979) J. Mol. Biol. 133, 249-266.
75   Grosschedl, R. and Schwarz, E. (1979) Nucl. Acids Res. 6,
     867-881.
76   Szybalski, E. and Szybalski, W. (1979) Gene 7, 217-270.
77   Rabbits, T.H. (1976) Nature 260, 221-225.
78   Becker, A. and Gold, M. (1975) Proc. Nat. Acad. Sci. U.S.A. 72,
     581-585.
79   Hohn, B. and Murray, K. (1977) Proc. Nat. Acad. Sci. U.S.A. 74,
     3259-3263.
80   Sternberg, N., Tiemeier, D. and Enquist, L. (1977) Gene 1,
     255-280.
81   Loenen, W. and Brammar, W.J. (1980) Gene (in press).
82   Williams, B.G. and Blattner, F.R. (1979) J. Virol. 29, 555-575.
83   Gottesman, M.E. and Yaramolinsky, M.B. (1968) 31, 487-505.
84   Lawn, R.M., Fritsch, E.F., Parker, R.C., Blake, G. and Maniatis,
     T. (1978) Cell 15, 1157-1174.
85   Tucker, P.W., Marcu, K.B., Newell, N., Richards, J. and
     Blattner, F.R. (1979) Science 206, 1303-1306.
86   Schroeder, J.L. and Blattner, F.R. (1978) Gene 4, 167-174.
87   Daniels, D.L., deWet, J.R. and Blattner, F.R. (1980) J. Virol.
     (in press).

88    deWet, J.R., Daniels, D.L., Schroeder, J.L., Williams, B.G.,
      Denniston-Thompson, K., Moore, D.D. and Blattner, F.R. (1980) J.
      Virol. (in press).
89    Davison, J., Brunel, F. and Merchez, M. (1979) Gene 8, 69-80.

Figure 3.  Restriction maps of the HV2 certified lambda DNA cloning
vectors.  The horizontal line at the level of the name of the vector
indicates the presence of wild-type lambda DNA.  The endpoints of
simple deletions are indicated by parentheses.  A line drawn above
the wild-type line indicates the presence of a substituted DNA, the
length of the line indicates the size of the substituted DNA and the
size of the lambda DNA removed is given by the length on the wild-
type line indicated to be missing by the dashed vertical lines.  This
convention allows the size of the phage DNA to be determined by
summing the measured solid horizontal lines or simply by estimating
at a glance.  The notations $\lambda$B and $\lambda$C refer to the EcoRI fragments of
lambda wild-type, named A, B, C, D, E and F, from left to right (6).
$\lambda$B' indicates that the $\lambda$B fragment is inverted.  $W^-$, $E^-$, $S^-$, $Z^-$, $J^-$,
$A^-$ and $B^-$ all refer to amber mutations in these essential genes.  The
restriction enzyme sites are shown by vertical lines.  The sites for
all the following enzymes are shown on each vector using the indicated abbreviations: EcoRI = R and RI, SstI = T and Sst, HindIII = H
and Hin, XbaI = X and Xba, SalI = S and Sal, XhoI = O and Xho.  A
small circle interrupting the vector diagram indicates that the restriction enzyme site at that position has been removed by mutation.

The size of the vector genome is indicated by the Net kb Deleted
column.  Adding 1.5 kb to the net deletion estimates the largest
piece of cloned DNA that the given vector can propagate as an insertion into a single restriction site.  The size capacities of each
vector for all insertions, and all allowable substitutions using
enzymes singly or in combination, are given in Table 2.  The certified vectors Charon 23A and 24A (not shown) are $W^-E^-$ versions of
Charons 3A and 4A, respectively.

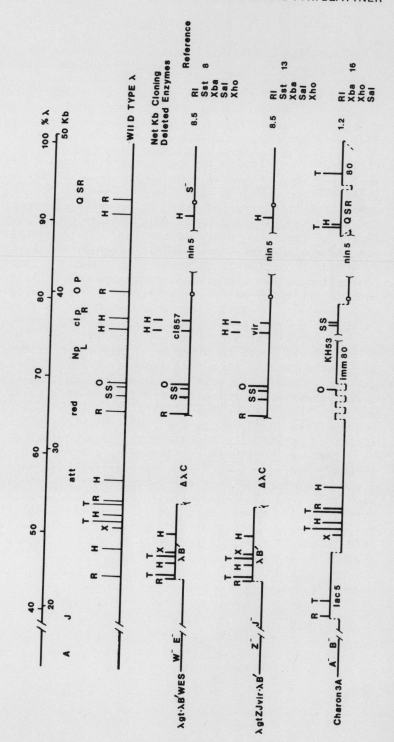

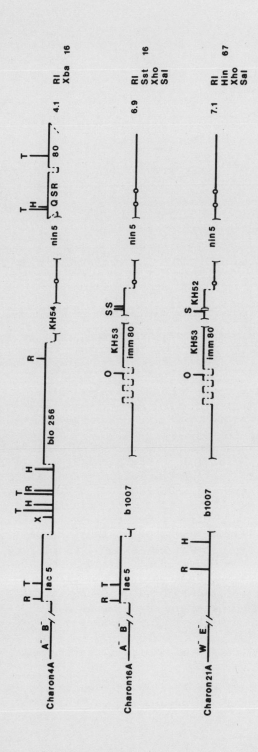

Figure 4. Restriction maps of bacteriophage lambda DNA cloning
vectors. The positions of the restriction sites of the various
enzymes are indicated by vertical lines and the nonstandard abbreviations are keyed to their standard designations and cleavage
sequences as follows:

    Bg = BglII A/GATCT
    Ba = BamHI G/GATCC
    Kp = KpnI GGTAC/C
    RI = EcoRI G/AATTC
    H3 = HindIII A/AGCTT
    Xa = XbaI T/CTAGA
    St = SstI GAGCT/C
    Sa = SalI G/TCGAC
    Xh = XhoI C/TCGAG

Deletions are indicated by parentheses, which are joined by a gabled
symbol when the deletion was formed by the action of restriction
enzymes and resulted in the restoration of a single site. Substitutions are indicated by boxes containing the name of the source of
substituted DNA. The substitution endpoints indicate the extent of
the lambda DNA removed, and therefore not the size of the substituted
fragment. The small double box at 58% lambda represents the attachment site att, and each half-box designates one of the two phage
halves, p or p', of that site. Insertions are shown as looped-out
ellipses. A small circle on the genome indicates that a restriction
site at that position has been removed by mutation. Shaded-in substitutions or insertions indicate that they have not been mapped in
the sited references for the enzymes listed above. The Charon phages
all originated in the Blattner laboratory and have been extensively
checked to verify their restriction maps for their enzymes. The
estimated position of each restriction site in each phage is given in
Table 1, as base pairs from the left end of the mature DNA molecule.
These estimates come from the averaging computer program (86), which
was used to produce the improved map of lambda by Daniels et al. (87)
and the maps of the Charon phages by deWet et al. (88). The maps of
phages from other laboratories have been derived by extrapolating the
maps of known substitutions to the structures deduced from the literature references. These are subject to error both from incomplete
published data and from the heterogeneity of substitutions (for example, the immunity substitution of φ21 carrying the cIts allele in the
N. Murray φ16 lacks the SalI site found in the other φ21 immunity
substitutions).

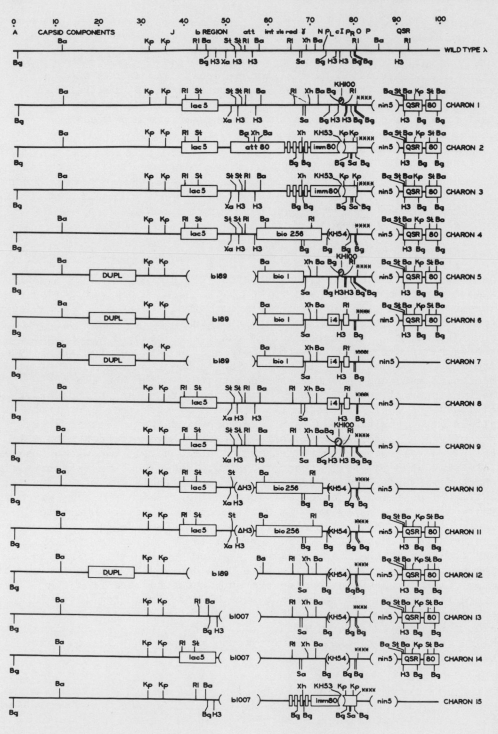

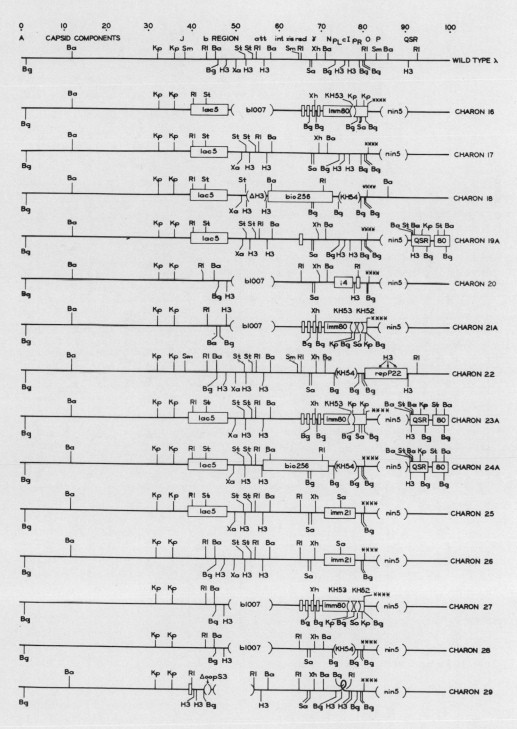

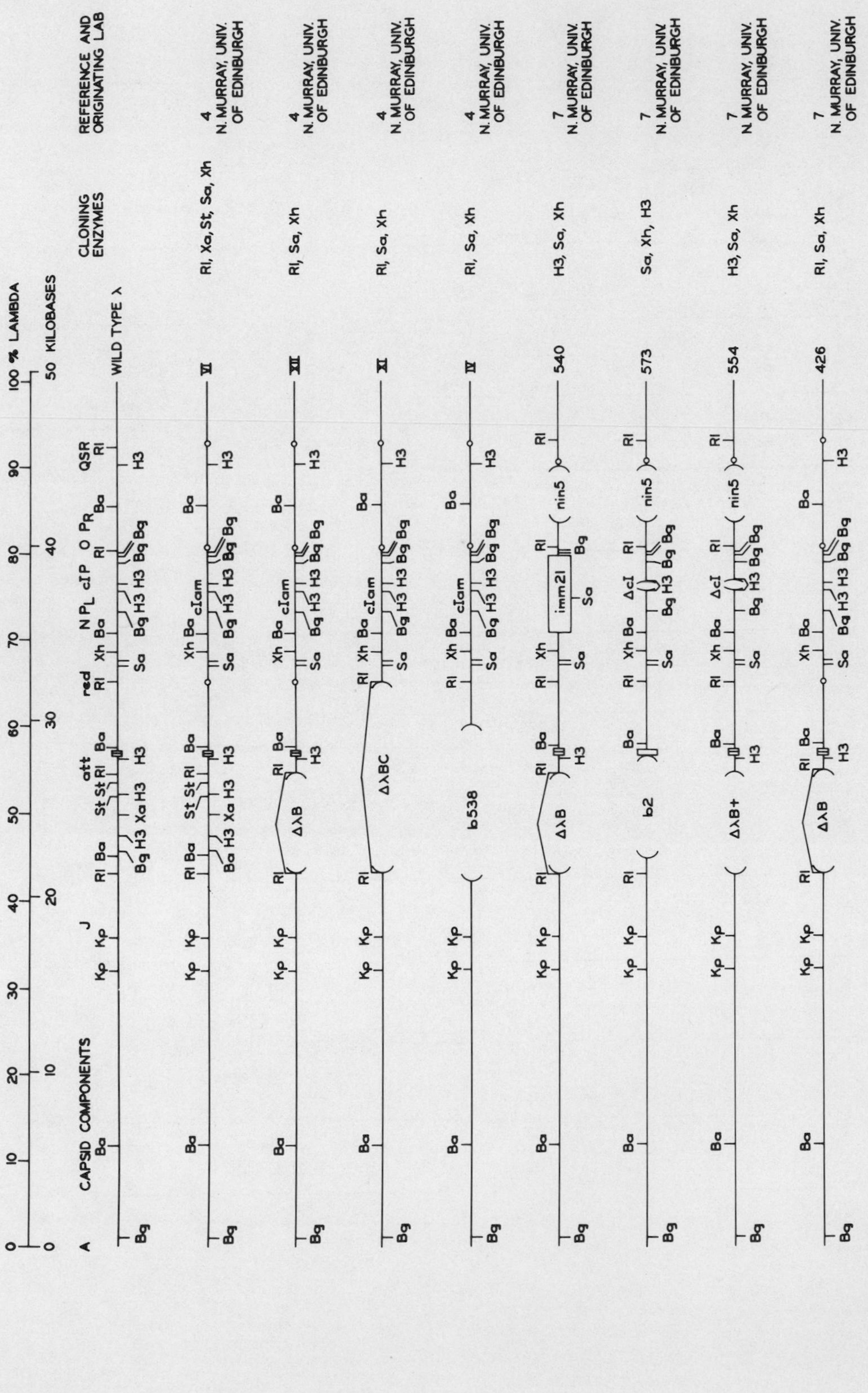

A

REFERENCE AND ORIGINATING LAB

CLONING ENZYMES

100 % LAMBDA

50 KILOBASES

CAPSID COMPONENTS

WILD TYPE λ

RI, Xa, St, Sa, Xh — 4, N. MURRAY, UNIV. OF EDINBURGH — VI

RI, Sa, Xh — 4, N. MURRAY, UNIV. OF EDINBURGH — XII

RI, Sa, Xh — 4, N. MURRAY, UNIV. OF EDINBURGH — XI

RI, Sa, Xh — 4, N. MURRAY, UNIV. OF EDINBURGH — IV

H3, Sa, Xh — 7, N. MURRAY, UNIV. OF EDINBURGH — 540

Sa, Xh, H3 — 7, N. MURRAY, UNIV. OF EDINBURGH — 573

H3, Sa, Xh — 7, N. MURRAY, UNIV. OF EDINBURGH — 554

RI, Sa, Xh — 7, N. MURRAY, UNIV. OF EDINBURGH — 426

nin5, imm21, ΔcI, b2, ΔλB, ΔλBC, ΔλB+, b538

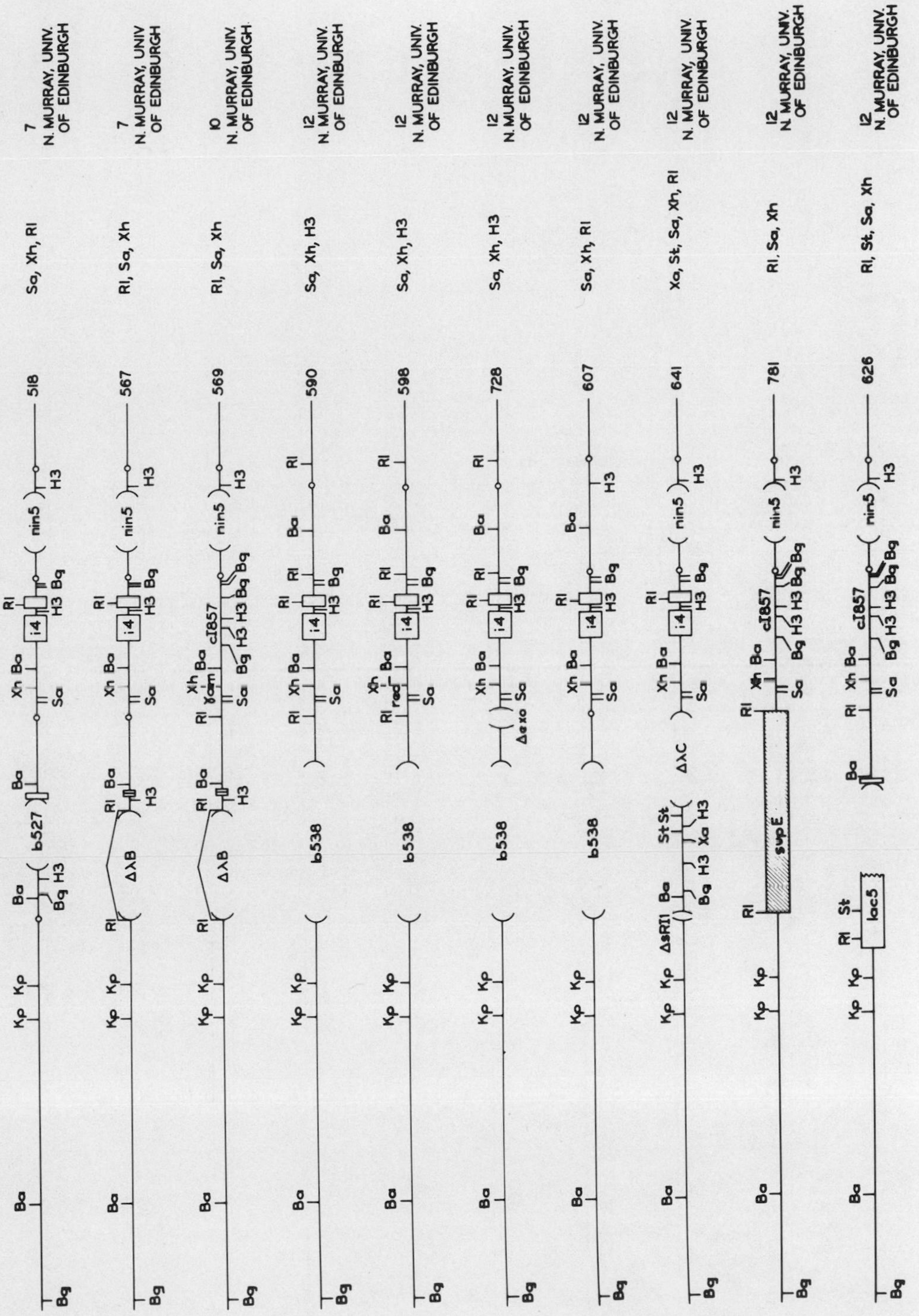
518
Sa, Xh, RI
7
N. MURRAY, UNIV. OF EDINBURGH
567
RI, Sa, Xh
7
N. MURRAY, UNIV OF EDINBURGH
569
RI, Sa, Xh
10
N. MURRAY, UNIV. OF EDINBURGH
590
Sa, Xh, H3
12
N. MURRAY, UNIV. OF EDINBURGH
598
Sa, Xh, H3
12
N. MURRAY, UNIV. OF EDINBURGH
728
Sa, Xh, H3
12
N. MURRAY, UNIV. OF EDINBURGH
607
Sa, Xh, RI
12
N. MURRAY, UNIV. OF EDINBURGH
641
Xa, St, Sa, Xh, RI
12
N. MURRAY, UNIV. OF EDINBURGH
781
RI, Sa, Xh
12
N. MURRAY, UNIV. OF EDINBURGH
626
RI, St, Sa, Xh
12
N. MURRAY, UNIV. OF EDINBURGH
b527
ΔλB
b538
ΔsRI1
ΔλC
supE
lac5
nin5
i4
cI857
Δexo
Ba
Bg
Kp
RI
Xh
Sa
H3
St
Xa

0  10  20  30  40  50  60  70  80  90  100 % LAMBDA

0  10  20  30  40  50 KILOBASES

A  CAPSID COMPONENTS  J  att  red  N P L cI P R O P  QSR  CLONING ENZYMES  REFERENCE AND ORIGINATING LAB

Ba  Kp Kp  RI Ba  St St RI Ba  RI Xh Ba  RI  Ba  RI  WILD TYPE λ

Bg  Bg H3 Xa H3  H3  Sa  Bg H3 H3 Bg Bg  H3

Ba  Kp Kp  RI St  Ba  RI Xh Ba KH54  (ΔcI)  (nin5)  631  RI, St, Sa, Xh  12 N. MURRAY, UNIV. OF EDINBURGH

Bg  lac5  Sa  Bg  Bg Bg  H3

Ba  Kp Kp  RI St  Ba  RI Xh  imm 21  (nin5)  647  RI, St, Sa, Xh  12 N. MURRAY, UNIV. OF EDINBURGH

Bg  lac5  Sa  Sa  Bg  H3

Ba  Kp Kp  RI St  Ba  RI Xh Ba KH54  (ΔcI)  (nin5)  791  RI, St, Sa, Xh  12 N. MURRAY, UNIV. OF EDINBURGH

Bg  Z ΔRI  lac5  Sa  Bg  Bg Bg  H3

Ba  Kp Kp  RI St  St St RI Ba  Xh  i21 cts  (nin5)  816  RI, St, Xa, Sa, Xh  71 N. MURRAY, UNIV. OF EDINBURGH

Bg  lac5  Xa H3  H3  Sa  Bg  H3

Ba  Kp Kp  RI  trpE  RI Ba  RI Xh  imm21  (nin5)  RI  678  H3, Sa, Xh  12 N. MURRAY, UNIV. OF EDINBURGH

Bg  ΔλB  H3  Sa  Sa  Bg

Ba  Kp Kp  RI  trpE  RI Ba  Xh Ba  γam  RI  (nin5)  RI  679  H3, Sa, Xh  12 N. MURRAY, UNIV. OF EDINBURGH

Bg  ΔλB  H3  Sa Bg H3 H3 Bg Bg

Ba  Kp Kp  RI  trpE att- redΔ  RI  Xh Ba  γam  RI  (nin5)  RI  709  H3, Sa, Xh  12 N. MURRAY, UNIV. OF EDINBURGH

Bg  ΔλB  H3  Sa Bg H3 H3 Bg Bg

Ba  Kp Kp  RI  trpE att- xisΔ  RI  Xh Ba  γam  RI  (nin5)  RI  696  H3, Sa, Xh  12 N. MURRAY, UNIV. OF EDINBURGH

Bg  ΔλB  H3  Sa Bg H3 H3 Bg Bg

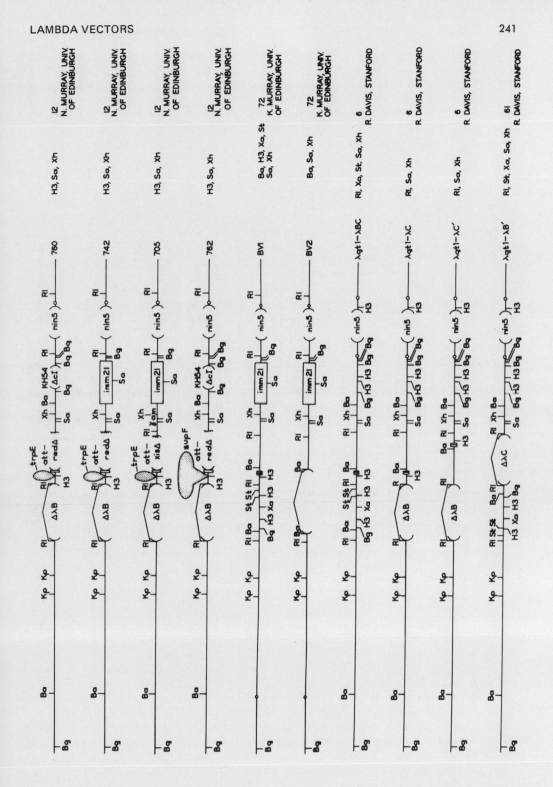

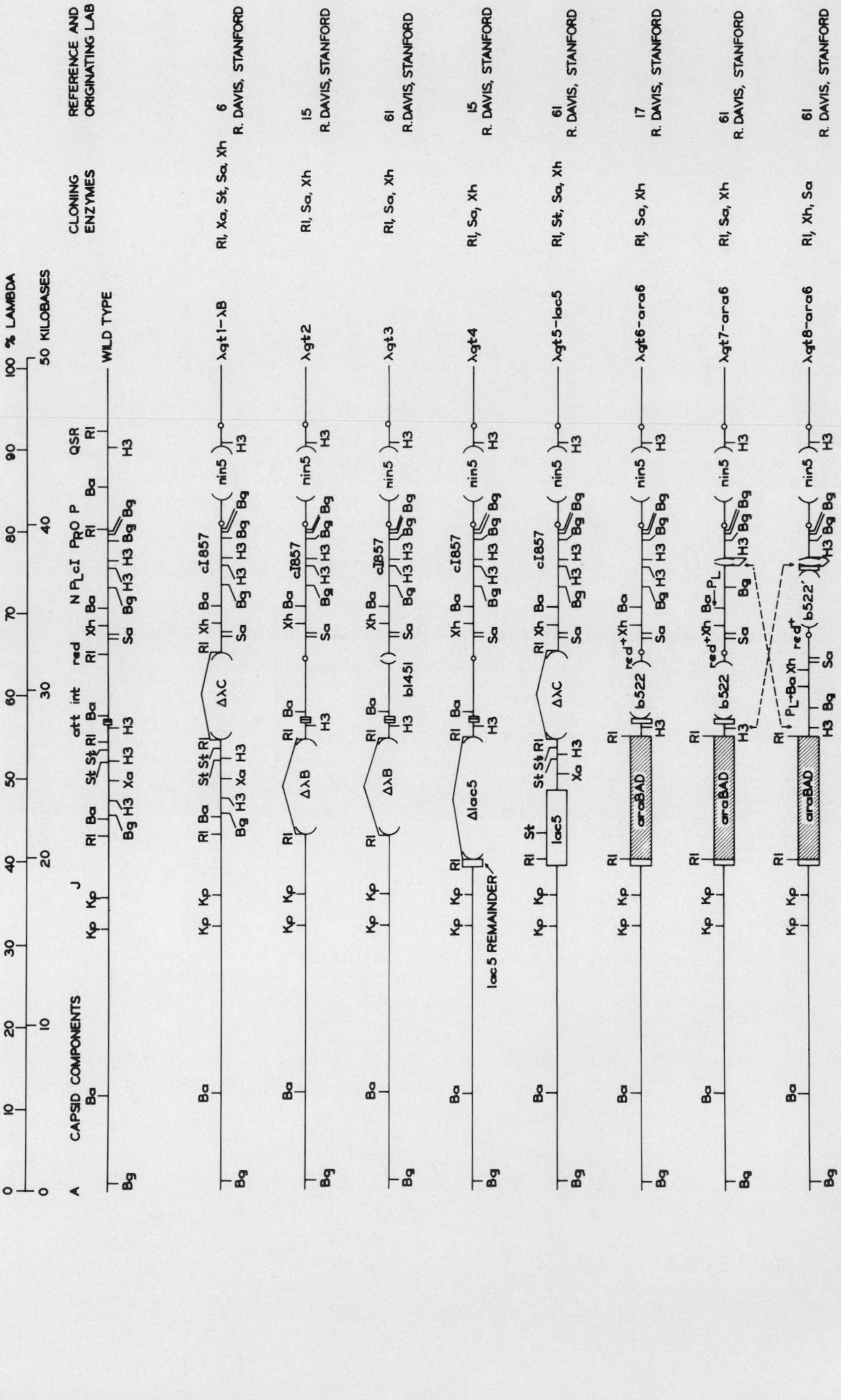

REFERENCE AND ORIGINATING LAB

CLONING ENZYMES

100 % LAMBDA

50 KILOBASES

WILD TYPE

A    CAPSID COMPONENTS

λgt1–λB    RI, Xa, St, Sa, Xh    6    R. DAVIS, STANFORD

λgt2    RI, Sa, Xh    I5    R. DAVIS, STANFORD

λgt3    RI, Sa, Xh    6I    R. DAVIS, STANFORD

λgt4    RI, Sa, Xh    I5    R. DAVIS, STANFORD

λgt5–lac5    RI, St, Sa, Xh    6I    R. DAVIS, STANFORD

λgt6–ara6    RI, Sa, Xh    I7    R. DAVIS, STANFORD

λgt7–ara6    RI, Sa, Xh    6I    R. DAVIS, STANFORD

λgt8–ara6    RI, Xh, Sa    6I    R. DAVIS, STANFORD

Ba Bg | Kp Kp | RI Xh | i21 CLEAR / Sa | Bg / nin5 H3 | λgt30 | RI, Sa, Xh | 61 R. DAVIS, STANFORD

E. coli 6Kb | b221

Ba Bg | Kp Kp | RI Ba / Bg H3 Xa | St / RI / ΔλC | RI Xh Ba / Sa / Bg H3 H3 Bg Bg / nin5 H3 | cI857 | λgt40 | RI, Xa, St, Sa, Xh | 18 R. DAVIS, STANFORD

Warm Ba Eam Bg | Kp Kp | RI / 1.8 kb T5 | RI / 1.8 kb T5 | RI Xh Ba / Sa / Bg H3 H3 Bg Bg / nin5 Sam H3 | cI857 | λgtWES·T5622 | RI, Sa, Xh | 89 J. DAVISON, INSTITUTE OF CELLULAR PATHOLOGY, BRUSSELS

Ba Bg | Kp Kp | RI St / lac5 | St St RI Ba / Xa H3 / H3 | Xh RI Ba / Sa / Ba / Bg H3 H3 Bg Bg / H3 | cI857 | "B" | RI, St, Xa, Sa, Xh | 5 P. TIOLLAIS INSTITUTE PASTEUR

Ba Bg | Kp Kp | RI St / lac5 | St St RI Ba / Xa H3 / H3 | Xh Ba / Sa / Ba / Bg H3 H3 Bg Bg / H3 | cI857 | "C" | RI, St, Xa, Sa, Xh | 5 P. TIOLLAIS INSTITUTE PASTEUR

Ba Bg | Kp Kp | RI St / lac5 | St St RI Ba / Bg H3 Xa H3 / H3 | RI Xh / Sa / Ba / Bg Bg / nin5 H3 | i21 ts | "2d" | RI, Xa, St, Sa, Xh | 14 P. TIOLLAIS INSTITUTE PASTEUR

Ba Bg | Kp Kp | RI St / lac5 | St St RI / Xa H3 / H3 | RI Xh / Sa / Ba / Bg Bg / nin5 H3 | i21 ts | "2e" | RI, St, Xa, Sa, Xh | 14 P. TIOLLAIS INSTITUTE PASTEUR

Ba Bg | Kp Kp | RI St / lac5 | RI Ba / H3 b508 / H3 | RI Xh / Sa / Ba / Bg Bg / nin5 H3 | i21 ts | "2g" | RI, Xa, St, Sa, Xh | 14 P. TIOLLAIS INSTITUTE PASTEUR

Ba Bg | Kp Kp | RI St / b515 | RI Ba / H3 / H3 | RI Xh / Sa / Ba / Bg Bg / nin5 H3 | i21 ts | "2i" | RI, St, Sa, Xh | 14 P. TIOLLAIS INSTITUTE PASTEUR

Ba Zam Bg | Kp Kp Jam | RI Ba / Bg H3 H3 / H3 | St St RI Ba / Xa H3 / H3 | RI Xh Ba / Sa / Ba / Bg Bg / nin5 vir H3 | λgt vir JZ-λBC | RI, Xa, St, Sa, Xh | 13 P. SHARP MIT

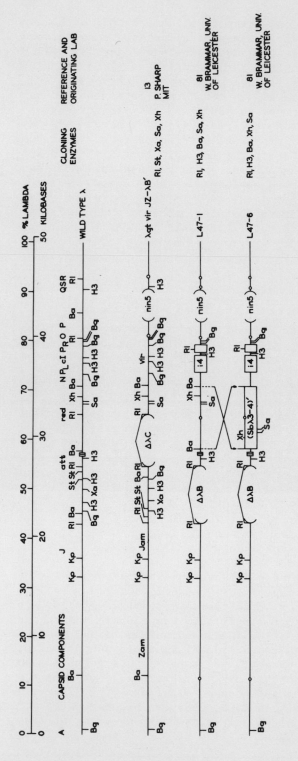

Table 1.

Coordinates of Restriction Enzyme Sites Listed as Base Pairs from
the Left End of the Mature DNA of the Listed Phages

---

The following abbreviations have been used: LEND = left end, REND =
right end, LVL = left viability limit, RVL = right viability limit
(used to demarcate the replaceable region for the computer generation
of Table 2), BAM1 = BamHI, BGL2 = BglII, HIN3 = HindIII, KPN1 = KpnI,
R1 = EcoRI, SAL1 = SalI, SST1 = SstI, XBA1 = XbaI, XHO1 = XhoI.  The
names of each file correspond as closely as possible to the names of
the vectors given in Figure 4, using the additional abbreviations CH
for Charon and L for lambda.

---

CH-1

| Site | Position |
|------|----------|
| LEND | 0 |
| BGL2 | 471 |
| BAM1 | 5559 |
| KPN1 | 17430 |
| KPN1 | 18950 |
| R1 | 19939 |
| SST1 | 21043 |
| XBA1 | 24959 |
| SST1 | 25230 |
| HIN3 | 25618 |
| SST1 | 26344 |
| R1 | 26565 |
| HIN3 | 27916 |
| BAM1 | 28511 |
| R1 | 32268 |
| SAL1 | 33309 |
| SAL1 | 33817 |
| XHO1 | 34075 |
| BAM1 | 35073 |
| BGL2 | 36282 |
| HIN3 | 37504 |
| BGL2 | 38165 |
| R1 | 39020 |
| HIN3 | 39264 |
| HIN3 | 39389 |
| BGL2 | 39908 |
| BGL2 | 40559 |
| BGL2 | 40619 |
| **** | 40973 |
| BAM1 | 43020 |
| SST1 | 43170 |
| HIN3 | 43212 |
| BAM1 | 43526 |
| KPN1 | 45412 |
| BGL2 | 45476 |
| SST1 | 47032 |
| BAM1 | 47454 |
| BGL2 | 47631 |
| REND | 48944 |

CH-2

| Site | Position |
|------|----------|
| LEND | 0 |
| BGL2 | 471 |
| BAM1 | 5559 |
| KPN1 | 17430 |
| KPN1 | 18950 |
| R1 | 19939 |
| SST1 | 21043 |
| BAM1 | 27061 |
| XHO1 | 27366 |
| BAM1 | 30136 |
| BGL2 | 32130 |
| XHO1 | 32886 |
| BGL2 | 33012 |
| KPN1 | 35757 |
| BGL2 | 36488 |
| SAL1 | 36775 |
| SAL1 | 36928 |
| KPN1 | 37567 |
| BGL2 | 38516 |
| **** | 38530 |
| BAM1 | 40577 |
| SST1 | 40727 |
| HIN3 | 40769 |
| BAM1 | 41083 |
| KPN1 | 42969 |
| BGL2 | 43033 |
| SST1 | 44589 |
| BAM1 | 45011 |
| BGL2 | 45188 |
| REND | 46501 |

CH-3

| Site | Position |
|------|----------|
| LEND | 0 |
| BGL2 | 471 |
| BAM1 | 5559 |
| KPN1 | 17430 |
| KPN1 | 18950 |
| R1 | 19939 |
| SST1 | 21043 |
| XBA1 | 24959 |
| SST1 | 25230 |
| HIN3 | 25618 |
| SST1 | 26344 |
| R1 | 26565 |
| HIN3 | 27916 |
| BAM1 | 28511 |
| BGL2 | 33808 |
| XHO1 | 34564 |
| BGL2 | 34690 |
| KPN1 | 37435 |
| BGL2 | 38166 |
| SAL1 | 38453 |
| SAL1 | 38606 |
| KPN1 | 39245 |
| BGL2 | 40194 |
| **** | 40208 |
| BAM1 | 42255 |
| SST1 | 42405 |
| HIN3 | 42447 |
| BAM1 | 42761 |
| KPN1 | 44647 |
| BGL2 | 44711 |
| SST1 | 46267 |
| BAM1 | 46689 |
| BGL2 | 46866 |
| REND | 48179 |

CH-4

| Site | Position |
|------|----------|
| LEND | 0 |
| BGL2 | 471 |
| BAM1 | 5559 |
| KPN1 | 17430 |
| KPN1 | 18950 |
| R1 | 19939 |
| SST1 | 21043 |
| XBA1 | 24959 |
| SST1 | 25230 |
| HIN3 | 25618 |
| SST1 | 26344 |
| R1 | 26565 |
| HIN3 | 27916 |
| BAM1 | 28919 |
| BGL2 | 32506 |
| BGL2 | 32640 |
| R1 | 34317 |
| BGL2 | 35963 |
| BGL2 | 36323 |
| BGL2 | 36974 |
| BGL2 | 37034 |
| **** | 37388 |
| BAM1 | 39435 |
| SST1 | 39585 |
| HIN3 | 39627 |
| BAM1 | 39941 |
| KPN1 | 41827 |
| BGL2 | 41891 |
| SST1 | 43447 |
| BAM1 | 43869 |
| BGL2 | 44046 |
| REND | 45359 |

CH-5

| Site | Position |
|------|----------|
| LEND | 0 |
| BGL2 | 471 |
| BAM1 | 5559 |
| KPN1 | 23954 |
| KPN1 | 25474 |
| BAM1 | 27475 |
| SAL1 | 30241 |
| XHO1 | 30499 |
| BAM1 | 31497 |
| BGL2 | 32706 |
| HIN3 | 33928 |
| BGL2 | 34589 |
| R1 | 35444 |
| HIN3 | 35688 |
| HIN3 | 35813 |
| BGL2 | 36332 |
| BGL2 | 36983 |
| BGL2 | 37043 |
| **** | 37397 |
| BAM1 | 39444 |
| SST1 | 39594 |
| HIN3 | 39636 |
| BAM1 | 39950 |
| KPN1 | 41836 |
| BGL2 | 41900 |
| SST1 | 43456 |
| BAM1 | 43878 |
| BGL2 | 44055 |
| REND | 45368 |

## CH-6

| | |
|---|---|
| LEND | 0 |
| BGL2 | 471 |
| BAM1 | 5559 |
| KPN1 | 23954 |
| KPN1 | 25474 |
| BAM1 | 27475 |
| SAL1 | 30241 |
| XHO1 | 30499 |
| BAM1 | 31497 |
| HIN3 | 33507 |
| R1 | 33743 |
| BGL2 | 34645 |
| BGL2 | 34705 |
| **** | 35059 |
| BAM1 | 37106 |
| SST1 | 37256 |
| HIN3 | 37298 |
| BAM1 | 37612 |
| KPN1 | 39498 |
| BGL2 | 39562 |
| SST1 | 41118 |
| BAM1 | 41540 |
| BGL2 | 41717 |
| REND | 43030 |

## CH-7

| | |
|---|---|
| LEND | 0 |
| BGL2 | 471 |
| BAM1 | 5559 |
| KPN1 | 23954 |
| KPN1 | 25474 |
| BAM1 | 27475 |
| SAL1 | 30241 |
| XHO1 | 30499 |
| BAM1 | 31497 |
| HIN3 | 33507 |
| R1 | 33743 |
| BGL2 | 34645 |
| BGL2 | 34705 |
| **** | 35059 |
| REND | 41666 |

## CH-8

| | |
|---|---|
| LEND | 0 |
| BGL2 | 471 |
| BAM1 | 5559 |
| KPN1 | 17430 |
| KPN1 | 18950 |
| R1 | 19939 |
| SST1 | 21043 |
| XBA1 | 24959 |
| SST1 | 25230 |
| HIN3 | 25618 |
| SST1 | 26344 |
| R1 | 26565 |
| HIN3 | 27916 |
| BAM1 | 28511 |
| R1 | 32268 |
| SAL1 | 33309 |
| SAL1 | 33817 |
| XHO1 | 34075 |
| BAM1 | 35073 |
| HIN3 | 37083 |
| R1 | 37319 |
| BGL2 | 38221 |
| BGL2 | 38281 |
| **** | 38635 |
| REND | 45242 |

## CH-9

| | |
|---|---|
| LEND | 0 |
| BGL2 | 471 |
| BAM1 | 5559 |
| KPN1 | 17430 |
| KPN1 | 18950 |
| R1 | 19939 |
| SST1 | 21043 |
| XBA1 | 24959 |
| SST1 | 25230 |
| HIN3 | 25618 |
| SST1 | 26344 |
| R1 | 26565 |
| HIN3 | 27916 |
| BAM1 | 28511 |
| R1 | 32268 |
| SAL1 | 33309 |
| SAL1 | 33817 |
| XHO1 | 34075 |
| BAM1 | 35073 |
| BGL2 | 36282 |
| HIN3 | 37504 |
| BGL2 | 38165 |
| R1 | 39020 |
| HIN3 | 39264 |
| HIN3 | 39389 |
| BGL2 | 39908 |
| BGL2 | 40559 |
| BGL2 | 40619 |
| **** | 40973 |
| REND | 47580 |

## CH-10

| | |
|---|---|
| LEND | 0 |
| BGL2 | 471 |
| BAM1 | 5559 |
| KPN1 | 17430 |
| KPN1 | 18950 |
| R1 | 19939 |
| SST1 | 21043 |
| XBA1 | 24959 |
| SST1 | 25230 |
| HIN3 | 25618 |
| BAM1 | 26621 |
| BGL2 | 30208 |
| BGL2 | 30342 |
| R1 | 32019 |
| BGL2 | 33665 |
| BGL2 | 34025 |
| BGL2 | 34676 |
| BGL2 | 34736 |
| **** | 35090 |
| REND | 41697 |

## CH-11

| | |
|---|---|
| LEND | 0 |
| BGL2 | 471 |
| BAM1 | 5559 |
| KPN1 | 17430 |
| KPN1 | 18950 |
| R1 | 19939 |
| SST1 | 21043 |
| XBA1 | 24959 |
| SST1 | 25230 |
| HIN3 | 25618 |
| BAM1 | 26621 |
| BGL2 | 30208 |
| BGL2 | 30342 |
| R1 | 32019 |
| BGL2 | 33665 |
| BGL2 | 34025 |
| BGL2 | 34676 |
| BGL2 | 34736 |
| **** | 35090 |
| BAM1 | 37137 |
| SST1 | 37287 |
| HIN3 | 37329 |
| BAM1 | 37643 |
| KPN1 | 39529 |
| BGL2 | 39593 |
| SST1 | 41149 |
| BAM1 | 41571 |
| BGL2 | 41748 |
| REND | 43061 |

## CH-12

| | |
|---|---|
| LEND | 0 |
| BGL2 | 471 |
| BAM1 | 5559 |
| KPN1 | 23954 |
| KPN1 | 25474 |
| BAM1 | 26955 |
| R1 | 30712 |
| SAL1 | 31753 |
| SAL1 | 32261 |
| XHO1 | 32519 |
| BAM1 | 33517 |
| BGL2 | 34726 |
| BGL2 | 35086 |
| BGL2 | 35737 |
| BGL2 | 35797 |
| **** | 36151 |
| BAM1 | 38198 |
| SST1 | 38348 |
| HIN3 | 38390 |
| BAM1 | 38704 |
| KPN1 | 40590 |
| BGL2 | 40654 |
| SST1 | 42210 |
| BAM1 | 42632 |
| BGL2 | 42809 |
| REND | 44122 |

CH-13

| | |
|---|---|
| LEND | 0 |
| BGL2 | 471 |
| BAM1 | 5559 |
| KPN1 | 17430 |
| KPN1 | 18950 |
| R1 | 21708 |
| BAM1 | 22843 |
| BGL2 | 22932 |
| HIN3 | 23644 |
| R1 | 27491 |
| SAL1 | 28532 |
| SAL1 | 29040 |
| XHO1 | 29298 |
| BAM1 | 30296 |
| BGL2 | 31505 |
| BGL2 | 31865 |
| BGL2 | 32516 |
| BGL2 | 32576 |
| **** | 32930 |
| BAM1 | 34977 |
| SST1 | 35127 |
| HIN3 | 35169 |
| BAM1 | 35483 |
| KPN1 | 37369 |
| BGL2 | 37433 |
| SST1 | 38989 |
| BAM1 | 39411 |
| BGL2 | 39588 |
| REND | 40901 |

CH-14

| | |
|---|---|
| LEND | 0 |
| BGL2 | 471 |
| BAM1 | 5559 |
| KPN1 | 17430 |
| KPN1 | 18950 |
| R1 | 19939 |
| SST1 | 21043 |
| R1 | 27491 |
| SAL1 | 28532 |
| SAL1 | 29040 |
| XHO1 | 29298 |
| BAM1 | 30296 |
| BGL2 | 31505 |
| BGL2 | 31865 |
| BGL2 | 32516 |

| | |
|---|---|
| BGL2 | 32576 |
| **** | 32930 |
| BAM1 | 34977 |
| SST1 | 35127 |
| HIN3 | 35169 |
| BAM1 | 35483 |
| KPN1 | 37369 |
| BGL2 | 37433 |
| SST1 | 38989 |
| BAM1 | 39411 |
| BGL2 | 39588 |
| REND | 40901 |

CH-15

| | |
|---|---|
| LEND | 0 |
| BGL2 | 471 |
| BAM1 | 5559 |
| KPN1 | 17430 |
| KPN1 | 18950 |
| R1 | 21708 |
| BAM1 | 22843 |
| BGL2 | 22932 |
| HIN3 | 23644 |
| BGL2 | 29031 |
| XHO1 | 29787 |
| BGL2 | 29913 |
| KPN1 | 32658 |
| BGL2 | 33389 |
| SAL1 | 33676 |
| SAL1 | 33829 |
| KPN1 | 34468 |
| BGL2 | 35417 |
| **** | 35431 |
| REND | 42038 |

CH-16

| | |
|---|---|
| LEND | 0 |
| BGL2 | 471 |
| BAM1 | 5559 |
| KPN1 | 17430 |
| KPN1 | 18950 |
| R1 | 19939 |
| SST1 | 21043 |
| BGL2 | 28780 |
| XHO1 | 29536 |
| BGL2 | 29662 |
| KPN1 | 32407 |

| | |
|---|---|
| BGL2 | 33138 |
| SAL1 | 33425 |
| SAL1 | 33578 |
| KPN1 | 34217 |
| BGL2 | 35166 |
| **** | 35180 |
| REND | 41787 |

CH-17

| | |
|---|---|
| LEND | 0 |
| BGL2 | 471 |
| BAM1 | 5559 |
| KPN1 | 17430 |
| KPN1 | 18950 |
| R1 | 19939 |
| SST1 | 21043 |
| XBA1 | 24959 |
| SST1 | 25230 |
| HIN3 | 25618 |
| SST1 | 26344 |
| R1 | 26565 |
| HIN3 | 27916 |
| BAM1 | 28511 |
| SAL1 | 33309 |
| SAL1 | 33817 |
| XHO1 | 34075 |
| BAM1 | 35073 |
| BGL2 | 36282 |
| HIN3 | 37504 |
| HIN3 | 38084 |
| BGL2 | 38728 |
| BGL2 | 39379 |
| BGL2 | 39439 |
| **** | 39793 |
| REND | 46400 |

CH-18

| | |
|---|---|
| LEND | 0 |
| BGL2 | 471 |
| BAM1 | 5559 |
| KPN1 | 17430 |
| KPN1 | 18950 |
| R1 | 19939 |
| SST1 | 21043 |
| XBA1 | 24959 |
| SST1 | 25230 |
| HIN3 | 25618 |

| | |
|---|---|
| BAM1 | 26621 |
| BGL2 | 30208 |
| BGL2 | 30342 |
| R1 | 32019 |
| BGL2 | 33665 |
| BGL2 | 34025 |
| BGL2 | 34676 |
| BGL2 | 34736 |
| **** | 35090 |
| BAM1 | 37699 |
| REND | 44590 |

CH-19A

| | |
|---|---|
| LEND | 0 |
| BGL2 | 471 |
| BAM1 | 5559 |
| KPN1 | 17430 |
| KPN1 | 18950 |
| R1 | 19939 |
| SST1 | 21043 |
| XBA1 | 24959 |
| SST1 | 25230 |
| HIN3 | 25618 |
| SST1 | 26344 |
| R1 | 26565 |
| HIN3 | 27916 |
| BAM1 | 28511 |
| SAL1 | 33309 |
| SAL1 | 33817 |
| XHO1 | 34075 |
| BAM1 | 35073 |
| BGL2 | 36282 |
| HIN3 | 37504 |
| HIN3 | 38084 |
| BGL2 | 38728 |
| BGL2 | 39379 |
| BGL2 | 39439 |
| **** | 39793 |
| BAM1 | 41840 |
| SST1 | 41990 |
| HIN3 | 42032 |
| BAM1 | 42346 |
| KPN1 | 44232 |
| BGL2 | 44296 |
| SST1 | 45852 |
| BAM1 | 46274 |
| BGL2 | 46451 |
| REND | 47764 |

CH-20

| | |
|---|---|
| LEND | 0 |
| BGL2 | 471 |
| BAM1 | 5559 |
| KPN1 | 17430 |
| KPN1 | 18950 |
| R1 | 21708 |
| BAM1 | 22843 |
| BGL2 | 22932 |
| HIN3 | 23644 |
| R1 | 27491 |
| SAL1 | 28532 |
| SAL1 | 29040 |
| XHO1 | 29298 |
| BAM1 | 30296 |
| HIN3 | 32306 |
| R1 | 32542 |
| BGL2 | 33444 |
| BGL2 | 33504 |
| **** | 33858 |
| REND | 40465 |

CH-21A

| | |
|---|---|
| LEND | 0 |
| BGL2 | 471 |
| BAM1 | 5559 |
| KPN1 | 17430 |
| KPN1 | 18950 |
| R1 | 21708 |
| BAM1 | 22843 |
| BGL2 | 22932 |
| HIN3 | 23644 |
| BGL2 | 29031 |
| XHO1 | 29787 |
| BGL2 | 29913 |
| KPN1 | 32658 |
| BGL2 | 33389 |
| SAL1 | 33676 |
| KPN1 | 34315 |
| BGL2 | 35264 |
| **** | 35278 |
| REND | 41885 |

CH-22

| | |
|---|---|
| LEND | 0 |
| BGL2 | 471 |
| BAM1 | 5559 |
| KPN1 | 17430 |
| KPN1 | 18950 |
| SMA1 | 19842 |
| R1 | 21708 |
| BAM1 | 22843 |
| BGL2 | 22932 |
| HIN3 | 23644 |
| XBA1 | 24959 |
| SST1 | 25230 |
| HIN3 | 25618 |
| SST1 | 26344 |
| R1 | 26565 |
| HIN3 | 27916 |
| BAM1 | 28511 |
| SMA1 | 32140 |
| R1 | 32268 |
| SAL1 | 33309 |
| SAL1 | 33817 |
| XHO1 | 34075 |
| BAM1 | 35073 |
| BGL2 | 36282 |
| BGL2 | 36642 |
| BGL2 | 37293 |
| BGL2 | 37353 |
| P22? | 37706 |
| P22? | 40483 |
| HIN3 | 41697 |
| R1 | 42592 |
| REND | 46164 |

CH-23A

| | |
|---|---|
| LEND | 0 |
| BGL2 | 471 |
| BAM1 | 5559 |
| KPN1 | 17430 |
| KPN1 | 18950 |

| | |
|---|---|
| R1 | 19939 |
| SST1 | 21043 |
| XBA1 | 24959 |
| SST1 | 25230 |
| HIN3 | 25618 |
| SST1 | 26344 |
| R1 | 26565 |
| HIN3 | 27916 |
| BAM1 | 28511 |
| BGL2 | 33808 |
| XHO1 | 34564 |
| BGL2 | 34690 |
| KPN1 | 37435 |
| BGL2 | 38166 |
| SAL1 | 38453 |
| SAL1 | 38606 |
| KPN1 | 39245 |
| BGL2 | 40194 |
| **** | 40208 |
| BAM1 | 42255 |
| SST1 | 42405 |
| HIN3 | 42447 |
| BAM1 | 42761 |
| KPN1 | 44647 |
| BGL2 | 44711 |
| SST1 | 46267 |
| BAM1 | 46689 |
| BGL2 | 46866 |
| REND | 48179 |

CH-24A

| | |
|---|---|
| LEND | 0 |
| BGL2 | 471 |
| BAM1 | 5559 |
| KPN1 | 17430 |
| KPN1 | 18950 |
| R1 | 19939 |
| SST1 | 21043 |
| XBA1 | 24959 |
| SST1 | 25230 |
| HIN3 | 25618 |
| SST1 | 26344 |
| R1 | 26565 |
| HIN3 | 27916 |

| | |
|---|---|
| BAM1 | 28919 |
| BGL2 | 32506 |
| BGL2 | 32640 |
| R1 | 34317 |
| BGL2 | 35963 |
| BGL2 | 36323 |
| BGL2 | 36974 |
| BGL2 | 37034 |
| **** | 37388 |
| BAM1 | 39435 |
| SST1 | 39585 |
| HIN3 | 39627 |
| BAM1 | 39941 |
| KPN1 | 41827 |
| BGL2 | 41891 |
| SST1 | 43447 |
| BAM1 | 43869 |
| BGL2 | 44046 |
| REND | 45359 |

CH-25

| | |
|---|---|
| LEND | 0 |
| BGL2 | 471 |
| BAM1 | 5559 |
| KPN1 | 17430 |
| KPN1 | 18950 |
| R1 | 19939 |
| SST1 | 21043 |
| XBA1 | 24959 |
| SST1 | 25230 |
| HIN3 | 25618 |
| SST1 | 26344 |
| R1 | 26565 |
| HIN3 | 27916 |
| BAM1 | 28511 |
| R1 | 32268 |
| SAL1 | 33309 |
| SAL1 | 33817 |
| XHO1 | 34075 |
| SAL1 | 35756 |
| BGL2 | 37044 |
| BGL2 | 37104 |
| **** | 37458 |
| REND | 44065 |

## CH-26

| | |
|---|---|
| LEND | 0 |
| BGL2 | 471 |
| BAM1 | 5559 |
| KPN1 | 17430 |
| KPN1 | 18950 |
| R1 | 21708 |
| BAM1 | 22843 |
| BGL2 | 22932 |
| HIN3 | 23644 |
| XBA1 | 24959 |
| SST1 | 25230 |
| HIN3 | 25618 |
| SST1 | 26344 |
| R1 | 26565 |
| HIN3 | 27916 |
| BAM1 | 28511 |
| R1 | 32268 |
| SAL1 | 33309 |
| SAL1 | 33817 |
| XHO1 | 34075 |
| SAL1 | 35756 |
| BGL2 | 37044 |
| BGL2 | 37104 |
| **** | 37458 |
| REND | 44065 |

| | |
|---|---|
| SAL1 | 33676 |
| KPN1 | 34315 |
| BGL2 | 35264 |
| **** | 35278 |
| REND | 41885 |

## CH-28

| | |
|---|---|
| LEND | 0 |
| BGL2 | 471 |
| KPN1 | 17430 |
| KPN1 | 18950 |
| R1 | 21708 |
| BAM1 | 22843 |
| BGL2 | 22932 |
| HIN3 | 23644 |
| R1 | 27491 |
| SAL1 | 28532 |
| SAL1 | 29040 |
| XHO1 | 29298 |
| BAM1 | 30296 |
| BGL2 | 31505 |
| BGL2 | 31865 |
| BGL2 | 32516 |
| BGL2 | 32576 |
| **** | 32930 |
| REND | 39537 |

## CH-27

| | |
|---|---|
| LEND | 0 |
| BGL2 | 471 |
| KPN1 | 17430 |
| KPN1 | 18950 |
| R1 | 21708 |
| BAM1 | 22843 |
| BGL2 | 22932 |
| HIN3 | 23644 |
| BGL2 | 29031 |
| XHO1 | 29787 |
| BGL2 | 29913 |
| KPN1 | 32658 |
| BGL2 | 33389 |

## CH-29

| | |
|---|---|
| LEND | 0 |
| BGL2 | 471 |
| BAM1 | 5559 |
| KPN1 | 17430 |
| KPN1 | 18950 |
| R1 | 19939 |
| HIN3 | 20183 |
| HIN3 | 20308 |
| BGL2 | 20827 |
| R1 | 21181 |
| HIN3 | 22532 |
| BAM1 | 23127 |
| R1 | 26884 |

| | |
|---|---|
| SAL1 | 27925 |
| SAL1 | 28433 |
| XHO1 | 28691 |
| BAM1 | 29689 |
| BGL2 | 30898 |
| HIN3 | 32120 |
| BGL2 | 32781 |
| R1 | 33636 |
| HIN3 | 33880 |
| HIN3 | 34005 |
| BGL2 | 34524 |
| BGL2 | 35175 |
| BGL2 | 35235 |
| **** | 35589 |
| REND | 42196 |

## L47-1

| | |
|---|---|
| LEND | 0 |
| BGL2 | 471 |
| KPN1 | 17430 |
| KPN1 | 18950 |
| R1 | 21708 |
| HIN3 | 23059 |
| BAM1 | 23654 |
| SAL1 | 28452 |
| SAL1 | 28960 |
| XHO1 | 29218 |
| BAM1 | 30216 |
| HIN3 | 32226 |
| R1 | 32462 |
| BGL2 | 33364 |
| BGL2 | 33424 |
| **** | 33778 |
| REND | 40385 |

## L47-6

| | |
|---|---|
| LEND | 0 |
| BGL2 | 471 |
| KPN1 | 17430 |
| KPN1 | 18950 |
| R1 | 21708 |

| | |
|---|---|
| HIN3 | 23059 |
| BAM1 | 23654 |
| XHO1 | 24652 |
| SAL1 | 24910 |
| SAL1 | 25418 |
| BAM1 | 30216 |
| HIN3 | 32226 |
| R1 | 32462 |
| BGL2 | 33364 |
| BGL2 | 33424 |
| **** | 33778 |
| REND | 40385 |

## VI

| | |
|---|---|
| LEND | 0 |
| BGL2 | 471 |
| BAM1 | 5559 |
| KPN1 | 17430 |
| KPN1 | 18950 |
| R1 | 21708 |
| BAM1 | 22843 |
| BGL2 | 22932 |
| HIN3 | 23644 |
| XBA1 | 24959 |
| SST1 | 25230 |
| HIN3 | 25618 |
| SST1 | 26344 |
| R1 | 26565 |
| HIN3 | 27916 |
| BAM1 | 28511 |
| SAL1 | 33309 |
| SAL1 | 33817 |
| XHO1 | 34075 |
| BAM1 | 35073 |
| BGL2 | 36282 |
| HIN3 | 37504 |
| HIN3 | 38084 |
| BGL2 | 38728 |
| BGL2 | 39379 |
| BGL2 | 39439 |
| **** | 39793 |
| BAM1 | 42402 |
| HIN3 | 44826 |
| REND | 49293 |

### XII

| | |
|---|---|
| LEND | 0 |
| BGL2 | 471 |
| BAM1 | 5559 |
| KPN1 | 17430 |
| KPN1 | 18950 |
| R1 | 21708 |
| HIN3 | 23059 |
| BAM1 | 23654 |
| SAL1 | 28452 |
| SAL1 | 28960 |
| XHO1 | 29218 |
| BAM1 | 30216 |
| BGL2 | 31425 |
| HIN3 | 32647 |
| HIN3 | 33227 |
| BGL2 | 33871 |
| BGL2 | 34522 |
| BGL2 | 34582 |
| **** | 34936 |
| BAM1 | 37545 |
| HIN3 | 39969 |
| REND | 44436 |

### IV

| | |
|---|---|
| LEND | 0 |
| BGL2 | 471 |
| BAM1 | 5559 |
| KPN1 | 17430 |
| KPN1 | 18950 |
| R1 | 23846 |
| SAL1 | 24887 |
| SAL1 | 25395 |
| XHO1 | 25653 |
| BAM1 | 26651 |
| BGL2 | 27860 |
| HIN3 | 29082 |
| HIN3 | 29662 |
| BGL2 | 30306 |
| BGL2 | 30957 |
| BGL2 | 31017 |
| **** | 31371 |
| BAM1 | 33980 |
| HIN3 | 36404 |
| REND | 40871 |

### 573

| | |
|---|---|
| LEND | 0 |
| BGL2 | 471 |
| BAM1 | 5559 |
| KPN1 | 17430 |
| KPN1 | 18950 |
| R1 | 21708 |
| BAM1 | 22591 |
| R1 | 26348 |
| SAL1 | 27389 |
| SAL1 | 27897 |
| XHO1 | 28155 |
| BAM1 | 29153 |
| BGL2 | 30362 |
| HIN3 | 31584 |
| BGL2 | 32228 |
| BGL2 | 32879 |
| BGL2 | 32939 |
| R1 | 33293 |
| R1 | 36328 |
| REND | 39900 |

### 426

| | |
|---|---|
| LEND | 0 |
| BGL2 | 471 |
| BAM1 | 5559 |
| KPN1 | 17430 |
| KPN1 | 18950 |
| R1 | 21708 |
| HIN3 | 23059 |
| BAM1 | 23654 |
| SAL1 | 28452 |
| SAL1 | 28960 |
| XHO1 | 29218 |
| BAM1 | 30216 |
| BGL2 | 31425 |
| HIN3 | 32647 |
| HIN3 | 33227 |
| BGL2 | 33871 |
| BGL2 | 34522 |
| BGL2 | 34582 |
| **** | 34936 |
| BAM1 | 37545 |
| HIN3 | 39969 |
| REND | 44436 |

### XI

| | |
|---|---|
| LEND | 0 |
| BGL2 | 471 |
| BAM1 | 5559 |
| KPN1 | 17430 |
| KPN1 | 18950 |
| R1 | 21708 |
| SAL1 | 22749 |
| SAL1 | 23257 |
| XHO1 | 23515 |
| BAM1 | 24513 |
| BGL2 | 25722 |
| HIN3 | 26944 |
| HIN3 | 27524 |
| BGL2 | 28168 |
| BGL2 | 28819 |
| BGL2 | 28879 |
| **** | 29233 |
| BAM1 | 31842 |
| HIN3 | 34266 |
| REND | 38733 |

### 540

| | |
|---|---|
| LEND | 0 |
| BGL2 | 471 |
| BAM1 | 5559 |
| KPN1 | 17430 |
| KPN1 | 18950 |
| R1 | 21708 |
| HIN3 | 23059 |
| BAM1 | 23654 |
| R1 | 27411 |
| SAL1 | 28452 |
| SAL1 | 28960 |
| XHO1 | 29218 |
| SAL1 | 30899 |
| BGL2 | 32187 |
| BGL2 | 32247 |
| R1 | 32601 |
| R1 | 35636 |
| REND | 39208 |

### 554

| | |
|---|---|
| LEND | 0 |
| BGL2 | 471 |
| BAM1 | 5559 |
| KPN1 | 17430 |
| KPN1 | 18950 |
| HIN3 | 22862 |
| BAM1 | 23457 |
| R1 | 27214 |
| SAL1 | 28255 |
| SAL1 | 28763 |
| XHO1 | 29021 |
| BAM1 | 30019 |
| BGL2 | 31228 |
| HIN3 | 32450 |
| BGL2 | 33094 |
| BGL2 | 33745 |
| BGL2 | 33805 |
| R1 | 34159 |
| R1 | 37194 |
| REND | 40766 |

### 518

| | |
|---|---|
| LEND | 0 |
| BGL2 | 471 |
| BAM1 | 5559 |
| KPN1 | 17430 |
| KPN1 | 18950 |
| BAM1 | 22843 |
| BGL2 | 22932 |
| HIN3 | 23644 |
| BAM1 | 24056 |
| SAL1 | 28854 |
| SAL1 | 29362 |
| XHO1 | 29620 |
| BAM1 | 30618 |
| HIN3 | 32628 |
| R1 | 32864 |
| BGL2 | 33766 |
| BGL2 | 33826 |
| **** | 34180 |
| HIN3 | 36320 |
| REND | 40787 |

REND 41543

### 567

LEND 0
BGL2 471
BAM1 5559
KPN1 17430
KPN1 18950
R1 21708
HIN3 23059
BAM1 23654
SAL1 28452
SAL1 28960
XHO1 29218
BAM1 30216
HIN3 32226
R1 32462
BGL2 33364
BGL2 33424
**** 33778
HIN3 35918
REND 40385

### 569

LEND 0
BGL2 471
BAM1 5559
KPN1 17430
KPN1 18950
R1 21708
HIN3 23059
BAM1 23654
SAL1 28452
SAL1 28960
XHO1 29218
BAM1 30216
BGL2 31425
HIN3 32647
HIN3 33227
BGL2 33871
BGL2 34522
BGL2 34582
**** 34936
HIN3 37076

### 590

LEND 0
BGL2 471
BAM1 5559
KPN1 17430
KPN1 18950
R1 23846
SAL1 24887
SAL1 25395
XHO1 25653
BAM1 26651
HIN3 28661
R1 28897
BGL2 29799
BGL2 29859
R1 30213
BAM1 32822
R1 36141
REND 39713

### 598

LEND 0
BGL2 471
BAM1 5559
KPN1 17430
KPN1 18950
R1 23846
SAL1 24887
SAL1 25395
XHO1 25653
BAM1 26651
HIN3 28661
R1 28897
BGL2 29799
BGL2 29859
R1 30213
BAM1 32822
R1 36141
REND 39713

### 728

LEND 0
BGL2 471
BAM1 5559
KPN1 17430
KPN1 18950
SAL1 23901
SAL1 24409
XHO1 24667
BAM1 25665
HIN3 27675
R1 27911
BGL2 28813
BGL2 28873
R1 29227
BAM1 31836
R1 35155
REND 38727

### 607

LEND 0
BGL2 471
BAM1 5559
KPN1 17430
KPN1 18950
SAL1 24888
SAL1 25396
XHO1 25654
BAM1 26652
HIN3 28662
R1 28898
BGL2 29800
BGL2 29860
**** 30214
BAM1 32823
HIN3 35247
REND 39714

### 641

LEND 0
BGL2 471

BAM1 5559
KPN1 17430
KPN1 18950
BAM1 22646
BGL2 22735
HIN3 23447
XBA1 24762
SST1 25033
HIN3 25421
SST1 26147
R1 26368
SAL1 27409
SAL1 27917
XHO1 28175
BAM1 29173
HIN3 31183
R1 31419
BGL2 32321
BGL2 32381
**** 32735
HIN3 34875
REND 39342

### 781

LEND 0
BGL2 471
BAM1 5559
KPN1 17430
KPN1 18950
R1 21708
SUPE? 21709
SUPE? 31805
R1 31806
SAL1 32847
SAL1 33355
XHO1 33613
BAM1 34611
BGL2 35820
HIN3 37042
HIN3 37622
BGL2 38266
BGL2 38917
BGL2 38977
**** 39331
HIN3 41471
REND 45938

**626**

| | |
|---|---|
| LEND | 0 |
| BGL2 | 471 |
| BAM1 | 5559 |
| KPN1 | 17430 |
| KPN1 | 18950 |
| R1 | 19939 |
| SST1 | 21043 |
| BAM1 | 25201 |
| R1 | 28958 |
| SAL1 | 29999 |
| SAL1 | 30507 |
| XHO1 | 30765 |
| BAM1 | 31763 |
| BGL2 | 32972 |
| HIN3 | 34194 |
| HIN3 | 34774 |
| BGL2 | 35418 |
| BGL2 | 36069 |
| BGL2 | 36129 |
| **** | 36483 |
| HIN3 | 38623 |
| REND | 43090 |

**631**

| | |
|---|---|
| LEND | 0 |
| BGL2 | 471 |
| BAM1 | 5559 |
| KPN1 | 17430 |
| KPN1 | 18950 |
| R1 | 19939 |
| SST1 | 21043 |
| BAM1 | 25201 |
| R1 | 28958 |
| SAL1 | 29999 |
| SAL1 | 30507 |
| XHO1 | 30765 |
| BAM1 | 31763 |
| BGL2 | 32972 |
| BGL2 | 33332 |
| BGL2 | 33983 |
| BGL2 | 34043 |
| **** | 34397 |
| HIN3 | 36537 |
| REND | 41004 |

**647**

| | |
|---|---|
| LEND | 0 |
| BGL2 | 471 |
| BAM1 | 5559 |
| KPN1 | 17430 |
| KPN1 | 18950 |
| R1 | 19939 |
| SST1 | 21043 |
| BAM1 | 25201 |
| R1 | 28958 |
| SAL1 | 29999 |
| SAL1 | 30507 |
| XHO1 | 30765 |
| SAL1 | 32446 |
| BGL2 | 33734 |
| BGL2 | 33794 |
| **** | 34148 |
| HIN3 | 36288 |
| REND | 40755 |

**791**

| | |
|---|---|
| LEND | 0 |
| BGL2 | 471 |
| BAM1 | 5559 |
| KPN1 | 17430 |
| KPN1 | 18950 |
| R1 | 21708 |
| SST1 | 22812 |
| BAM1 | 26970 |
| R1 | 30727 |
| SAL1 | 31768 |
| SAL1 | 32276 |
| XHO1 | 32534 |
| BAM1 | 33532 |
| BGL2 | 34741 |
| BGL2 | 35101 |
| BGL2 | 35752 |
| BGL2 | 35812 |
| **** | 36166 |
| HIN3 | 38306 |
| REND | 42773 |

**816**

| | |
|---|---|
| LEND | 0 |
| BGL2 | 471 |
| BAM1 | 5559 |

(continuation of 816)

| | |
|---|---|
| KPN1 | 17430 |
| KPN1 | 18950 |
| R1 | 19939 |
| SST1 | 21043 |
| XBA1 | 25209 |
| SST1 | 25480 |
| HIN3 | 25868 |
| SST1 | 26594 |
| R1 | 26815 |
| HIN3 | 28166 |
| BAM1 | 28761 |
| R1 | 32518 |
| SAL1 | 33559 |
| SAL1 | 34067 |
| XHO1 | 34325 |
| BGL2 | 37294 |
| BGL2 | 37354 |
| **** | 37708 |
| HIN3 | 39848 |
| REND | 44315 |

**678**

| | |
|---|---|
| LEND | 0 |
| BGL2 | 471 |
| BAM1 | 5559 |
| KPN1 | 17430 |
| KPN1 | 18950 |
| R1 | 21708 |
| HIN3 | 23059 |
| TRPE? | 23060 |
| TRPE? | 28798 |
| HIN3 | 28799 |
| BAM1 | 29394 |
| R1 | 33151 |
| SAL1 | 34192 |
| SAL1 | 34700 |
| XHO1 | 34958 |
| SAL1 | 36639 |
| BGL2 | 37927 |
| BGL2 | 37987 |
| R1 | 38341 |
| R1 | 41376 |
| REND | 44948 |

**679**

| | |
|---|---|
| LEND | 0 |
| BGL2 | 471 |

(continuation of 679)

| | |
|---|---|
| BAM1 | 5559 |
| KPN1 | 17430 |
| KPN1 | 18950 |
| R1 | 21708 |
| HIN3 | 23059 |
| TRPE? | 23060 |
| TRPE? | 28798 |
| HIN3 | 28799 |
| BAM1 | 29394 |
| R1 | 33151 |
| SAL1 | 34192 |
| SAL1 | 34700 |
| XHO1 | 34958 |
| BAM1 | 35956 |
| BGL2 | 37165 |
| HIN3 | 38387 |
| HIN3 | 38967 |
| BGL2 | 39611 |
| BGL2 | 40262 |
| BGL2 | 40322 |
| R1 | 40676 |
| R1 | 43711 |
| REND | 47283 |

**709**

| | |
|---|---|
| LEND | 0 |
| BGL2 | 471 |
| BAM1 | 5559 |
| KPN1 | 17430 |
| KPN1 | 18950 |
| R1 | 21708 |
| HIN3 | 23059 |
| TRPE? | 23060 |
| TRPE? | 28798 |
| HIN3 | 28799 |
| SAL1 | 30161 |
| SAL1 | 30669 |
| XHO1 | 30927 |
| BAM1 | 31925 |
| BGL2 | 33134 |
| HIN3 | 34356 |
| HIN3 | 34936 |
| BGL2 | 35580 |
| BGL2 | 36231 |
| BGL2 | 36291 |
| R1 | 36645 |
| R1 | 39680 |
| REND | 43252 |

**696**

| | |
|---|---|
| LEND | 0 |
| BGL2 | 471 |
| BAM1 | 5559 |
| KPN1 | 17430 |
| KPN1 | 18950 |
| R1 | 21708 |
| HIN3 | 23059 |
| TRPE? | 23060 |
| TRPE? | 28798 |
| HIN3 | 28799 |
| R1 | 29740 |
| SAL1 | 30781 |
| SAL1 | 31289 |
| XHO1 | 31547 |
| BAM1 | 32545 |
| BGL2 | 33754 |
| HIN3 | 34976 |
| HIN3 | 35556 |
| BGL2 | 36200 |
| BGL2 | 36851 |
| BGL2 | 36911 |
| R1 | 37265 |
| R1 | 40300 |
| REND | 43872 |

**760**

| | |
|---|---|
| LEND | 0 |
| BGL2 | 471 |
| BAM1 | 5559 |
| KPN1 | 17430 |
| KPN1 | 18950 |
| R1 | 21708 |
| HIN3 | 23059 |
| TRPE? | 23060 |
| TRPE? | 28798 |
| HIN3 | 28799 |
| SAL1 | 30161 |

| | |
|---|---|
| SAL1 | 30669 |
| XHO1 | 30927 |
| BAM1 | 31925 |
| BGL2 | 33134 |
| BGL2 | 33494 |
| BGL2 | 34145 |
| BGL2 | 34205 |
| R1 | 34559 |
| R1 | 37594 |
| REND | 41166 |

**742**

| | |
|---|---|
| LEND | 0 |
| BGL2 | 471 |
| BAM1 | 5559 |
| KPN1 | 17430 |
| KPN1 | 18950 |
| R1 | 21708 |
| HIN3 | 23059 |
| TRPE? | 23060 |
| TRPE? | 28798 |
| HIN3 | 28799 |
| SAL1 | 30161 |
| SAL1 | 30669 |
| XHO1 | 30927 |
| SAL1 | 32608 |
| BGL2 | 33896 |
| BGL2 | 33956 |
| R1 | 34310 |
| R1 | 37345 |
| REND | 40917 |

**705**

| | |
|---|---|
| LEND | 0 |
| BGL2 | 471 |
| BAM1 | 5559 |
| KPN1 | 17430 |
| KPN1 | 18950 |
| R1 | 21708 |

| | |
|---|---|
| HIN3 | 23059 |
| TRPE? | 23060 |
| TRPE? | 28798 |
| HIN3 | 28799 |
| R1 | 29740 |
| SAL1 | 30781 |
| SAL1 | 31289 |
| XHO1 | 31547 |
| SAL1 | 33228 |
| BGL2 | 34516 |
| BGL2 | 34576 |
| R1 | 34930 |
| R1 | 37965 |
| REND | 41537 |

**762**

| | |
|---|---|
| LEND | 0 |
| BGL2 | 471 |
| BAM1 | 5559 |
| KPN1 | 17430 |
| KPN1 | 18950 |
| R1 | 21708 |
| HIN3 | 23059 |
| SUPF? | 23060 |
| SUPF? | 30978 |
| HIN3 | 30979 |
| SAL1 | 32341 |
| SAL1 | 32849 |
| XHO1 | 33107 |
| BAM1 | 34105 |
| BGL2 | 35314 |
| BGL2 | 35674 |
| BGL2 | 36325 |
| BGL2 | 36385 |
| R1 | 36739 |
| R1 | 39774 |
| REND | 43346 |

**BV1**

| | |
|---|---|
| LEND | 0 |
| BGL2 | 471 |
| KPN1 | 17430 |
| KPN1 | 18950 |
| R1 | 21708 |
| BAM1 | 22843 |
| BGL2 | 22932 |
| HIN3 | 23644 |
| XBA1 | 24959 |
| SST1 | 25230 |
| HIN3 | 25618 |
| SST1 | 26344 |
| R1 | 26565 |
| HIN3 | 27916 |
| BAM1 | 28511 |
| R1 | 32268 |
| SAL1 | 33309 |
| SAL1 | 33817 |
| XHO1 | 34075 |
| SAL1 | 35756 |
| BGL2 | 37044 |
| BGL2 | 37104 |
| R1 | 37458 |
| R1 | 40493 |
| REND | 44065 |

**BV2**

| | |
|---|---|
| LEND | 0 |
| BGL2 | 471 |
| KPN1 | 17430 |
| KPN1 | 18950 |
| R1 | 21708 |
| BAM1 | 22843 |
| R1 | 26600 |
| SAL1 | 27641 |
| SAL1 | 28149 |
| XHO1 | 28407 |
| SAL1 | 30088 |
| BGL2 | 31376 |
| BGL2 | 31436 |
| R1 | 31790 |
| R1 | 34825 |
| REND | 38397 |

## LGT1-LBC

| | |
|---|---|
| LEND | 0 |
| BGL2 | 471 |
| BAM1 | 5559 |
| KPN1 | 17430 |
| KPN1 | 18950 |
| R1 | 21708 |
| BAM1 | 22843 |
| BGL2 | 22932 |
| HIN3 | 23644 |
| XBA1 | 24959 |
| SST1 | 25230 |
| HIN3 | 25618 |
| SST1 | 26344 |
| R1 | 26565 |
| HIN3 | 27916 |
| BAM1 | 28511 |
| R1 | 32268 |
| SAL1 | 33309 |
| SAL1 | 33817 |
| XHO1 | 34075 |
| BAM1 | 35073 |
| BGL2 | 36282 |
| HIN3 | 37504 |
| HIN3 | 38084 |
| BGL2 | 38728 |
| BGL2 | 39379 |
| BGL2 | 39439 |
| **** | 39793 |
| HIN3 | 41961 |
| REND | 46428 |

## LGT1-LC

| | |
|---|---|
| LEND | 0 |
| BGL2 | 471 |
| BAM1 | 5559 |
| KPN1 | 17430 |
| KPN1 | 18950 |
| R1 | 21708 |
| HIN3 | 23059 |
| BAM1 | 23654 |
| R1 | 27411 |
| SAL1 | 28452 |
| SAL1 | 28960 |
| XHO1 | 29218 |
| BAM1 | 30216 |
| BGL2 | 31425 |
| HIN3 | 32647 |
| HIN3 | 33227 |
| BGL2 | 33871 |
| BGL2 | 34522 |
| BGL2 | 34582 |
| **** | 34936 |
| HIN3 | 37104 |
| REND | 41571 |

## LGT1-LC'

| | |
|---|---|
| LEND | 0 |
| BGL2 | 471 |
| BAM1 | 5559 |
| KPN1 | 17430 |
| KPN1 | 18950 |
| R1 | 21708 |
| BAM1 | 25465 |
| HIN3 | 26060 |
| R1 | 27411 |
| SAL1 | 28452 |
| SAL1 | 28960 |
| XHO1 | 29218 |
| BAM1 | 30216 |
| BGL2 | 31425 |
| HIN3 | 32647 |
| HIN3 | 33227 |
| BGL2 | 33871 |
| BGL2 | 34522 |
| BGL2 | 34582 |
| **** | 34936 |
| HIN3 | 37104 |
| REND | 41571 |

## LGT1-LB'

| | |
|---|---|
| LEND | 0 |
| BGL2 | 471 |
| BAM1 | 5559 |
| KPN1 | 17430 |
| KPN1 | 18950 |
| R1 | 21708 |
| SST1 | 21929 |
| HIN3 | 22655 |
| SST1 | 23043 |
| XBA1 | 23314 |
| HIN3 | 24629 |
| BGL2 | 25341 |
| BAM1 | 25430 |
| R1 | 26565 |
| SAL1 | 27606 |
| SAL1 | 28114 |
| XHO1 | 28372 |
| BAM1 | 29370 |
| BGL2 | 30579 |
| HIN3 | 31801 |
| HIN3 | 32381 |
| BGL2 | 33025 |
| BGL2 | 33676 |
| BGL2 | 33736 |
| **** | 34090 |
| HIN3 | 36258 |
| REND | 40725 |

## LGT1-LB

| | |
|---|---|
| LEND | 0 |
| BGL2 | 471 |
| BAM1 | 5559 |
| KPN1 | 17430 |
| KPN1 | 18950 |
| R1 | 21708 |
| BAM1 | 22843 |
| BGL2 | 22932 |
| HIN3 | 23644 |
| XBA1 | 24959 |
| SST1 | 25230 |
| HIN3 | 25618 |
| SST1 | 26344 |
| R1 | 26565 |
| SAL1 | 27606 |
| SAL1 | 28114 |
| XHO1 | 28372 |
| BAM1 | 29370 |
| BGL2 | 30579 |
| HIN3 | 31801 |
| HIN3 | 32381 |
| BGL2 | 33025 |
| BGL2 | 33676 |
| BGL2 | 33736 |
| **** | 34090 |
| HIN3 | 36258 |
| REND | 40725 |

## LGT2

| | |
|---|---|
| LEND | 0 |
| BGL2 | 471 |
| BAM1 | 5559 |
| KPN1 | 17430 |
| KPN1 | 18950 |
| R1 | 21708 |
| HIN3 | 23059 |
| BAM1 | 23654 |
| SAL1 | 28452 |
| SAL1 | 28960 |
| XHO1 | 29218 |
| BAM1 | 30216 |
| BGL2 | 31425 |
| HIN3 | 32647 |
| HIN3 | 33227 |
| BGL2 | 33871 |
| BGL2 | 34522 |
| BGL2 | 34582 |
| **** | 34936 |
| HIN3 | 37104 |
| REND | 41571 |

## LGT3

| | |
|---|---|
| LEND | 0 |
| BGL2 | 471 |
| BAM1 | 5559 |
| KPN1 | 17430 |
| KPN1 | 18950 |
| R1 | 21708 |
| HIN3 | 23059 |
| BAM1 | 23654 |
| SAL1 | 27862 |
| SAL1 | 28370 |
| XHO1 | 28628 |
| BAM1 | 29626 |
| BGL2 | 30835 |
| HIN3 | 32057 |
| HIN3 | 32637 |
| BGL2 | 33281 |
| BGL2 | 33932 |
| BGL2 | 33992 |
| **** | 34346 |
| HIN3 | 36514 |
| REND | 40981 |

| LGT4 | | | |
|---|---|---|---|
| LEND | 0 |
| BGL2 | 471 |
| BAM1 | 5559 |
| KPN1 | 17430 |
| KPN1 | 18950 |
| R1 | 19939 |
| HIN3 | 21290 |
| BAM1 | 21885 |
| SAL1 | 26683 |
| SAL1 | 27191 |
| XHO1 | 27449 |
| BAM1 | 28447 |
| BGL2 | 29656 |
| HIN3 | 30878 |
| HIN3 | 31458 |
| BGL2 | 32102 |
| BGL2 | 32753 |
| BGL2 | 32813 |
| **** | 33167 |
| HIN3 | 35335 |
| REND | 39802 |

| LGT5 | |
|---|---|
| LEND | 0 |
| BGL2 | 471 |
| BAM1 | 5559 |
| KPN1 | 17430 |
| KPN1 | 18950 |
| R1 | 19939 |
| SST1 | 21043 |
| XBA1 | 25209 |
| SST1 | 25480 |
| HIN3 | 25868 |
| SST1 | 26594 |
| R1 | 26815 |
| SAL1 | 27856 |
| SAL1 | 28364 |
| XHO1 | 28622 |
| BAM1 | 29620 |
| BGL2 | 30829 |
| HIN3 | 32051 |

| | |
|---|---|
| HIN3 | 32631 |
| BGL2 | 33275 |
| BGL2 | 33926 |
| BGL2 | 33986 |
| **** | 34340 |
| HIN3 | 36508 |
| REND | 40975 |

| LGT6-AR6 | |
|---|---|
| LEND | 0 |
| BGL2 | 471 |
| BAM1 | 5559 |
| KPN1 | 17430 |
| KPN1 | 18950 |
| R1 | 19939 |
| ARA6? | 19940 |
| ARA6? | 25638 |
| R1 | 25639 |
| HIN3 | 26990 |
| SAL1 | 28977 |
| SAL1 | 29485 |
| XHO1 | 29743 |
| BAM1 | 30741 |
| BGL2 | 31950 |
| HIN3 | 33172 |
| HIN3 | 33752 |
| BGL2 | 34396 |
| BGL2 | 35047 |
| BGL2 | 35107 |
| **** | 35461 |
| HIN3 | 37629 |
| REND | 42096 |

| LGT7-AR6 | |
|---|---|
| LEND | 0 |
| BGL2 | 471 |
| BAM1 | 5559 |
| KPN1 | 17430 |
| KPN1 | 18950 |
| R1 | 19939 |
| ARA6? | 19940 |

| | |
|---|---|
| ARA6? | 25638 |
| R1 | 25639 |
| HIN3 | 26990 |
| SAL1 | 28977 |
| SAL1 | 29485 |
| XHO1 | 29743 |
| BAM1 | 30741 |
| BGL2 | 31950 |
| HIN3 | 33172 |
| BGL2 | 33816 |
| BGL2 | 34467 |
| BGL2 | 34527 |
| **** | 34881 |
| HIN3 | 37049 |
| REND | 41516 |

| LGT8-AR6 | |
|---|---|
| LEND | 0 |
| BGL2 | 471 |
| BAM1 | 5559 |
| KPN1 | 17430 |
| KPN1 | 18950 |
| R1 | 19939 |
| ARA6? | 19940 |
| ARA6? | 25638 |
| R1 | 25639 |
| HIN3 | 26990 |
| BGL2 | 28212 |
| BAM1 | 29421 |
| XHO1 | 30419 |
| SAL1 | 30677 |
| SAL1 | 31185 |
| HIN3 | 33172 |
| BGL2 | 33816 |
| BGL2 | 34467 |
| BGL2 | 34527 |
| **** | 34881 |
| HIN3 | 37049 |
| REND | 41516 |

| LGT30 | |
|---|---|

| | |
|---|---|
| LEND | 0 |
| BGL2 | 471 |
| BAM1 | 5559 |
| KPN1 | 17430 |
| KPN1 | 18950 |
| R1 | 21304 |
| SAL1 | 22345 |
| SAL1 | 22853 |
| COL1? | 22854 |
| COL1? | 28852 |
| SAL1 | 28853 |
| XHO1 | 29111 |
| BGL2 | 32080 |
| BGL2 | 32140 |
| **** | 32494 |
| HIN3 | 34662 |
| REND | 39129 |

| LGT40 | |
|---|---|
| LEND | 0 |
| BGL2 | 471 |
| BAM1 | 5559 |
| KPN1 | 17430 |
| KPN1 | 18950 |
| R1 | 21708 |
| BAM1 | 22843 |
| BGL2 | 22932 |
| HIN3 | 23644 |
| XBA1 | 24959 |
| SST1 | 25230 |
| R1 | 25451 |
| SAL1 | 26492 |
| SAL1 | 27000 |
| XHO1 | 27258 |
| BAM1 | 28256 |
| BGL2 | 29465 |
| HIN3 | 30687 |
| HIN3 | 31267 |
| BGL2 | 31911 |
| BGL2 | 32562 |
| BGL2 | 32622 |
| **** | 32976 |
| HIN3 | 35144 |
| REND | 39611 |

LGTT5622

| | | | | |
|---|---|---|---|---|
| | | BAM1 | 35323 | KPN1 | 18950 |
| | | BGL2 | 36532 | R1 | 21708 |

| | | | | | | |
|---|---|---|---|---|---|
| LGTT5622 | | BAM1 | 35323 | KPN1 | 18950 | 2-G |
| | | BGL2 | 36532 | R1 | 21708 | |
| | | HIN3 | 37754 | BAM1 | 22843 | |
| LEND | 0 | HIN3 | 38334 | BGL2 | 22932 | LEND | 0 |
| BGL2 | 471 | BGL2 | 38978 | HIN3 | 23644 | BGL2 | 471 |
| BAM1 | 5559 | BGL2 | 39629 | XBA1 | 24959 | BAM1 | 5559 |
| KPN1 | 17430 | BGL2 | 39689 | SST1 | 25230 | KPN1 | 17430 |
| KPN1 | 18950 | **** | 40043 | HIN3 | 25618 | KPN1 | 18950 |
| R1 | 21708 | BAM1 | 42652 | SST1 | 26344 | R1 | 19939 |
| T5? | 21709 | HIN3 | 45076 | R1 | 26565 | SST1 | 21043 |
| T5? | 23507 | REND | 49543 | HIN3 | 27916 | XBA1 | 25209 |
| R1 | 23508 | | | BAM1 | 28511 | SST1 | 25480 |
| T5? | 23509 | C | | R1 | 32268 | HIN3 | 25868 |
| T5? | 25307 | | | SAL1 | 33309 | SST1 | 26594 |
| R1 | 25308 | LEND | 0 | SAL1 | 33817 | R1 | 26815 |
| SAL1 | 26349 | BGL2 | 471 | XHO1 | 34075 | HIN3 | 28166 |
| SAL1 | 26857 | BAM1 | 5559 | SAL1 | 35756 | R1 | 30474 |
| XHO1 | 27115 | KPN1 | 17430 | BGL2 | 37044 | SAL1 | 31515 |
| BAM1 | 28113 | KPN1 | 18950 | BGL2 | 37104 | SAL1 | 32023 |
| BGL2 | 29322 | R1 | 19939 | **** | 37458 | XHO1 | 32281 |
| HIN3 | 30544 | SST1 | 21043 | HIN3 | 39626 | SAL1 | 33962 |
| HIN3 | 31124 | XBA1 | 25209 | REND | 44093 | BGL2 | 35250 |
| BGL2 | 31768 | SST1 | 25480 | | | BGL2 | 35310 |
| BGL2 | 32419 | HIN3 | 25868 | 2-E | | **** | 35664 |
| BGL2 | 32479 | SST1 | 26594 | | | HIN3 | 37832 |
| **** | 32833 | R1 | 26815 | LEND | 0 | REND | 42299 |
| HIN3 | 35001 | HIN3 | 28166 | BGL2 | 471 | | |
| REND | 39468 | BAM1 | 28761 | BAM1 | 5559 | 2-I | |
| | | SAL1 | 33559 | KPN1 | 17430 | | |
| B | | SAL1 | 34067 | KPN1 | 18950 | LEND | 0 |
| | | XHO1 | 34325 | R1 | 19939 | BGL2 | 471 |
| LEND | 0 | BAM1 | 35323 | SST1 | 21043 | BAM1 | 5559 |
| BGL2 | 471 | BGL2 | 36532 | XBA1 | 25209 | KPN1 | 17430 |
| BAM1 | 5559 | HIN3 | 37754 | SST1 | 25480 | KPN1 | 18950 |
| KPN1 | 17430 | HIN3 | 38334 | HIN3 | 25868 | R1 | 19939 |
| KPN1 | 18950 | BGL2 | 38978 | SST1 | 26594 | SST1 | 21043 |
| R1 | 19939 | BGL2 | 39629 | R1 | 26815 | R1 | 24940 |
| SST1 | 21043 | BGL2 | 39689 | HIN3 | 28166 | HIN3 | 26291 |
| XBA1 | 25209 | **** | 40043 | BAM1 | 28761 | BAM1 | 26886 |
| SST1 | 25480 | BAM1 | 42652 | R1 | 32518 | R1 | 30643 |
| HIN3 | 25868 | HIN3 | 45076 | SAL1 | 33559 | SAL1 | 31684 |
| SST1 | 26594 | REND | 49543 | SAL1 | 34067 | SAL1 | 32192 |
| R1 | 26815 | | | XHO1 | 34325 | XHO1 | 32450 |
| HIN3 | 28166 | 2-D | | SAL1 | 36006 | SAL1 | 34131 |
| BAM1 | 28761 | | | BGL2 | 37294 | BGL2 | 35419 |
| R1 | 32518 | LEND | 0 | BGL2 | 37354 | BGL2 | 35479 |
| SAL1 | 33559 | BGL2 | 471 | **** | 37708 | **** | 35833 |
| SAL1 | 34067 | BAM1 | 5559 | HIN3 | 39876 | HIN3 | 38001 |
| XHO1 | 34325 | KPN1 | 17430 | REND | 44343 | REND | 42468 |

| JZ-LBC | | | | REND | 46428 | BGL2 | 25341 |
|---|---|---|---|---|---|---|---|
| | | R1 | 26565 | | | BAM1 | 25430 |
| | | HIN3 | 27916 | | | R1 | 26565 |
| | | BAM1 | 28511 | JZ-LB' | | SAL1 | 27606 |
| LEND | 0 | R1 | 32268 | | | SAL1 | 28114 |
| BGL2 | 471 | SAL1 | 33309 | | | XHO1 | 28372 |
| BAM1 | 5559 | SAL1 | 33817 | LEND | 0 | BAM1 | 29370 |
| KPN1 | 17430 | XHO1 | 34075 | BGL2 | 471 | BGL2 | 30579 |
| KPN1 | 18950 | BAM1 | 35073 | BAM1 | 5559 | HIN3 | 31801 |
| R1 | 21708 | BGL2 | 36282 | KPN1 | 17430 | HIN3 | 32381 |
| BAM1 | 22843 | HIN3 | 37504 | KPN1 | 18950 | BGL2 | 33025 |
| BGL2 | 22932 | HIN3 | 38084 | R1 | 21708 | BGL2 | 33676 |
| HIN3 | 23644 | BGL2 | 38728 | SST1 | 21929 | BGL2 | 33736 |
| XBA1 | 24959 | BGL2 | 39379 | HIN3 | 22655 | **** | 34090 |
| SST1 | 25230 | BGL2 | 39439 | SST1 | 23043 | HIN3 | 36258 |
| HIN3 | 25618 | **** | 39793 | XBA1 | 23314 | REND | 40725 |
| SST1 | 26344 | HIN3 | 41961 | HIN3 | 24629 | | |

Table 2

The Cloning Capacities of Each Vector Given as the Minimum and
Maximum Number of Base Pairs of DNA that Can Be Cloned by the Use
of Each Enzyme or Enzyme Combination

---

The minimum size limit of 38 kb was chosen because phages with shorter genomes grow poorly and select spontaneous duplications.  The maximum of 51 kb was chosen because that is the size of a well-characterized clone of Charon 3.  This is a conservative estimate since 53
kb phages have been successfully and stably propagated.  Furthermore,
these limits may differ between host strains, different transfection
and packaging systems, and different growth conditions.

---

CH-1

| ENZYMES | | MAX SIZE | MIN SIZE |
|---|---|---|---|
| R1 | ALONE | 21137 | 8137 |
| SAL1 | ALONE | 2564 | 0 |
| XBA1 | ALONE | 2056 | 0 |
| XHO1 | ALONE | 2056 | 0 |
| XBA1 | SAL1 | 10914 | 0 |
| SAL1 | XHO1 | 2822 | 0 |
| XBA1 | XHO1 | 11172 | 0 |

CH-2

| ENZYMES | | MAX SIZE | MIN SIZE |
|---|---|---|---|
| R1 | ALONE | 4499 | 0 |
| SAL1 | ALONE | 4652 | 0 |
| XHO1 | ALONE | 10019 | 0 |
| R1 | SAL1 | 21488 | 8488 |
| R1 | XHO1 | 17446 | 4446 |
| XHO1 | SAL1 | 14061 | 1061 |

CH-3

| ENZYMES | | MAX SIZE | MIN SIZE |
|---|---|---|---|
| R1 | ALONE | 9447 | 0 |
| SAL1 | ALONE | 2974 | 0 |
| XBA1 | ALONE | 2821 | 0 |
| XHO1 | ALONE | 2821 | 0 |
| R1 | SAL1 | 21488 | 8488 |
| R1 | XHO1 | 17446 | 4446 |
| XBA1 | SAL1 | 16468 | 3468 |
| XHO1 | SAL1 | 6863 | 0 |
| XBA1 | XHO1 | 12426 | 0 |

CH-4

| ENZYMES | | MAX SIZE | MIN SIZE |
|---|---|---|---|
| R1 | ALONE | 20019 | 7019 |
| XBA1 | ALONE | 5641 | 0 |

CH-5

| ENZYMES | | MAX SIZE | MIN SIZE |
|---|---|---|---|
| R1 | ALONE | 5632 | 0 |
| SAL1 | ALONE | 5632 | 0 |
| XHO1 | ALONE | 5632 | 0 |
| SAL1 | R1 | 10835 | 0 |
| XHO1 | R1 | 10577 | 0 |
| SAL1 | XHO1 | 5890 | 0 |

CH-6

| ENZYMES | | MAX SIZE | MIN SIZE |
|---|---|---|---|
| R1 | ALONE | 7970 | 0 |
| SAL1 | ALONE | 7970 | 0 |
| XHO1 | ALONE | 7970 | 0 |
| SAL1 | R1 | 11472 | 0 |
| XHO1 | R1 | 11214 | 0 |
| SAL1 | XHO1 | 8228 | 0 |

CH-7

| ENZYMES | | MAX SIZE | MIN SIZE |
|---|---|---|---|
| HIN3 | ALONE | 9334 | 0 |
| R1 | ALONE | 9334 | 0 |
| SAL1 | ALONE | 9334 | 0 |
| XHO1 | ALONE | 9334 | 0 |
| HIN3 | R1 | 9570 | 0 |
| SAL1 | HIN3 | 12600 | 0 |
| XHO1 | HIN3 | 12342 | 0 |
| SAL1 | R1 | 12836 | 0 |
| XHO1 | R1 | 12578 | 0 |
| SAL1 | XHO1 | 9592 | 0 |

CH-8

| ENZYMES | | MAX SIZE | MIN SIZE |
|---|---|---|---|
| HIN3 | ALONE | 17223 | 4223 |
| R1 | ALONE | 23138 | 10138 |
| SAL1 | ALONE | 6266 | 0 |
| SST1 | ALONE | 11059 | 0 |
| XBA1 | ALONE | 5758 | 0 |
| XHO1 | ALONE | 5758 | 0 |
| SST1 | HIN3 | 21798 | 8798 |
| XBA1 | HIN3 | 17882 | 4882 |
| SST1 | SAL1 | 18532 | 5532 |
| XBA1 | SAL1 | 14616 | 1616 |
| SAL1 | XHO1 | 6524 | 0 |
| SST1 | XHO1 | 18790 | 5790 |
| XBA1 | XHO1 | 14874 | 1874 |

## CH-9

| ENZYMES | | MAX SIZE | MIN SIZE |
|---|---|---|---|
| HIN3 | ALONE | 17191 | 4191 |
| R1 | ALONE | 22501 | 9501 |
| SAL1 | ALONE | 3928 | 0 |
| SST1 | ALONE | 8721 | 0 |
| XBA1 | ALONE | 3420 | 0 |
| XHO1 | ALONE | 3420 | 0 |
| R1 | HIN3 | 22870 | 9870 |
| SST1 | HIN3 | 21766 | 8766 |
| XBA1 | HIN3 | 17850 | 4850 |
| SST1 | SAL1 | 16194 | 3194 |
| XBA1 | SAL1 | 12278 | 0 |
| SAL1 | XHO1 | 4186 | 0 |
| SST1 | XHO1 | 16452 | 3452 |
| XBA1 | XHO1 | 12536 | 0 |

## CH-10

| ENZYMES | | MAX SIZE | MIN SIZE |
|---|---|---|---|
| HIN3 | ALONE | 9303 | 0 |
| R1 | ALONE | 21383 | 8383 |
| SST1 | ALONE | 13490 | 490 |
| XBA1 | ALONE | 9303 | 0 |
| SST1 | HIN3 | 13878 | 878 |
| XBA1 | HIN3 | 9962 | 0 |

## CH-11

| ENZYMES | | MAX SIZE | MIN SIZE |
|---|---|---|---|
| R1 | ALONE | 20019 | 7019 |
| XBA1 | ALONE | 7939 | 0 |

## CH-12

| ENZYMES | | MAX SIZE | MIN SIZE |
|---|---|---|---|
| R1 | ALONE | 6878 | 0 |
| SAL1 | ALONE | 7386 | 0 |
| XHO1 | ALONE | 6878 | 0 |
| R1 | SAL1 | 8427 | 0 |
| R1 | XHO1 | 8685 | 0 |
| SAL1 | XHO1 | 7644 | 0 |

## CH-13

| ENZYMES | | MAX SIZE | MIN SIZE |
|---|---|---|---|
| R1 | ALONE | 15882 | 2882 |
| SAL1 | ALONE | 10607 | 0 |
| XHO1 | ALONE | 10099 | 0 |
| R1 | SAL1 | 17431 | 4431 |
| R1 | XHO1 | 17689 | 4689 |
| SAL1 | XHO1 | 10865 | 0 |

CH-14

| ENZYMES | | MAX SIZE | MIN SIZE |
|---------|---------|----------|----------|
| R1 | ALONE | 17651 | 4651 |
| SAL1 | ALONE | 10607 | 0 |
| XHO1 | ALONE | 10099 | 0 |
| R1 | SAL1 | 19200 | 6200 |
| R1 | XHO1 | 19458 | 6458 |
| SAL1 | XHO1 | 10865 | 0 |

CH-15

| ENZYMES | | MAX SIZE | MIN SIZE |
|---------|---------|----------|----------|
| HIN3 | ALONE | 8962 | 0 |
| R1 | ALONE | 8962 | 0 |
| SAL1 | ALONE | 9115 | 0 |
| XHO1 | ALONE | 8962 | 0 |
| R1 | HIN3 | 10898 | 0 |
| HIN3 | SAL1 | 19147 | 6147 |
| HIN3 | XHO1 | 15105 | 2105 |
| R1 | SAL1 | 21083 | 8083 |
| R1 | XHO1 | 17041 | 4041 |
| XHO1 | SAL1 | 13004 | 4 |

CH-16

| ENZYMES | | MAX SIZE | MIN SIZE |
|---------|---------|----------|----------|
| R1 | ALONE | 9213 | 0 |
| SAL1 | ALONE | 9366 | 0 |
| SST1 | ALONE | 9213 | 0 |
| XHO1 | ALONE | 9213 | 0 |
| R1 | SAL1 | 22852 | 9852 |
| R1 | SST1 | 10317 | 0 |
| R1 | XHO1 | 18810 | 5810 |
| SST1 | SAL1 | 21748 | 8748 |
| XHO1 | SAL1 | 13255 | 255 |
| SST1 | XHO1 | 17706 | 4706 |

## CH-17

| ENZYMES | | MAX SIZE | MIN SIZE |
|---|---|---|---|
| HIN3 | ALONE | 17066 | 4066 |
| R1 | ALONE | 11226 | 0 |
| SAL1 | ALONE | 5108 | 0 |
| SST1 | ALONE | 9901 | 0 |
| XBA1 | ALONE | 4600 | 0 |
| XHO1 | ALONE | 4600 | 0 |
| R1 | HIN3 | 22745 | 9745 |
| SST1 | HIN3 | 21641 | 8641 |
| XBA1 | HIN3 | 17725 | 4725 |
| R1 | SAL1 | 18478 | 5478 |
| R1 | XHO1 | 18736 | 5736 |
| SST1 | SAL1 | 17374 | 4374 |
| XBA1 | SAL1 | 13458 | 458 |
| SAL1 | XHO1 | 5366 | 0 |
| SST1 | XHO1 | 17632 | 4632 |
| XBA1 | XHO1 | 13716 | 716 |

## CH-18

| ENZYMES | | MAX SIZE | MIN SIZE |
|---|---|---|---|
| HIN3 | ALONE | 6410 | 0 |
| R1 | ALONE | 18490 | 5490 |
| SST1 | ALONE | 10597 | 0 |
| XBA1 | ALONE | 6410 | 0 |
| SST1 | HIN3 | 10985 | 0 |
| XBA1 | HIN3 | 7069 | 0 |

## CH-19A

| ENZYMES | | MAX SIZE | MIN SIZE |
|---|---|---|---|
| R1 | ALONE | 9862 | 0 |
| SAL1 | ALONE | 3744 | 0 |
| XBA1 | ALONE | 3236 | 0 |
| XHO1 | ALONE | 3236 | 0 |
| R1 | SAL1 | 17114 | 4114 |
| R1 | XHO1 | 17372 | 4372 |
| XBA1 | SAL1 | 12094 | 0 |
| SAL1 | XHO1 | 4002 | 0 |
| XBA1 | XHO1 | 12352 | 0 |

## CH-20

| ENZYMES | | MAX SIZE | MIN SIZE |
|---|---|---|---|
| HIN3 | ALONE | 19197 | 6197 |
| R1 | ALONE | 21369 | 8369 |
| SAL1 | ALONE | 11043 | 0 |
| XHO1 | ALONE | 10535 | 0 |
| SAL1 | XHO1 | 11301 | 0 |

## CH-21A

| ENZYMES | | MAX SIZE | MIN SIZE |
|---|---|---|---|
| HIN3 | ALONE | 9115 | 0 |
| R1 | ALONE | 9115 | 0 |
| SAL1 | ALONE | 9115 | 0 |
| XHO1 | ALONE | 9115 | 0 |
| R1 | HIN3 | 11051 | 0 |
| HIN3 | SAL1 | 19147 | 6147 |
| HIN3 | XHO1 | 15258 | 2258 |
| R1 | SAL1 | 21083 | 8083 |
| R1 | XHO1 | 17194 | 4194 |
| XHO1 | SAL1 | 13004 | 4 |

## CH-22

| ENZYMES | | MAX SIZE | MIN SIZE |
|---|---|---|---|
| SAL1 | ALONE | 5344 | 0 |
| SST1 | ALONE | 5950 | 0 |
| XBA1 | ALONE | 4836 | 0 |
| XHO1 | ALONE | 4836 | 0 |
| SST1 | SAL1 | 13423 | 423 |
| XBA1 | SAL1 | 13694 | 694 |
| SAL1 | XHO1 | 5602 | 0 |
| XBA1 | SST1 | 6221 | 0 |
| SST1 | XHO1 | 13681 | 681 |
| XBA1 | XHO1 | 13952 | 952 |

## CH-23A

| ENZYMES | | MAX SIZE | MIN SIZE |
|---|---|---|---|
| R1 | ALONE | 9447 | 0 |
| SAL1 | ALONE | 2974 | 0 |
| XBA1 | ALONE | 2821 | 0 |
| XHO1 | ALONE | 2821 | 0 |
| R1 | SAL1 | 21488 | 8488 |
| R1 | XHO1 | 17446 | 4446 |
| XBA1 | SAL1 | 16468 | 3468 |
| XHO1 | SAL1 | 6863 | 0 |
| XBA1 | XHO1 | 12426 | 0 |

## CH-24A

| ENZYMES | | MAX SIZE | MIN SIZE |
|---|---|---|---|
| R1 | ALONE | 20019 | 7019 |
| XBA1 | ALONE | 5641 | 0 |

CH-25

| ENZYMES | | MAX SIZE | MIN SIZE |
|---|---|---|---|
| HIN3 | ALONE | 9233 | 0 |
| R1 | ALONE | 19264 | 6264 |
| SAL1 | ALONE | 9382 | 0 |
| SST1 | ALONE | 12236 | 0 |
| XBA1 | ALONE | 6935 | 0 |
| XHO1 | ALONE | 6935 | 0 |
| HIN3 | SAL1 | 17073 | 4073 |
| SST1 | HIN3 | 13808 | 808 |
| XBA1 | HIN3 | 9892 | 0 |
| HIN3 | XHO1 | 15392 | 2392 |
| R1 | SAL1 | 22752 | 9752 |
| R1 | XHO1 | 21071 | 8071 |
| SST1 | SAL1 | 21648 | 8648 |
| XBA1 | SAL1 | 17732 | 4732 |
| SST1 | XHO1 | 19967 | 6967 |
| XBA1 | XHO1 | 16051 | 3051 |

CH-26

| ENZYMES | | MAX SIZE | MIN SIZE |
|---|---|---|---|
| HIN3 | ALONE | 11207 | 0 |
| R1 | ALONE | 17495 | 4495 |
| SAL1 | ALONE | 9382 | 0 |
| SST1 | ALONE | 8049 | 0 |
| XBA1 | ALONE | 6935 | 0 |
| XHO1 | ALONE | 6935 | 0 |
| HIN3 | SAL1 | 19047 | 6047 |
| HIN3 | XHO1 | 17366 | 4366 |
| R1 | SAL1 | 20983 | 7983 |
| R1 | XHO1 | 19302 | 6302 |
| SST1 | SAL1 | 17461 | 4461 |
| XBA1 | SAL1 | 17732 | 4732 |
| XBA1 | SST1 | 8320 | 0 |
| SST1 | XHO1 | 15780 | 2780 |
| XBA1 | XHO1 | 16051 | 3051 |

## CH-27

| ENZYMES | | MAX SIZE | MIN SIZE |
|---|---|---|---|
| BAM1 | ALONE | 9115 | 0 |
| HIN3 | ALONE | 9115 | 0 |
| R1 | ALONE | 9115 | 0 |
| SAL1 | ALONE | 9115 | 0 |
| XHO1 | ALONE | 9115 | 0 |
| BAM1 | HIN3 | 9916 | 0 |
| R1 | BAM1 | 10250 | 0 |
| BAM1 | SAL1 | 19948 | 6948 |
| BAM1 | XHO1 | 16059 | 3059 |
| R1 | HIN3 | 11051 | 0 |
| HIN3 | SAL1 | 19147 | 6147 |
| HIN3 | XHO1 | 15258 | 2258 |
| R1 | SAL1 | 21083 | 8083 |
| R1 | XHO1 | 17194 | 4194 |
| XHO1 | SAL1 | 13004 | 4 |

## CH-28

| ENZYMES | | MAX SIZE | MIN SIZE |
|---|---|---|---|
| BAM1 | ALONE | 18916 | 5916 |
| HIN3 | ALONE | 11463 | 0 |
| R1 | ALONE | 17246 | 4246 |
| SAL1 | ALONE | 11971 | 0 |
| XHO1 | ALONE | 11463 | 0 |
| R1 | BAM1 | 20051 | 7051 |
| HIN3 | SAL1 | 16859 | 3859 |
| HIN3 | XHO1 | 17117 | 4117 |
| R1 | SAL1 | 18795 | 5795 |
| R1 | XHO1 | 19053 | 6053 |
| SAL1 | XHO1 | 12229 | 0 |

## CH-29

| ENZYMES | | MAX SIZE | MIN SIZE |
|---|---|---|---|
| HIN3 | ALONE | 22626 | 9626 |
| R1 | ALONE | 22501 | 9501 |
| SAL1 | ALONE | 9312 | 0 |
| XHO1 | ALONE | 8804 | 0 |
| R1 | HIN3 | 22870 | 9870 |
| SAL1 | XHO1 | 9570 | 0 |

## L47-1

| ENZYMES | | MAX SIZE | MIN SIZE |
|---|---|---|---|
| BAM1 | ALONE | 17177 | 4177 |
| HIN3 | ALONE | 19782 | 6782 |
| R1 | ALONE | 21369 | 8369 |
| SAL1 | ALONE | 11123 | 0 |
| XHO1 | ALONE | 10615 | 0 |
| SAL1 | XHO1 | 11381 | 0 |

## L47-6

| ENZYMES | | MAX SIZE | MIN SIZE |
|---------|-------|----------|----------|
| BAM1 | ALONE | 17177 | 4177 |
| HIN3 | ALONE | 19782 | 6782 |
| R1 | ALONE | 21369 | 8369 |
| SAL1 | ALONE | 11123 | 0 |
| XHO1 | ALONE | 10615 | 0 |
| XHO1 | SAL1 | 11381 | 0 |

## VI

| ENZYMES | | MAX SIZE | MIN SIZE |
|---------|-------|----------|----------|
| R1 | ALONE | 6564 | 0 |
| SAL1 | ALONE | 2215 | 0 |
| SST1 | ALONE | 2821 | 0 |
| XBA1 | ALONE | 1707 | 0 |
| XHO1 | ALONE | 1707 | 0 |
| R1 | SAL1 | 13816 | 816 |
| R1 | XHO1 | 14074 | 1074 |
| SST1 | SAL1 | 10294 | 0 |
| XBA1 | SAL1 | 10565 | 0 |
| SAL1 | XHO1 | 2473 | 0 |
| XBA1 | SST1 | 3092 | 0 |
| SST1 | XHO1 | 10552 | 0 |
| XBA1 | XHO1 | 10823 | 0 |

## XII

| ENZYMES | | MAX SIZE | MIN SIZE |
|---------|-------|----------|----------|
| R1 | ALONE | 6564 | 0 |
| SAL1 | ALONE | 7072 | 0 |
| XHO1 | ALONE | 6564 | 0 |
| R1 | SAL1 | 13816 | 816 |
| R1 | XHO1 | 14074 | 1074 |
| SAL1 | XHO1 | 7330 | 0 |

## XI

| ENZYMES | | MAX SIZE | MIN SIZE |
|---------|-------|----------|----------|
| R1 | ALONE | 12267 | 0 |
| SAL1 | ALONE | 12775 | 0 |
| XHO1 | ALONE | 12267 | 0 |
| R1 | SAL1 | 13816 | 816 |
| R1 | XHO1 | 14074 | 1074 |
| SAL1 | XHO1 | 13033 | 33 |

IV

| ENZYMES | | MAX SIZE | MIN SIZE |
|---------|---------|----------|----------|
| R1   | ALONE | 10129 | 0 |
| SAL1 | ALONE | 10637 | 0 |
| XHO1 | ALONE | 10129 | 0 |
| R1   | SAL1  | 11678 | 0 |
| R1   | XHO1  | 11936 | 0 |
| SAL1 | XHO1  | 10895 | 0 |

540

| ENZYMES | | MAX SIZE | MIN SIZE |
|---------|---------|----------|----------|
| HIN3 | ALONE | 11792 | 0 |
| SAL1 | ALONE | 14239 | 1239 |
| XHO1 | ALONE | 11792 | 0 |
| HIN3 | SAL1  | 19632 | 6632 |
| HIN3 | XHO1  | 17951 | 4951 |

573

| ENZYMES | | MAX SIZE | MIN SIZE |
|---------|---------|----------|----------|
| HIN3 | ALONE | 11100 | 0 |
| SAL1 | ALONE | 11608 | 0 |
| XHO1 | ALONE | 11100 | 0 |
| SAL1 | HIN3  | 15295 | 2295 |
| XHO1 | HIN3  | 14529 | 1529 |
| SAL1 | XHO1  | 11866 | 0 |

554

| ENZYMES | | MAX SIZE | MIN SIZE |
|---------|---------|----------|----------|
| HIN3 | ALONE | 19822 | 6822 |
| SAL1 | ALONE | 10742 | 0 |
| XHO1 | ALONE | 10234 | 0 |
| SAL1 | XHO1  | 11000 | 0 |

426

| ENZYMES | | MAX SIZE | MIN SIZE |
|---------|---------|----------|----------|
| R1   | ALONE | 6564  | 0 |
| SAL1 | ALONE | 7072  | 0 |
| XHO1 | ALONE | 6564  | 0 |
| R1   | SAL1  | 13816 | 816 |
| R1   | XHO1  | 14074 | 1074 |
| SAL1 | XHO1  | 7330  | 0 |

### 518

| ENZYMES | | MAX SIZE | MIN SIZE |
|---------|---------|----------|----------|
| R1 | ALONE | 10213 | 0 |
| SAL1 | ALONE | 10721 | 0 |
| XHO1 | ALONE | 10213 | 0 |
| SAL1 | R1 | 14223 | 1223 |
| XHO1 | R1 | 13457 | 457 |
| SAL1 | XHO1 | 10979 | 0 |

### 567

| ENZYMES | | MAX SIZE | MIN SIZE |
|---------|---------|----------|----------|
| R1 | ALONE | 21369 | 8369 |
| SAL1 | ALONE | 11123 | 0 |
| XHO1 | ALONE | 10615 | 0 |
| SAL1 | XHO1 | 11381 | 0 |

### 569

| ENZYMES | | MAX SIZE | MIN SIZE |
|---------|---------|----------|----------|
| R1 | ALONE | 9457 | 0 |
| SAL1 | ALONE | 9965 | 0 |
| XHO1 | ALONE | 9457 | 0 |
| R1 | SAL1 | 16709 | 3709 |
| R1 | XHO1 | 16967 | 3967 |
| SAL1 | XHO1 | 10223 | 0 |

### 590

| ENZYMES | | MAX SIZE | MIN SIZE |
|---------|---------|----------|----------|
| HIN3 | ALONE | 11287 | 0 |
| SAL1 | ALONE | 11795 | 0 |
| XHO1 | ALONE | 11287 | 0 |
| SAL1 | HIN3 | 15061 | 2061 |
| XHO1 | HIN3 | 14295 | 1295 |
| SAL1 | XHO1 | 12053 | 0 |

### 598

| ENZYMES | | MAX SIZE | MIN SIZE |
|---------|---------|----------|----------|
| HIN3 | ALONE | 11287 | 0 |
| SAL1 | ALONE | 11795 | 0 |
| XHO1 | ALONE | 11287 | 0 |
| SAL1 | HIN3 | 15061 | 2061 |
| XHO1 | HIN3 | 14295 | 1295 |
| SAL1 | XHO1 | 12053 | 0 |

### 728

| ENZYMES | | MAX SIZE | MIN SIZE |
|---|---|---|---|
| HIN3 | ALONE | 12273 | 0 |
| SAL1 | ALONE | 12781 | 0 |
| XHO1 | ALONE | 12273 | 0 |
| SAL1 | HIN3 | 16047 | 3047 |
| XHO1 | HIN3 | 15281 | 2281 |
| SAL1 | XHO1 | 13039 | 39 |

### 607

| ENZYMES | | MAX SIZE | MIN SIZE |
|---|---|---|---|
| R1 | ALONE | 11286 | 0 |
| SAL1 | ALONE | 11794 | 0 |
| XHO1 | ALONE | 11286 | 0 |
| SAL1 | R1 | 15296 | 2296 |
| XHO1 | R1 | 14530 | 1530 |
| SAL1 | XHO1 | 12052 | 0 |

### 641

| ENZYMES | | MAX SIZE | MIN SIZE |
|---|---|---|---|
| R1 | ALONE | 16709 | 3709 |
| SAL1 | ALONE | 12166 | 0 |
| SST1 | ALONE | 12772 | 0 |
| XBA1 | ALONE | 11658 | 0 |
| XHO1 | ALONE | 11658 | 0 |
| SST1 | R1 | 18044 | 5044 |
| XBA1 | R1 | 18315 | 5315 |
| SST1 | SAL1 | 14542 | 1542 |
| XBA1 | SAL1 | 14813 | 1813 |
| SAL1 | XHO1 | 12424 | 0 |
| XBA1 | SST1 | 13043 | 43 |
| SST1 | XHO1 | 14800 | 1800 |
| XBA1 | XHO1 | 15071 | 2071 |

### 781

| ENZYMES | | MAX SIZE | MIN SIZE |
|---|---|---|---|
| R1 | ALONE | 15160 | 2160 |
| SAL1 | ALONE | 5570 | 0 |
| XHO1 | ALONE | 5062 | 0 |
| R1 | SAL1 | 16709 | 3709 |
| R1 | XHO1 | 16967 | 3967 |
| SAL1 | XHO1 | 5828 | 0 |

626

| ENZYMES | | MAX SIZE | MIN SIZE |
|---------|-------|----------|----------|
| R1 | ALONE | 16929 | 3929 |
| SAL1 | ALONE | 8418 | 0 |
| SST1 | ALONE | 7910 | 0 |
| XHO1 | ALONE | 7910 | 0 |
| R1 | SAL1 | 18478 | 5478 |
| R1 | XHO1 | 18736 | 5736 |
| SST1 | SAL1 | 17374 | 4374 |
| SAL1 | XHO1 | 8676 | 0 |
| SST1 | XHO1 | 17632 | 4632 |

631

| ENZYMES | | MAX SIZE | MIN SIZE |
|---------|-------|----------|----------|
| R1 | ALONE | 19015 | 6015 |
| SAL1 | ALONE | 10504 | 0 |
| SST1 | ALONE | 9996 | 0 |
| XHO1 | ALONE | 9996 | 0 |
| R1 | SAL1 | 20564 | 7564 |
| R1 | XHO1 | 20822 | 7822 |
| SST1 | SAL1 | 19460 | 6460 |
| SAL1 | XHO1 | 10762 | 0 |
| SST1 | XHO1 | 19718 | 6718 |

647

| ENZYMES | | MAX SIZE | MIN SIZE |
|---------|-------|----------|----------|
| R1 | ALONE | 19264 | 6264 |
| SAL1 | ALONE | 12692 | 0 |
| SST1 | ALONE | 10245 | 0 |
| XHO1 | ALONE | 10245 | 0 |
| R1 | SAL1 | 22752 | 9752 |
| R1 | XHO1 | 21071 | 8071 |
| SST1 | SAL1 | 21648 | 8648 |
| SST1 | XHO1 | 19967 | 6967 |

791

| ENZYMES | | MAX SIZE | MIN SIZE |
|---------|-------|----------|----------|
| R1 | ALONE | 17246 | 4246 |
| SAL1 | ALONE | 8735 | 0 |
| SST1 | ALONE | 8227 | 0 |
| XHO1 | ALONE | 8227 | 0 |
| R1 | SAL1 | 18795 | 5795 |
| R1 | XHO1 | 19053 | 6053 |
| SST1 | SAL1 | 17691 | 4691 |
| SAL1 | XHO1 | 8993 | 0 |
| SST1 | XHO1 | 17949 | 4949 |

816

| ENZYMES | | MAX SIZE | MIN SIZE |
|---------|---------|----------|----------|
| R1 | ALONE | 19264 | 6264 |
| SAL1 | ALONE | 7193 | 0 |
| SST1 | ALONE | 12236 | 0 |
| XBA1 | ALONE | 6685 | 0 |
| XHO1 | ALONE | 6685 | 0 |
| R1 | SAL1 | 20813 | 7813 |
| R1 | XHO1 | 21071 | 8071 |
| SST1 | SAL1 | 19709 | 6709 |
| XBA1 | SAL1 | 15543 | 2543 |
| SAL1 | XHO1 | 7451 | 0 |
| SST1 | XHO1 | 19967 | 6967 |
| XBA1 | XHO1 | 15801 | 2801 |

678

| ENZYMES | | MAX SIZE | MIN SIZE |
|---------|---------|----------|----------|
| HIN3 | ALONE | 11792 | 0 |
| SAL1 | ALONE | 8499 | 0 |
| XHO1 | ALONE | 6052 | 0 |
| HIN3 | SAL1 | 19632 | 6632 |
| HIN3 | XHO1 | 17951 | 4951 |

679

| ENZYMES | | MAX SIZE | MIN SIZE |
|---------|---------|----------|----------|
| HIN3 | ALONE | 19625 | 6625 |
| SAL1 | ALONE | 4225 | 0 |
| XHO1 | ALONE | 3717 | 0 |
| SAL1 | XHO1 | 4483 | 0 |

709

| ENZYMES | | MAX SIZE | MIN SIZE |
|---------|---------|----------|----------|
| HIN3 | ALONE | 19625 | 6625 |
| SAL1 | ALONE | 8256 | 0 |
| XHO1 | ALONE | 7748 | 0 |
| SAL1 | XHO1 | 8514 | 0 |

696

| ENZYMES | | MAX SIZE | MIN SIZE |
|---------|---------|----------|----------|
| HIN3 | ALONE | 19625 | 6625 |
| SAL1 | ALONE | 7636 | 0 |
| XHO1 | ALONE | 7128 | 0 |
| SAL1 | XHO1 | 7894 | 0 |

### 760

| ENZYMES | | MAX SIZE | MIN SIZE |
|---------|---------|----------|----------|
| HIN3 | ALONE | 15574 | 2574 |
| SAL1 | ALONE | 10342 | 0 |
| XHO1 | ALONE | 9834 | 0 |
| HIN3 | SAL1 | 17444 | 4444 |
| HIN3 | XHO1 | 17702 | 4702 |
| SAL1 | XHO1 | 10600 | 0 |

### 742

| ENZYMES | | MAX SIZE | MIN SIZE |
|---------|---------|----------|----------|
| HIN3 | ALONE | 15823 | 2823 |
| SAL1 | ALONE | 12530 | 0 |
| XHO1 | ALONE | 10083 | 0 |
| HIN3 | SAL1 | 19632 | 6632 |
| HIN3 | XHO1 | 17951 | 4951 |

### 705

| ENZYMES | | MAX SIZE | MIN SIZE |
|---------|---------|----------|----------|
| HIN3 | ALONE | 15203 | 2203 |
| SAL1 | ALONE | 11910 | 0 |
| XHO1 | ALONE | 9463 | 0 |
| HIN3 | SAL1 | 19632 | 6632 |
| HIN3 | XHO1 | 17951 | 4951 |

### 762

| ENZYMES | | MAX SIZE | MIN SIZE |
|---------|---------|----------|----------|
| HIN3 | ALONE | 15574 | 2574 |
| SAL1 | ALONE | 8162 | 0 |
| XHO1 | ALONE | 7654 | 0 |
| HIN3 | SAL1 | 17444 | 4444 |
| HIN3 | XHO1 | 17702 | 4702 |
| SAL1 | XHO1 | 8420 | 0 |

## BV1

| ENZYMES | | MAX SIZE | MIN SIZE |
|---------|------|----------|----------|
| BAM1 | ALONE | 12603 | 0 |
| HIN3 | ALONE | 11207 | 0 |
| SAL1 | ALONE | 9382 | 0 |
| SST1 | ALONE | 8049 | 0 |
| XBA1 | ALONE | 6935 | 0 |
| XHO1 | ALONE | 6935 | 0 |
| BAM1 | SAL1 | 19848 | 6848 |
| BAM1 | XHO1 | 18167 | 5167 |
| HIN3 | SAL1 | 19047 | 6047 |
| HIN3 | XHO1 | 17366 | 4366 |
| SST1 | SAL1 | 17461 | 4461 |
| XBA1 | SAL1 | 17732 | 4732 |
| XBA1 | SST1 | 8320 | 0 |
| SST1 | XHO1 | 15780 | 2780 |
| XBA1 | XHO1 | 16051 | 3051 |

## BV2

| ENZYMES | | MAX SIZE | MIN SIZE |
|---------|------|----------|----------|
| BAM1 | ALONE | 12603 | 0 |
| SAL1 | ALONE | 15050 | 2050 |
| XHO1 | ALONE | 12603 | 0 |
| BAM1 | SAL1 | 19848 | 6848 |
| BAM1 | XHO1 | 18167 | 5167 |

## LGT1-LBC

| ENZYMES | | MAX SIZE | MIN SIZE |
|---------|------|----------|----------|
| R1 | ALONE | 15132 | 2132 |
| SAL1 | ALONE | 5080 | 0 |
| SST1 | ALONE | 5686 | 0 |
| XBA1 | ALONE | 4572 | 0 |
| XHO1 | ALONE | 4572 | 0 |
| R1 | SAL1 | 16681 | 3681 |
| R1 | XHO1 | 16939 | 3939 |
| SST1 | SAL1 | 13159 | 159 |
| XBA1 | SAL1 | 13430 | 430 |
| SAL1 | XHO1 | 5338 | 0 |
| XBA1 | SST1 | 5957 | 0 |
| SST1 | XHO1 | 13417 | 417 |
| XBA1 | XHO1 | 13688 | 688 |

## LGT1-LC

| ENZYMES | | MAX SIZE | MIN SIZE |
|---|---|---|---|
| R1 | ALONE | 15132 | 2132 |
| SAL1 | ALONE | 9937 | 0 |
| XHO1 | ALONE | 9429 | 0 |
| R1 | SAL1 | 16681 | 3681 |
| R1 | XHO1 | 16939 | 3939 |
| SAL1 | XHO1 | 10195 | 0 |

## LGT1-LC'

| ENZYMES | | MAX SIZE | MIN SIZE |
|---|---|---|---|
| R1 | ALONE | 15132 | 2132 |
| SAL1 | ALONE | 9937 | 0 |
| XHO1 | ALONE | 9429 | 0 |
| R1 | SAL1 | 16681 | 3681 |
| R1 | XHO1 | 16939 | 3939 |
| SAL1 | XHO1 | 10195 | 0 |

## LGT1-LB'

| ENZYMES | | MAX SIZE | MIN SIZE |
|---|---|---|---|
| R1 | ALONE | 15132 | 2132 |
| SAL1 | ALONE | 10783 | 0 |
| SST1 | ALONE | 11389 | 0 |
| XBA1 | ALONE | 10275 | 0 |
| XHO1 | ALONE | 10275 | 0 |
| R1 | SAL1 | 16681 | 3681 |
| R1 | XHO1 | 16939 | 3939 |
| SST1 | SAL1 | 16460 | 3460 |
| XBA1 | SAL1 | 15075 | 2075 |
| SAL1 | XHO1 | 11041 | 0 |
| SST1 | XBA1 | 11660 | 0 |
| SST1 | XHO1 | 16718 | 3718 |
| XBA1 | XHO1 | 15333 | 2333 |

## LGT1-LB

| ENZYMES | | MAX SIZE | MIN SIZE |
|---|---|---|---|
| R1 | ALONE | 15132 | 2132 |
| SAL1 | ALONE | 10783 | 0 |
| SST1 | ALONE | 11389 | 0 |
| XBA1 | ALONE | 10275 | 0 |
| XHO1 | ALONE | 10275 | 0 |
| R1 | SAL1 | 16681 | 3681 |
| R1 | XHO1 | 16939 | 3939 |
| SST1 | SAL1 | 13159 | 159 |
| XBA1 | SAL1 | 13430 | 430 |
| SAL1 | XHO1 | 11041 | 0 |
| XBA1 | SST1 | 11660 | 0 |
| SST1 | XHO1 | 13417 | 417 |
| XBA1 | XHO1 | 13688 | 688 |

## LGT2

| ENZYMES | | MAX SIZE | MIN SIZE |
|---|---|---|---|
| R1 | ALONE | 9429 | 0 |
| SAL1 | ALONE | 9937 | 0 |
| XHO1 | ALONE | 9429 | 0 |
| R1 | SAL1 | 16681 | 3681 |
| R1 | XHO1 | 16939 | 3939 |
| SAL1 | XHO1 | 10195 | 0 |

## LGT3

| ENZYMES | | MAX SIZE | MIN SIZE |
|---|---|---|---|
| R1 | ALONE | 10019 | 0 |
| SAL1 | ALONE | 10527 | 0 |
| XHO1 | ALONE | 10019 | 0 |
| R1 | SAL1 | 16681 | 3681 |
| R1 | XHO1 | 16939 | 3939 |
| SAL1 | XHO1 | 10785 | 0 |

## LGT4

| ENZYMES | | MAX SIZE | MIN SIZE |
|---|---|---|---|
| R1 | ALONE | 11198 | 0 |
| SAL1 | ALONE | 11706 | 0 |
| XHO1 | ALONE | 11198 | 0 |
| R1 | SAL1 | 18450 | 5450 |
| R1 | XHO1 | 18708 | 5708 |
| SAL1 | XHO1 | 11964 | 0 |

## LGT5

| ENZYMES | | MAX SIZE | MIN SIZE |
|---|---|---|---|
| R1 | ALONE | 16901 | 3901 |
| SAL1 | ALONE | 10533 | 0 |
| SST1 | ALONE | 15576 | 2576 |
| XBA1 | ALONE | 10025 | 0 |
| XHO1 | ALONE | 10025 | 0 |
| R1 | SAL1 | 18450 | 5450 |
| R1 | XHO1 | 18708 | 5708 |
| SST1 | SAL1 | 17346 | 4346 |
| XBA1 | SAL1 | 13180 | 180 |
| SAL1 | XHO1 | 10791 | 0 |
| SST1 | XHO1 | 17604 | 4604 |
| XBA1 | XHO1 | 13438 | 438 |

## LGT6-AR6

| ENZYMES | | MAX SIZE | MIN SIZE |
|---|---|---|---|
| R1 | ALONE | 14604 | 1604 |
| SAL1 | ALONE | 9412 | 0 |
| XHO1 | ALONE | 8904 | 0 |
| R1 | SAL1 | 18450 | 5450 |
| R1 | XHO1 | 18708 | 5708 |
| SAL1 | XHO1 | 9670 | 0 |

## LGT7-AR6

| ENZYMES | | MAX SIZE | MIN SIZE |
|---|---|---|---|
| R1 | ALONE | 15184 | 2184 |
| SAL1 | ALONE | 9992 | 0 |
| XHO1 | ALONE | 9484 | 0 |
| R1 | SAL1 | 19030 | 6030 |
| R1 | XHO1 | 19288 | 6288 |
| SAL1 | XHO1 | 10250 | 0 |

## LGT8-AR6

| ENZYMES | | MAX SIZE | MIN SIZE |
|---|---|---|---|
| R1 | ALONE | 15184 | 2184 |
| SAL1 | ALONE | 9992 | 0 |
| XHO1 | ALONE | 9484 | 0 |
| R1 | SAL1 | 20730 | 7730 |
| R1 | XHO1 | 19964 | 6964 |
| XHO1 | SAL1 | 10250 | 0 |

## LGT30

| ENZYMES | | MAX SIZE | MIN SIZE |
|---|---|---|---|
| R1 | ALONE | 11871 | 0 |
| SAL1 | ALONE | 20318 | 7318 |
| XHO1 | ALONE | 11871 | 0 |
| R1 | SAL1 | 21359 | 8359 |
| R1 | XHO1 | 19678 | 6678 |

## LGT40

| ENZYMES | | MAX SIZE | MIN SIZE |
|---|---|---|---|
| R1 | ALONE | 15132 | 2132 |
| SAL1 | ALONE | 11897 | 0 |
| SST1 | ALONE | 11389 | 0 |
| XBA1 | ALONE | 11389 | 0 |
| XHO1 | ALONE | 11389 | 0 |
| R1 | SAL1 | 16681 | 3681 |
| R1 | XHO1 | 16939 | 3939 |
| SST1 | SAL1 | 13159 | 159 |
| XBA1 | SAL1 | 13430 | 430 |
| SAL1 | XHO1 | 12155 | 0 |
| XBA1 | SST1 | 11660 | 0 |
| SST1 | XHO1 | 13417 | 417 |
| XBA1 | XHO1 | 13688 | 688 |

LGTT5622

| ENZYMES | | MAX SIZE | MIN SIZE |
|---|---|---|---|
| R1 | ALONE | 15132 | 2132 |
| SAL1 | ALONE | 12040 | 0 |
| XHO1 | ALONE | 11532 | 0 |
| R1 | SAL1 | 16681 | 3681 |
| R1 | XHO1 | 16939 | 3939 |
| SAL1 | XHO1 | 12298 | 0 |

B

| ENZYMES | | MAX SIZE | MIN SIZE |
|---|---|---|---|
| R1 | ALONE | 14036 | 1036 |
| SAL1 | ALONE | 1965 | 0 |
| SST1 | ALONE | 7008 | 0 |
| XBA1 | ALONE | 1457 | 0 |
| XHO1 | ALONE | 1457 | 0 |
| R1 | SAL1 | 15585 | 2585 |
| R1 | XHO1 | 15843 | 2843 |
| SST1 | SAL1 | 14481 | 1481 |
| XBA1 | SAL1 | 10315 | 0 |
| SAL1 | XHO1 | 2223 | 0 |
| SST1 | XHO1 | 14739 | 1739 |
| XBA1 | XHO1 | 10573 | 0 |

C

| ENZYMES | | MAX SIZE | MIN SIZE |
|---|---|---|---|
| R1 | ALONE | 8333 | 0 |
| SAL1 | ALONE | 1965 | 0 |
| SST1 | ALONE | 7008 | 0 |
| XBA1 | ALONE | 1457 | 0 |
| XHO1 | ALONE | 1457 | 0 |
| R1 | SAL1 | 15585 | 2585 |
| R1 | XHO1 | 15843 | 2843 |
| SST1 | SAL1 | 14481 | 1481 |
| XBA1 | SAL1 | 10315 | 0 |
| SAL1 | XHO1 | 2223 | 0 |
| SST1 | XHO1 | 14739 | 1739 |
| XBA1 | XHO1 | 10573 | 0 |

2-D

| ENZYMES | | MAX SIZE | MIN SIZE |
|---------|-------|----------|----------|
| R1      | ALONE | 17467    | 4467     |
| SAL1    | ALONE | 9354     | 0        |
| SST1    | ALONE | 8021     | 0        |
| XBA1    | ALONE | 6907     | 0        |
| XHO1    | ALONE | 6907     | 0        |
| R1      | SAL1  | 20955    | 7955     |
| R1      | XHO1  | 19274    | 6274     |
| SST1    | SAL1  | 17433    | 4433     |
| XBA1    | SAL1  | 17704    | 4704     |
| XBA1    | SST1  | 8292     | 0        |
| SST1    | XHO1  | 15752    | 2752     |
| XBA1    | XHO1  | 16023    | 3023     |

2-E

| ENZYMES | | MAX SIZE | MIN SIZE |
|---------|-------|----------|----------|
| R1      | ALONE | 19236    | 6236     |
| SAL1    | ALONE | 9104     | 0        |
| SST1    | ALONE | 12208    | 0        |
| XBA1    | ALONE | 6657     | 0        |
| XHO1    | ALONE | 6657     | 0        |
| R1      | SAL1  | 22724    | 9724     |
| R1      | XHO1  | 21043    | 8043     |
| SST1    | SAL1  | 21620    | 8620     |
| XBA1    | SAL1  | 17454    | 4454     |
| SST1    | XHO1  | 19939    | 6939     |
| XBA1    | XHO1  | 15773    | 2773     |

2-G

| ENZYMES | | MAX SIZE | MIN SIZE |
|---------|-------|----------|----------|
| R1      | ALONE | 19236    | 6236     |
| SAL1    | ALONE | 11148    | 0        |
| SST1    | ALONE | 14252    | 1252     |
| XBA1    | ALONE | 8701     | 0        |
| XHO1    | ALONE | 8701     | 0        |
| R1      | SAL1  | 22724    | 9724     |
| R1      | XHO1  | 21043    | 8043     |
| SST1    | SAL1  | 21620    | 8620     |
| XBA1    | SAL1  | 17454    | 4454     |
| SST1    | XHO1  | 19939    | 6939     |
| XBA1    | XHO1  | 15773    | 2773     |

### 2-1

| ENZYMES | | MAX SIZE | MIN SIZE |
|---------|---------|----------|----------|
| R1 | ALONE | 19236 | 6236 |
| SAL1 | ALONE | 10979 | 0 |
| SST1 | ALONE | 8532 | 0 |
| XHO1 | ALONE | 8532 | 0 |
| R1 | SAL1 | 22724 | 9724 |
| R1 | XHO1 | 21043 | 8043 |
| SST1 | SAL1 | 21620 | 8620 |
| SST1 | XHO1 | 19939 | 6939 |

### JZ-LBC

| ENZYMES | | MAX SIZE | MIN SIZE |
|---------|---------|----------|----------|
| R1 | ALONE | 15132 | 2132 |
| SAL1 | ALONE | 5080 | 0 |
| SST1 | ALONE | 5686 | 0 |
| XBA1 | ALONE | 4572 | 0 |
| XHO1 | ALONE | 4572 | 0 |
| R1 | SAL1 | 16681 | 3681 |
| R1 | XHO1 | 16939 | 3939 |
| SST1 | SAL1 | 13159 | 159 |
| XBA1 | SAL1 | 13430 | 430 |
| SAL1 | XHO1 | 5338 | 0 |
| XBA1 | SST1 | 5957 | 0 |
| SST1 | XHO1 | 13417 | 417 |
| XBA1 | XHO1 | 13688 | 688 |

### JZ-LB'

| ENZYMES | | MAX SIZE | MIN SIZE |
|---------|---------|----------|----------|
| R1 | ALONE | 15132 | 2132 |
| SAL1 | ALONE | 10783 | 0 |
| SST1 | ALONE | 11389 | 0 |
| XBA1 | ALONE | 10275 | 0 |
| XHO1 | ALONE | 10275 | 0 |
| R1 | SAL1 | 16681 | 3681 |
| R1 | XHO1 | 16939 | 3939 |
| SST1 | SAL1 | 16460 | 3460 |
| XBA1 | SAL1 | 15075 | 2075 |
| SAL1 | XHO1 | 11041 | 0 |
| SST1 | XBA1 | 11660 | 0 |
| SST1 | XHO1 | 16718 | 3718 |
| XBA1 | XHO1 | 15333 | 2333 |

# INDEX

AccI, 194
Adenovirus, 74
  capping, 105
  DNA sequencing, 190,197
  E3 promoter, 107
  early genes, 105
  genome map, 104
  late genes, 105
  poly(A), 105
  splicing, 105
  structural protein fiber gene,
    109
  SV40 hybrids, 103-114
  vectors, 103
ADP glycosyl transferase, 23
Affinity chromatography, 43
Aggammaglobulinemic horse serum,
  37,39
Agrobacterium plasmids, 157
Albumin mRNA, 61
Alcohol dehydrogenase-1 locus, 23
Aleurone, 22
Alkaline hydrolysis of DNA, 51
Alkaline phosphatase, 40,173,174
  206,207,209,210
AluI, 11,153,154,197
Alum in adjuvant, 34
Amplification of cosmids, 143
Amplification of plasmids,
  126,128,135
Annealing rate, 50
Anthocyanin mutations, 22
Anthranilate synthetase, 218,219
Antibiotic resistance genes on
  plasmids, 133
Antigen binding assays, 39
ara promoter, 144

Ascites fluid, 42
  induction, 32
AsuI, 186,191,194
att, 203,204
AvaI,137,141,142,176,177,186,187,
  194
AvaII, 194
8-Azaguanine resistance, 35
Azotobacter, plasmids in, 157

b2, 204
b region, 203
Bacillus licheniformis, 124,125
  cloning in, 159
Bacillus pumilis, 119
  cloning in, 159
Bacillus subtilis
  cloning in, 115-131,159,160
  competence, 115
  heterospecific barriers, 115,
    128
  minicells, 126-128
  modification, 117,118
  recE, 116,123,125,126
  restriction, 117,118
  sporulation, 115
  transformation with plasmid
    DNA, 116,117
  transformation of protoplasts,
    126
Bacteriocin genes on plasmids,
  133
Bacteriophage genes in mouse
  cells, 90
BalI, 194
BALB/c mice, 32

283